MATHEMATICS FOR ELECTRONICS

MATHEMATICS FOR ELECTRONICS

Donald P. Leach

Reston Publishing Company, Inc.
A Prentice-Hall Company
Reston, Virginia

Library of Congress Cataloging in Publication Data

Leach, Donald P
  Mathematics for electronics.

  Includes index.
  1. Electronics—Mathematics. I. Title.
TK7835.L42   512′.1   78-15317
ISBN 0-8359-4277-5

© 1979 by Reston Publishing Company, Inc.
A Prentice-Hall Company
Reston, Virginia 22090

10 9 8 7 6 5 4 3

Printed in the United States of America

CONTENTS

PREFACE

To the Student:

This text will help you learn the mathematics necessary to solve technical problems in electricity and electronics. It is not theoretical, but rather stresses the "how-to-apply-it" aspect of mathematics. You should have available an electronic hand calculator.*

The first few sections of the text provide the opportunity to become proficient at using a hand calculator to solve arithmetic problems (addition, subtraction, multiplication, division, square roots, and cube roots). In later sections, topics from algebra and trigonometry are introduced, along with a wide variety of problems from electricity and electronics involving algebra and trigonometry in their solutions. These sections provide the opportunity to become proficient in using the calculator to solve problems requiring the use of algebra and trigonometry.

There is a wide variety of electronic hand calculators available, and the selection of the one suitable for your own individual purposes involves a certain amount of *personal preference*. The mathematical operations that can be performed with a calculator are specified in terms of its *functions*. Generally speaking, it is desirable to have a unit described as a scientific calculator (as opposed to a business calculator), and it should as a minimum include the following functions: arithmetic ($+, -, \times, \div$), square root, sine, cosine, tangent, log (base 10), ln (base e), e^x (natural antilog), y^x (power of a number). Polar-to-rectangular and rectangular-to-polar conversions are highly desirable but not absolutely necessary.

This study of mathematics proceeds in a logical fashion, and you will find it worthwhile to understand each topic before proceeding to the next one. Doing daily homework is essential. As any athlete or performing artist knows, daily practice is the only way to develop skill and confidence.

The introduction of each new topic or technique is followed by an example to illustrate its application. Study every example in detail and be certain that you understand it completely. It may occasionally be necessary to reread the previous material or to seek help from the instructor, but stick to it until you have a clear understanding of the point(s) illustrated by the example.

The exercises at the end of each section provide the opportunity to apply the concepts presented in that section, and you should consider the assigned homework as the *minimum* amount of work necessary to develop skill and confidence. Working additional problems will almost certainly increase your knowledge and confidence. Answers to the odd-numbered problems are given in the back of the book and they should be used as a check.

After completing this text, you will have a good grasp of those principles of mathematics used in electricity and electronics.

---

*Although it is assumed that an electronic hand calculator is available, it is not absolutely essential. Instead of a calculator, it is certainly possible to use a slide rule provided it has the following scales, or their equivalents (A, B, C, D, S, T, L, LL1, LL2, LL3, LL01, LL02, LL03).

To the Instructor:

This text presents those topics of arithmetic, algebra, and trigonometry necessary to participate effectively in a two-year college-level program in electricity/electronics. The only prerequisite is a knowledge of the fundamentals of arithmetic. There is enough material for a full year of work, but it can be taught in less than a year with an appropriate increase in weekly class time.

The material is divided into three major sections: Fundamental Operations, Algebra, and Trigonometry. The text is written to help students to learn to use mathematics. As such, it does not concentrate on mathematical derivations and proofs, but rather teaches "how to apply it." Topics from electricity/electronics are introduced so that electrical problems can be used as "vehicles" to teach mathematical techniques. This forms a natural tie between the mathematics course that is commonly prerequisite to or corequisite with a first course in electric circuits. Since the laws and formulas borrowed from electricity are used as vehicles to teach mathematics, they are stated without proof, in the belief that their derivations and proofs are best discussed in detail in an appropriate electronics text (course).

The topics are presented in such a way that the text can be used in a mathematics course having no electronics prerequisite or corequisite. Furthermore, the pace and order of topics have been carefully keyed with a number of popular electronics circuits texts, such that this text can be used to teach mathematics concurrently with a beginning course in circuits. When offered this way, there must be careful coordination, and the necessary mathematical topics should be encountered in the mathematics course just before they are needed in the circuits course.

To achieve the proper pace, this text differs from most other math texts in at least two ways. First, a considerable number of topics from mathematics that seem to fall into the "nice-to-know-but-not-very-useful" category have been eliminated. For example, the interpolation between given values in a table of logarithms or trigonometric functions is not discussed since the calculator literally puts the values at your fingertips. Since the presentation is based on the use of hand calculators, a lengthy discussion of theory and operation of the slide rule has been omitted. Second, the typical order of topics is somewhat rearranged. For example, a limited introduction to exponents is given in the first part of Fundamental Operations, but a complete discussion of the laws of exponents is delayed until needed in Unit 12 of Algebra.

The first part, Fundamental Operations, is intended to be quite flexible in order to accommodate the wide range of student backgrounds and abilities encountered in this introductory course. An exceptionally well prepared student may have prior knowledge of all the topics presented in the first part, including the ability to use an electronic calculator to solve the problems. On the other hand, there may be students who are encountering this material for the first time. Most students will have a prior knowledge of some of the material, but their confidence and competence in solving the problems will vary widely. The task of the instructor is to determine the individual needs of each class and then tailor the material accordingly. Each of the units in this first part can then be expanded upon, or reduced, or perhaps skipped entirely depending on the specific needs and desired results.

Many students will have a prior knowledge of the mathematical topics presented in some or all of the sections, but they will be using an electronic hand calculator for the first time. These students can be accommodated by using the appropriate sections to achieve proficiency with the calculator.

The electronic hand calculator has made the slide rule virtually obsolete, and this text was written assuming that each student has a scientific calculator at his disposal. The use of a slide rule is certainly not prohibited, and most of the problems requiring the use of a calculator can be done with a slide rule instead. Thus a student may opt to use an appropriate slide rule in place of a calculator. In spite of the fact that calculators may have ten digit accuracy, the numerical problems in this text are limited to no more than four digit accuracy—a reasonable and practical compromise for most applications. Answers to the odd-numbered problems are given in the back of the text and the Instructor's Manual contains all the answers.

To err is human, and I would be most grateful to receive notice of any errors or discrepancies you might note in this text.

MATHEMATICAL SYMBOLS

$\times$ or $\cdot$	multiplied by
$\div$	divided by
$+$	positive, plus, add
$-$	negative, minus, subtract
$=$	equals
$\equiv$	is defined as
$\cong$	is approximately equal to
$\neq$	does not equal
$>$	is greater than
$<$	is less than
$\geqslant$	is greater than or equal to
$\leqslant$	is less than or equal to
$\therefore$	therefore
$\angle$	angle
$\perp$	is perpendicular to
$\parallel$	is parallel to
$\lvert n \rvert$	absolute value of n
$\%$	percent
$\propto$	is proportional to

ABBREVIATIONS

Term	Abbreviation	Term	Abbreviation
Alternating current	ac	Counterclockwise	ccw
Ampere	A	Cubic centimeter	cm^3
Ampere-hour	Ah	Cubic foot	ft^3
Amplitude modulation	AM	Cubic inch	in^3
Antilogarithm	antilog	Cubic meter	m^3
Audio frequency	AF	Cubic yard	yd^3
Average	avg	Cycles per second	Hz
Centimeter	cm	Decibel	dB
Circular	cir	Degrees Celsius	°C
Clockwise	cw	Degrees Fahrenheit	°F
Continuous wave	CW	Direct current	dc
Cosecant	csc	Electromotive force	emf
Cosine	cos	Equation	Eq.
Cotangent	cot or ctn	Farad	F
Coulomb	C		

TERM	ABBREVIATION	TERM	ABBREVIATION
Foot, feet	ft	Maximum	max
Feet per minute	ft/min	Mega	
Feet per second	ft/s	(prefix, $= 1 \times 10^6$)	M
Feet per second		Megacycles per second	MHz
squared	ft/s^2	Megahertz	MHz
Figure	Fig.	Megavolt	MV
Frequency	spell out	Megohm	MΩ
Frequency modulation	FM	Meter	m
Giga		Meter-kilogram-second	
(prefix, $= 1 \times 10^9$)	G	system	MKS
Gigahertz	GHz	Meters per second	m/s
Gram	g	Mho	S
Henry	H	Micro	
Hertz	Hz	(prefix, $= 1 \times 10^{-6}$)	μ
High frequency	HF	Microampere	μA
Hour	h	Microfarad	μF
Inch	in	Microhenry	μH
Inches per second	in/s	Micromho	μS
Intermediate		Micromicro	
frequency	IF	(prefix, $= 1 \times 10^{-12}$)	p
Kilo (prefix, $= 1 \times 10^3$)	k	Micromicrofarad	pF
Kilocycles per second	kHz	Microsecond	μs
Kilogram	kg	Microvolt	μV
Kilohertz	kHz	Microwatt	μW
Kilohm	kΩ	Mile	mi
Kilometer	km	Miles per hour	mi/h
Kilometers per hour	km/h	Miles per minute	mi/min
Kilovars	kvar	Miles per second	mi/s
Kilovolt	kV	Milli	
Kilovoltampere	kVA	(prefix, $= 1 \times 10^{-3}$)	m
Kilowatt	kW	Milliampere	mA
Kilowatthour	kWh	Millihenry	mH
Logarithm		Millimeter	mm
(common, base 10)	log	Millisecond	ms
Logarithm (any base)	log$_a$	Millivolt	mV
Logarithm		Milliwatt	mW
(natural base e)	log$_e$, ln	Minimum	min
Low frequency	LF	Nano	
Lowest common		(prefix $= 1 \times 10^{-9}$)	n
denominator	LCD	Nanoampere	nA
Lowest common		Nanofarad	nF
multiple	LCM	Nanosecond	ns

Term	Abbreviation	Term	Abbreviation
Nanowatt	nW	Second	s
Ohms	Ω	Sine	sin
Peak-to-peak	p-p	Square centimeter	cm^2
Pico		Square foot	ft^2
(prefix, $= 1 \times 10^{-12}$)	p	Square inch	in^2
Picoampere	pA	Square meter	m^2
Picofarad	pF	Square yard	yd^2
Picosecond	ps	Tangent	tan
Picowatt	pW	Ultrahigh frequency	UHF
Pound	1b	Var	
Power factor	PF	(reactive voltampere)	var
Radian	rad	Very high frequency	VHF
Radio frequency	RF	Volt	V
Revolutions per minute	rev/min	Voltampere	VA
Revolutions per second	rev/s	Watt	W
Root mean square	rms	Watthour	Wh
Secant	sec	Wattsecond	Ws

FUNDAMENTAL
OPERATIONS

UNIT 1
INTRODUCTION

1.1 OVERVIEW

This is a text in applied mathematics written for the technician who wants to learn how to use mathematics as a tool for solving problems. The topics are of course mathematically precise, but the emphasis is definitely on practical applications. The usual tedious derivations of formulas and proofs of theorems found in traditional mathematics texts have been omitted. Furthermore, a number of topics considered to be in the "nice to know, but not very useful" category have been eliminated. The guiding principle is always—"Is this topic necessary to solve problems in electricity/electronics?"

The trusty old slide rule has been thrust into the world of fond memories, along with the horse and buggy, by the availability of a wide variety of modern electronic hand calculators. Mastering the operation of an electronic calculator is imperative. Without this ability, the education of a capable technician is incomplete. To this end, any modern mathematics text must make the greatest possible utilization of the electronic hand calculator—this text does exactly that.

As in most endeavors, the benefit you derive from this book will be directly proportional to the effort you expend. If you always do only the minimum work required, you can expect your achievement to be minimal; on the other hand, a conscientious, concentrated effort on your part will certainly yield the maximum positive results for you. There is nothing magic about the text or the topics presented; the same topics can be found in numerous other books. However, it is hoped that the topics presented and the order of presentation will allow you to derive the maximum benefit, with the least effort, in the shortest possible time.

1.2 ARITHMETIC

Carefully defined terms and symbols are basic to the study of mathematics, and mathematical symbols form a universal language understood by people throughout the world. A clear understanding of such terms and symbols is essential for anyone seeking proficiency in solving mathematical problems.

Whole Numbers

The arabic numerals 0, 1, 2, 3, 4, 5, 6, 7, 8, and 9 are the *digits* of arithmetic used in counting. As such, each of these numerals (numbers) can be thought of as representing a quantity of things counted. Some specific types of numbers encountered in arithmetic problems include the following:

WHOLE NUMBER: Any integer or integral number formed by counting the basic digits. Thus 0, 3, 11, 21, and 243 are examples of whole numbers.

FACTOR: A factor of any whole number is any other whole number that will divide it exactly. As an illustration, 2 is a factor of 10, since 2 divides 10 into exactly 5 equal parts. Similarly, 5 is seen to be a factor of 10.

ODD NUMBER: An integer not exactly divisible by 2. For example, 3, 5, 7, 13, and 27 are all odd numbers.

EVEN NUMBER: An integer that is exactly divisible by 2. For example, 4, 8, 10, and 64 are all even numbers.

Signs of Operation

Generally speaking, arithmetic refers to the four basic operations of addition, subtraction, multiplication, and division applied to numbers. The signs representing these four operations are:

ADDITION ($+$): The result of adding two numbers is a *sum*.

SUBTRACTION ($-$): The result of subtracting one number from another is a *difference*.

MULTIPLICATION ($\times$) or ($\cdot$): The result of multiplying two numbers is a *product*.

DIVISION ($\div$): The result of dividing one number by another is a *quotient*.

From practical experience, it is clear that the order in which two numbers are written is unimportant when determining their sum. For example, the sum of 2 and 3 is always equal to 5, and it does not matter whether it is written as $2+3$ or $3+2$. Formally, this is known as the *Commutative Law of Addition*.

There is a similar law for multiplication known as the *Commutative Law of Multiplication*. It simply states that the order in which two numbers are written is unimportant in determining their product. For instance, 2 times 3 always equals 6, and it can be written as 2×3 or 3×2.

When performing a series of operations, multiplications are done first, then divisions, and finally additions and subtractions. This is known as the *order of operations*. For example, $7\times8+4\div2-6$ is evaluated as follows:
First, do the multiplication to obtain

$$56+4\div2-6$$

Then do the division

$$56+2-6$$

Then do the addition

$$58-6$$

And finally the subtraction

$$52$$

Notice that if either an addition or a subtraction is performed *before* the multiplication and division, the result will be incorrect. Try it.

Signs of Connection

A sign of connection is used in mathematics to express a relationship between two quantities. The seven basic signs of connection and the meaning of each are given below:

EQUAL SIGN ($=$): This sign is used to state the equality of two quantities. For example, the fact that "three plus four is equal to seven" can be written using mathematical signs as "$3+4=7$."

NOT EQUAL ($\neq$): Used to state the inequality of two quantities. For example, the fact that "three plus four is not equal to eight" can be written as "$3+4\neq8$."

APPROXIMATELY ($\cong$): Used to express the fact that two quantities are approximately (but not exactly) equal. For instance, "π is approximately equal to 3.14" can be written as "$\pi \cong 3.14$."

GREATER THAN ($>$): Used to state that a given quantity is greater than another. For example, the fact that "seven is greater than four" can be written as "$7>4$."

LESS THAN ($<$): Used to state that a given quantity is less than another. Thus, "five is less than thirteen" can be written as "$5<13$."

GREATER THAN OR EQUAL TO ($\geqslant$): Used to state that a certain quantity is greater than or equal to another quantity. For instance, "The temperature of water when in a liquid state must be greater than or equal to 0°C," (water freezes at 0°C) can be written as "Temperature $\geqslant 0$°C."

LESS THAN OR EQUAL TO ($\leqslant$): Used to state that a given quantity is less than or equal to another quantity. For example, "The quantity of liquid that can be added to a 25-liter container must certainly be less than or equal to 25 liters," can be written as "Quantity $\leqslant 25$ liters."

EXAMPLE 1.1. Use signs of operation and connection to express the following:
(a) The sum of 5 and 11 is equal to 16.
(b) The difference found by subtracting 85 from 180 is not equal to 85.
(c) The product of 3 and 17 is 51.
(d) The quotient found by dividing 150 by 6 is 25.
(e) The quantity of liquid in an 18-liter container is less than or equal to 18 liters.

SOLUTION. Note the appropriate signs of operation and connection:
(a) $5+11=16$
(b) $180-85\neq85$
(c) $3\times17=51$

or $\qquad 3\cdot17=51$

(d) $150\div6=\dfrac{150}{6}=150/6=25$

Notice that a division has been indicated in three different ways.
(e) *Quantity* $\leqslant 18$ liters

Fractions

A *common fraction* is an indicated division of two whole numbers. Thus 1/3, 2/7, and 19/87 are all common fractions. The number above the line in a fraction is called the *numerator*, while the number below the line is called the *denominator*. For instance, in the common fraction 7/8, the numerator is 7 and the denominator is 8. Notice carefully that a common fraction having zero for a denominator is not allowed since division by zero is undefined in mathematics. Thus a fraction such as 13/0 is undefined since it has a zero denominator.

Any common fraction having a numerator smaller than its denominator is called a *proper* fraction; if the numerator is greater than the denominator, it is called an *improper* fraction. For example, 7/13 is a proper fraction, while 7/4 is an improper fraction.

If the indicated division of a proper fraction is carried out, the result is a *decimal fraction*. For instance, 0.75 is the decimal fraction equivalent to 3/4. Similarly, 0.318 is the decimal fraction equivalent to the proper fraction 318/1000.

When finding the decimal equivalent of a common fraction such as 1/3, the division process will continue to generate one 3 after another for as long as the division is continued. This is indicated by using "three dots" following the decimal number. Thus,

$$\frac{1}{3} = 0.333\ldots$$

Mixed Numbers

A *mixed number* consists of a whole number and a fraction. For instance, $1\frac{3}{4}$, $21\frac{5}{8}$, 6.37, and 87.92 are all mixed numbers. Mixed numbers can be used to count "things" as well as "fractional parts of things." Thus quantities such a 1.5 watts and $17\frac{1}{2}$ kilometers can be specified.

Negative Numbers

Any number can be considered as a distance measured from a fixed point on a straight line as shown in Figure 1.1. On this line the beginning point is zero, and a number is located by measuring (ascending) to the right of zero. For example, $2\frac{1}{2}$ is located at point A, while the number at point B is 3.8.

A natural extension of the measuring line in Figure 1.1 is to add a second measuring line extending to the left of the reference, as shown in Figure 1.2. To distinguish between numbers measured to the *right* of reference and numbers measured to the *left* of reference, a negative sign is placed in front of any number measured to the left of the reference. Any number measured to the left of the

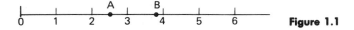

Figure 1.1

Figure 1.2

reference is called a *negative number*, while any number to the right is called a *positive number*. Notice that the positive number 3.7 could be written with a plus sign as $+3.7$, but the plus sign is generally omitted as a matter of convenience.

As an example, the positive number 3.7 is located at point A, while the negative number $-1\frac{3}{4}$ is located at point B, and $+1\frac{3}{4} = 1\frac{3}{4}$ is found at point C. In general, any number preceded by a negative sign is a negative number, and any number preceded by a positive sign (or no sign) is a positive number.

EXAMPLE 1.2.   The thermometer in Figure 1.3 is calibrated with a scale in degrees Celsius (°C). The reference temperature is 0°C, the temperature at which water freezes. Temperatures above 0°C are positive numbers, while temperatures below 0°C are negative numbers. Read from the scale the temperatures at points A and B.

SOLUTION.   Point A is less than 0°C and is thus a negative number. It is halfway between 0°C and -20°C, and it must therefore be -10°C. Point B is a temperature above 0°C and is thus a positive number. Clearly, point B is at $+20$°C, or simply 20°C.

Absolute Value

The *absolute value* or *magnitude* of any number is simply its numerical value without regard to sign. For instance, the absolute value of $+7.3$ is simply 7.3. The absolute value of -1.9 is 1.9. Both point B and point C in Figure 1.2 have the same absolute value—namely, $1\frac{3}{4}$. The absolute value can be thought of as the distance from the reference point without regard to the direction (left or right) on the number line.

Two vertical bars placed around a quantity is the mathematical sign meaning the absolute value. For instance, "the absolute value of 7" is written $|7|$. Similarly, $|-1.8|$ is read "the absolute value of minus 1.8."

EXAMPLE 1.3.   Write in words the meaning of:
(a) $|-71| = 71$
(b) $\dfrac{|-8|}{4} = 2$
(c) $-10° \leqslant$ temperature $\leqslant 10°$
(d) $|\text{Temperature}| \leqslant 10°$

SOLUTION.
(a) The absolute value of minus seventy-one is equal to seventy-one.
(b) The magnitude of negative eight, divided by four, is equal to two.
(c) The temperature is greater than or equal to minus 10°, but less than or equal to 10°.

(d) The magnitude of the temperature is less than or equal to 10°.

Notice that statements (c) and (d) provide the exact same information; namely, the temperature cannot be more than 10° removed from the 0° reference, either positive or negative.

Exercises 1.2

1. The standard resistance sizes for carbon-composition resistors from 10 ohms up to 100 ohms are: 10, 12, 15, 18, 22, 27, 33, 39, 47, 56, 68, and 82 ohms. Which values are even numbers? Which values are odd numbers? List them.

2. If the resistance values in Problem 1 are all doubled, how many of them will be even numbers? Odd numbers? List them.

3. If ten times each resistance value in Problem 1 is halved, how many of them will be even numbers? Odd numbers? List them.

4. List all of the factors for the resistance value 39 ohms. (*Hint*: The factors for 10 are 10, 5, 2, and 1.) What are the factors of the number 27?

5. Use signs of operation and connection to express the following:
   (a) The sum of 3 and 11 is equal to 14.
   (b) The difference found by subtracting 17.1 from 28.3 is equal to 11.2.
   (c) The product of 7 and 18 is equal to 126.
   (d) The quotient found by dividing 88 by 4 is 22.
   (e) The sum of 11 and 1/8 is less than 15.
   (f) The product of 9 and 11 is greater than 90.
   (g) The volume is less than or equal to 33 liters.
   (h) The resistance is greater than or equal to 4700 ohms, but less than or equal to 4900 ohms.

6. Write in words the meaning of the following:
   (a) $6 + 11 = 17$
   (b) $17 - 5 = 12$
   (c) $5 + 20 \neq 21$
   (d) $3 \times 7 = 21$
   (e) $3 \cdot 9 = 27$
   (f) $121 \div 11 = 11$
   (g) $144/12 > 9$
   (h) $1/2 < 5/8$
   (i) Length $\geqslant 35$ centimeters
   (j) Resistance $\leqslant 750$ ohms
   (k) Volume $\cong 2.63$ liters
   (l) Velocity $=$ distance/time

7. Find the numerical value of each of the following:
   (a) $81 \div 9 - 6$
   (b) $5 \times 7 + 4$
   (c) $48 - 6/3$
   (d) $2 \times 33 \div 11$
   (e) $5 \times 17 - 64 \div 8$
   (f) $225 \div 15 + 5 \times 7$

8. Write the number represented by each point labeled on the following number line. Don't forget the proper sign.

9. Draw a temperature scale (similar to Figure 1.3) using degrees Fahrenheit (°F). Locate the point where water boils (212°F), where water freezes (32°F), and the points 0°F and −20°F.

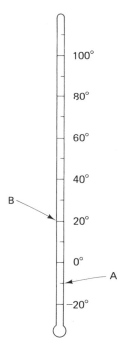

Figure 1.3

10. What is the absolute value (magnitude) of each point labeled on the number line in Problem 8?

11. Write in words the meaning of each of the following:

(a) $|3.7|$

(b) $|-16| = 16$

(c) $|-11| > 5$

(d) $\dfrac{|6|}{5} < 2$

(e) $\dfrac{|\text{Voltage}|}{\text{Resistance}} = \text{current}$

12. Demonstrate whether or not the following two expressions have the same meaning:

$$-15 \leqslant \text{voltage} \leqslant 15$$

$$|\text{Voltage}| \leqslant 15$$

(*Hint:* Use the number line.)

1.3 ALGEBRA

Literal Numbers

Occasionally when solving a problem, it is convenient to assign a letter or a symbol to represent a certain quantity whose actual numerical value is unknown. For example, the letter Q might represent the actual quantity of gasoline contained in a 20-gallon tank. Obviously, Q could then have a numerical value anywhere within the limits of 0 to 20; that is, Q must be *equal to or greater than* 0, but *less than or equal to* 20. A concise mathematical expression for this statement is

$$0 \leqslant Q \leqslant 20$$

Similarly, the quantity of electric current I through the resistor R in Figure 1.4 is dependent on the value of the battery voltage V. All of these letters that represent variable quantities (Q, I, R, V) are called *literal numbers*. Literal numbers are used in expressing the various laws and relationships of science in a compact mathematical form; they are widely used in all branches of science and engineering.

Sometimes the same letter is used to represent two different values of the same quantity, for instance, two different resistances or voltages in the same circuit. Therefore, to further distinguish these values, literal numbers (letters) are sometimes given primes or subscripts. For example, primes can be used with the letter V

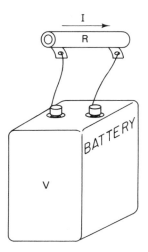

Figure 1.4

to represent two different voltages, such as, V' (read "vee prime") and V'' ("vee double prime"). Subscripts are also used in the same way, for instance, V_1 ("vee sub one"), or V_4 ("vee sub four"), or V_x ("vee sub x").

Terms and Expressions

Literal numbers can be combined with other numbers by using any of the four signs of operation. For example: $Q+3$, $V-4$, $P/7$, and $4R$ or $7IR$. Each of the quantities $Q+3$, $V-4$, $P/7$, and $4R$ or $7IR$ is known as an *algebraic expression*. An expression containing no $+$ or $-$ signs is an *algebraic term*. Thus $P/7$, $4R$, and $7IR$ are all algebraic terms. Notice that $4R$ really means "four times R," and it could be written as $4\times R$ or $4 \cdot R$ or simply $4R$. Similarly, $7IR = 7 \times I \times R = 7 \cdot I \cdot R$.

Each of the numbers in a term, or the product of any combination of the numbers, is called a *factor* of that term. For example, 7, I, R, $7I$, $7R$, and IR are all factors of the term $7IR$. The numerical part of a term is known as the *numerical coefficient*, or simply the *coefficient* of that term. Thus the term $7IR$ has a numerical coefficient of 7. In algebra, an expression consisting of a single term is called a *monomial*, while an expression containing two or more terms is a *polynomial*.

Formulas

Joining two algebraic expressions with a sign of connection will generate an *algebraic sentence*—sometimes referred to as a *formula*. For example, the area A of the rectangle in Figure 1.5 is given by the formula

$$A = hw \tag{1.1}$$

where $h=$ the height, $w=$ the width, and $A=$ the area. To find the value of an algebraic expression, or to evaluate a formula, simply means to find its numerical value. Thus a formula is evaluated by substituting the proper numbers for each letter and calculating the numerical value.

EXAMPLE 1.4. Find the area A of the rectangle in the figure below if $h=6$ in. and $w = 11$ in.

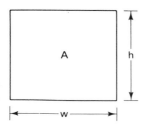

Figure 1.5

SOLUTION. Use the formula $A = hw$ and substitute the given values for h and w to obtain

$$A = hw = 6 \times 11 = 66 \text{ in.}^2$$

EXAMPLE 1.5. Evaluate the term $7aRz$ if $a=5$, $R=9$, and $z=0.1$.

SOLUTION. Substitute the given values for a, R, and z:

$$7aRz = 7 \times 5 \times 9 \times 0.1 = 31.5$$

EXAMPLE 1.6. For the rectangle in Figure 1.5, $w=3$ in., but h can have any value from 1 in. to 24 in. Write a sentence showing the limits of the area A.

SOLUTION. Use the formula $A = hw$ and calculate the minimum and maximum values of A.

For $h=1$ in.,

$$A_{min} = hw = 1 \times 3 = 3 \text{ in.}^2$$

For $h=24$ in.,

$$A_{max} = hw = 24 \times 3 = 72 \text{ in.}^2$$

Thus,

$$3 \leqslant A \leqslant 72$$

If the rectangle in Figure 1.5 has its height equal to its width (i.e., a square), then $h=w$ and Eq. (1.1) becomes $A = h \cdot w = h \cdot h$ (or $A = w \cdot w$). A short way of writing $h \cdot h$ is h^2—read "h squared," or "h to the second power." The small 2 is the *exponent* or *power* of h. The idea can be extended; for example,

$$i^2 = i \cdot i \qquad\qquad r^3 = r \cdot r \cdot r$$

$$v^5 = v \cdot v \cdot v \cdot v \cdot v \qquad 2^4 = 2 \cdot 2 \cdot 2 \cdot 2 = 16$$

$$10^2 = 10 \cdot 10 = 100 \qquad 10^4 = 10 \cdot 10 \cdot 10 \cdot 10 = 10,000$$

EXAMPLE 1.7. Find the area of the square in Figure 1.5 if $h = w = 7$ ft.

SOLUTION.

$$A = h^2 = 7 \times 7 = 49 \text{ ft}^2$$

Constants and Variables

The formula for the area A of a circle in terms of its radius r is $A = \pi r^2$. This formula contains three literal numbers: A, π, and r. Since the value of π is always equal to 3.14159..., it is said to be a *constant*. The radius r can be assigned any value, and thus r is said to be a *variable*; in fact, it is called the *independent variable*. The area A is a quantity whose value is dependent on the radius, and thus A is called the *dependent variable*. Thus literal numbers can be more precisely referred to as either *constants* or *variables*.

Signs of Grouping

Three different signs of grouping are:

$$\text{Parentheses} \quad (\ )$$

$$\text{Brackets} \quad [\ ]$$

$$\text{Braces} \quad \{\ \}$$

These signs are used to specify the order in which operations are to be performed. For instance, the standard order of operations specifies multiplication first and then addition, and thus the expression $3I+7$ must be evaluated by first multiplying the given value of I by 3 and then adding 7 to this product. On the other hand, the expression $3(I+7)$ must be evaluated by first adding 7 to the given value of I and then multiplying this sum by 3. The parentheses specify that *the quantity inside the parentheses must be evaluated first*. The same is true for the other signs of grouping.

When more than one sign of grouping is used, the evaluation proceeds by performing the operations indicated by the *innermost* groups first. For example,

$$[150 \div (3+2)3] \div 2 = [150 \div (5)3] \div 2 = [150 \div 15] \div 2 = 10 \div 2 = 5$$

EXAMPLE 1.8.   Evaluate the following expressions for $I=5$:

(a) $3I+7$

(b) $3(I+7)$

(c) $[65 \div (8+I)] \times (I+3)$

SOLUTION.

(a) Substitute $I=5$, and then evaluate according to the standard order.

$$3I+7 = 3 \cdot 5 + 7 = 15 + 7 = 22$$

Notice that the incorrect result of 36 is obtained if the addition is done first.

(b) Substitute $I=5$, and evaluate the quantity inside the parentheses first.

$$3(I+7) = 3(5+7) = 3(12) = 36$$

(c) Substitute $I=5$, and remember to evaluate the innermost groups first.

$$[65 \div (8+I)] \times (I+3) = [65 \div (8+5)]$$
$$\times (5+3)$$
$$= [65 \div 13] \times (8)$$
$$= [5] \times (8) = 40$$

Exercises 1.3

1. Use signs of connection to express the following ideas:
   (a) A voltage V is always less than 220 volts.
   (b) The power P dissipated in a resistor is always greater than or equal to zero.
   (c) The current I in a lamp is less than 1.5 amperes.
   (d) The resistance of a resistor R is between 450 and 550 ohms.

2. Write out in words the meanings of the following algebraic expressions:

   (a) $V \geqslant 20$ (b) $I \leqslant 21.5$ (c) $0 \leqslant P \leqslant 35$

   V is the letter used to designate a voltage in an electric circuit, I is used to represent electric current, and P is used to represent power.

3. Write an algebraic sentence (formula) that states that the volume V of a box is equal to the product of the height h, times the width w, times the depth d.

4. Electric energy is measured in kilowatthours (kWh). If energy costs 4¢ per kWh, write the formula used to calculate the monthly energy bill.

5. Evaluate the following if $I = 2$, $V = 15$, and $P = 7.5$:

   (a) $I + 3$ (b) $30 + V$ (c) $V - 15$ (d) $21P$

   (e) $\dfrac{4+P}{6}$ (f) $\dfrac{VI}{5}$ (g) $3V - \dfrac{I}{7}$ (h) $\dfrac{6VI - 1.5P}{2.2P}$

   (i) $550VI$

6. The formula for power dissipated in a resistance is $P = VI$, where P is the power in watts, V is the voltage across the resistor in volts, and I is the current through the resistor in amperes. Calculate the power if:
   (a) $V = 10$ volts and $I = 1.5$ amperes
   (b) $V = 220$ volts and $I = 30$ amperes
   (c) $V = 12$ volts and $I = 0.015$ ampere

7. Another formula for the power dissipated in a resistance can be written as $P = I^2 R$. Calculate the power if:
   (a) $I = 6$ amperes and $R = 100$ ohms
   (b) $I = 0.1$ ampere and $R = 1000$ ohms

8. The area A of a circle of radius r is given by $A = \pi r^2$. Find A if r is equal to:
   (a) 3 in.   (b) 10 ft   (c) 2.5 meters
   Use $\pi \cong 3.14$.

9. Evaluate the following expressions:

   (a) $8(7-2)$ (b) $3(7.5+1.5) - 6 \div 3$ (c) $\dfrac{(15.1 - 7.1)}{4} + \{5 + (6-3)2\}$

10. Evaluate the following expressions if $I_1=7$ and $I_2=4$:

   (a) $3I_1+7(I_2+I_1)$ (d) $4(I_1+I_2)+2(I_1-I_2)$

   (b) $\dfrac{(I_1-I_2)}{3}+7I_2$ (e) $\dfrac{2(I_1+2I_2)}{5}-\dfrac{(2I_1-3I_2)}{2}$

   (c) $[10(I_1-3)]+[4(I_2-1)\div6]$ (f) $\{[10-(I_1+2I_2)\div3]+\dfrac{(5I_1-5I_2)}{5}\}\div4$

11. Evaluate the following:

   (a) h^2 if $h=7$ (e) 10^2

   (b) v^3 if $v=2$ (f) 10^3

   (c) 2^4 (g) 10^4

   (d) 3^3

UNIT 2
BASIC HAND-CALCULATOR OPERATIONS

One of the most significant products resulting from the recent revolutionary advances made in the electronics industry is the small hand-held solid-state calculator. The hand calculator places within the grasp of every individual the power to solve mathematical problems that are impractical using paper and pencil alone; many such problems have been solved only recently using large-scale digital computers. It is certainly a worthwhile goal for every student of engineering and technology to become proficient in the use of an electronic hand calculator.

As mentioned in the Preface, there are a number of different calculators available. For use in solving the problems in this text, it is advisable to use a "scientific calculator" having at least the following functions: arithmetic ($+$, $-$, $\times$, $\div$), square root, sine, cosine, tangent, log (base 10), ln (base e), e^x (natural antilog), y^x (power of a number). Polar to rectangular and rectangular to polar conversions are desirable. Calculators providing these functions include the Hewlett-Packard HP-21 or the HP-25, the Rockwell International Unicom 202/SR, the Bowmar MX100, and the Texas Instruments SR-50.

2.1 SIGNIFICANT FIGURES

Accuracy

Although most hand calculators are capable of displaying numbers having either 8 or 10 digits, the electronics problems treated here do not require numerical solutions having this accuracy. The *accuracy* of a number is determined by the significant figures contained in that number. A *significant figure* can be defined as any digit that is considered reliable as a result of measurement or mathematical computation.

For example, the resistance of a certain carbon composition resistor is given as 18 ohms (two significant figures), but a more accurate measurement reveals that the resistance is 18.25 ohms (four significant figures). It should be clear that the accuracy of a number is increased as the quantity of significant figures in the number is increased.

The problems in this text will be treated with an accuracy specified by *no more than four significant figures*. This provides a very realistic accuracy for almost all practical problems encountered. Indeed, the use of more than four-digit accuracy is simply an unnecessary burden that is not justified in most real-life situations.

Measured Data

The practical aspects of electronics nearly always involve the collection of measured data, and measured data are inherently inexact. When recording measured data, it is customary to include all numbers represented by the graduations on the measuring device, and to include one *estimated* number that is actually equal to one-half of the smallest graduation. Every number read and recorded is considered to be a significant figure.

As an example, the thermometers in Figure 2.1 are calibrated to measure temperature in degrees Celsius (°C), and the smallest graduation is 1°C. This means reliable measurements can be made to the nearest 0.5°C (1/2 the smallest graduation). The procedure is to *imagine* one additional graduation line halfway between each existing line on the thermometer and to record the reading closest to the top of the mercury column.

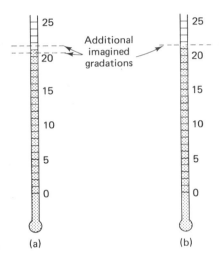

Figure 2.1.   Thermometers calibrated in degrees C.   (a)   (b)

In Figure 2.1(a), the proper reading is 21.0°C. Notice there are three significant figures in this reading. Since 1°C is the smallest graduation, the value 21°C is recorded, and this reading is then refined by estimating to the nearest half degree. The third significant figure (the zero after the decimal point) shows that the reading is closer to 21.0°C than it is to 21.5°C or 20.5°C.

As another example, the reading in Figure 2.1(b) is taken by first recording 21°C, and then estimating an additional 0.5°C. The proper recording must then be 21.5°C—a number having three significant figures. Notice that the third digit in this measurement is "doubtful," since the mercury column is not exactly on the imagined 1/2 degree line, but the measurement is reliable since it is possible to repeatedly estimate this third digit accurately. At the same time, an attempt to refine the measurement to say 21.4°C leads to a *false accuracy* since it implies the ability to imagine ten additional graduation lines in between each existing line; such estimation generally cannot be repeated reliably and should not be attempted.

EXAMPLE 2.1 Record the voltmeter reading shown in Figure 2.2.

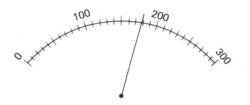

Figure 2.2

SOLUTION. The smallest graduation on the voltmeter scale is 10 volts, and thus readings can be made to the nearest 5 volts (one-half the smallest graduation). The needle is between 170 volts and 180 volts. It is possible to estimate the 175-volt graduation, but the needle is closer to 180 than it is to 175. The correct reading is then 180 volts.

Manipulated Data

When doing calculations using measured or given data, it is possible for *false accuracy* to occur in the solutions. For example, consider the following three measurements of weight, perhaps measured with different instruments.

Weight A	175 lb read to the nearest 1 lb
Weight B	11.5 lb read to the nearest 0.1 lb
Weight C	9.20 lb read to the nearest 0.01 lb

In finding the sum of these three weights, it is tempting to simply add all the numbers together and write the total as 195.70 lb. But this solution has five significant figures, and the last zero implies a measurement (calculation) accurate to the nearest 0.01 lb. This is certainly false accuracy since weight A is only accurate to the nearest 1 lb! Clearly it is not justifiable to obtain a result containing more significant figures than any one of the original readings. In this case, the sum should be reduced to three significant figures by dropping digits that are not significant. When dropping a digit having a value of 5 or more, the last digit retained should be increased by 1. Thus, the sum 195.70 becomes 195.7 (dropping the zero), and then 196 (dropping the 7 and adding 1). The correct sum is then 196 lb.

The following rules will be helpful while performing calculations:

1. When doing addition or subtraction, do not carry the result beyond the first column containing a doubtful figure.

2. When doing multiplication or division, carry the result to the same number of significant figures contained in the quantity entered into the calculation that has the fewest number of significant figures.

3. When dropping digits, the last digit retained is increased by one if the digit dropped is 5 or more.

EXAMPLE 2.2
(a) Find the sum of 17.5 and 0.701.
(b) Find the quotient given by 17.5/0.701.

SOLUTION.
(a) Adding the two numbers yields

$$17.5$$
$$\underline{0.701}$$
$$18.201$$

The first column containing a doubtful figure is the third column—the 5 in the reading 17.5 is doubtful. The result is thus properly reduced to 18.2.

(b) Carrying out the indicated division yields

$$\frac{17.5}{0.701} = 12.2675$$

Since both of the original numbers contain three significant figures, the quotient should be shortened to three significant figures also. Notice how a retained digit is increased when the digit dropped is 5 or more. Thus,

$$12.2675 \cong 12.268$$
$$12.268 \cong 12.27$$
$$12.27 \cong 12.3$$

The desired result is then 12.3.

Exercises 2.1

1. Use a ruler having a scale calibrated in centimeters (cm) to measure the length of each line drawn below. The measurements should be recorded to three significant figures.

   (a) _____

   (b) _____

   (c) _____

   (d) _____

   (e) ___

2. Record each voltmeter reading shown in Figure 2.3.

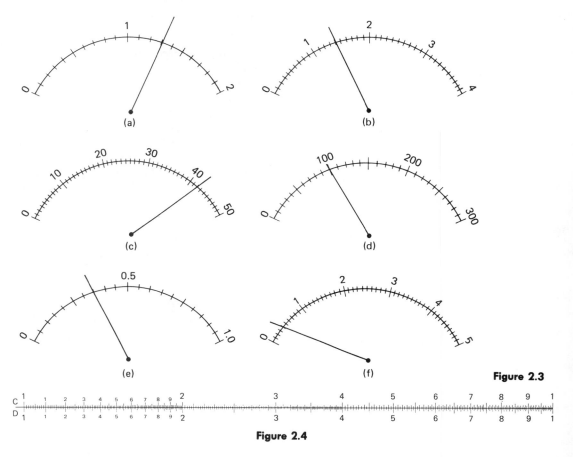

Figure 2.3

Figure 2.4

3. The scale shown in Figure 2.4 is known as a *logarithmic scale*. It is used on slide rules as well as on certain graph paper. Notice that the divisions are all of different lengths. Carefully locate the following points on the scale:

    (a) 2.31   (b) 4.25   (c) 6.15   (d) 8.30   (e) 9.95   (f) 1.855
    (g) 1.015

4. Perform the following arithmetic operations. Express the result in each case to the proper number of significant figures by dropping digits if necessary.

    (a)   11.5   (b)   21.3   (c)   107   (d)   0.713
          (+) 1.16   (−) 20.1   (+) 582.6   (+) 0.112

    (e) 1.22×3.7   (i) 42÷3.5
    (f) 11.0×1.22   (j) 3.74÷1.7
    (g) 722×633   (k) 384÷250
    (h) 0.3×8.8   (l) 650÷120

2.2   ARITHMETIC OPERATIONS

For solving mathematical problems, an electronic hand calculator is vastly superior to a slide rule in many different respects. In the first place, numbers are displayed on a calculator by means of 7-segment indicators,* and it is not necessary to learn how to read the graduations of a logarithmic scale as found on the slide rule (see Exercises 2.1, Problem 3). Furthermore, a calculator has much greater accuracy (8 or 10 digits); it is much faster and more convenient to operate; and it definitely has a much greater versatility. For instance, it is possible to do both addition and subtraction using a hand calculator.

Calculators can be placed in either of two major categories according to the manner in which they are designed to perform functions. The first group operates with "algebraic logic," for example, the Unicom 202/SR and the Texas Instruments SR-50. The second group uses "reverse polish notation" (abbreviated as RPN), for example, the Hewlett-Packard HP-21. There are advantages and disadvantages for each system. In general, it is possible to do problems with fewer *key strokes* when using a calculator of the RPN type than when using a calculator of the algebraic type.

It is neither desirable nor necessary to discuss here the relative merits of one calculator as opposed to another. Nor is it feasible to offer instruction in all of the various types of calculators available. Rather at this point it is necessary for you to utilize the owner's manual that accompanies *your* calculator and learn at the minimum the following operations:

1. Power on	5. Subtraction
2. Clear	6. Multiplication
3. Keying in numbers	7. Division
4. Addition	8. Combinations of the above operations

The purpose of the problems in this section is to provide the opportunity to learn the above arithmetic operations, and to offer the practice time necessary to gain confidence, speed, and accuracy in using the hand calculator. It is recommended that you do calculations to 4-digit accuracy unless otherwise stated, and that you use "fixed decimal point display." Scientific notation will be covered in a following section of this text. It is suggested that you manipulate the calculator with the hand *opposite* your writing hand. Thus if you are right-handed, manipulating the calculator with your left hand will leave your right hand free for writing.

---

*For example, see A. P. Malvin and D. P. Leach, *Digital Principles and Applications*, 2nd. ed. (New York: McGraw-Hill, 1975); or D. P. Leach, *Experiments in Digital Principles* (New York: McGraw-Hill, 1976).

Exercises 2.2

1. Enter each of the following numbers on your calculator:
   (a) 41.2
   (b) 0.445
   (c) 0.000531
   (d) 873,192
   (e) 7,863.773
   (f) 67.054321
   (g) -0.10203
   (h) π
   (i) 100.02
   (j) -567.38
   (k) -31.62
   (l) -0.0000785

2. To find the reciprocal of a number means to find the value of that number divided into unity. Thus, the reciprocal of the number N is

$$\text{Reciprocal of } N = 1 \div N = \frac{1}{N}$$

   For example, the reciprocal of 2 is $1/2 = 0.5000$. Find the reciprocal of the following numbers (remember, use 4-digit accuracy):
   (a) 3
   (b) 4
   (c) 5
   (d) 6
   (e) 7
   (f) 8
   (g) 9
   (h) 10
   (i) 15
   (j) 33
   (k) π
   (l) 27
   (m) 83
   (n) 65
   (o) 43
   (p) 56
   (q) 94
   (r) 123
   (s) 987
   (t) 333
   (u) 500
   (v) 111
   (w) 0.72
   (x) 0.135
   (y) 0.021
   (z) 3.72
   (aa) 1.99
   (bb) 15.33

3. Perform the following additions:
   (a) $3+7$
   (b) $8+9$
   (c) $6+3$
   (d) $\pi+7$
   (e) $11+38$
   (f) $19+27$
   (g) $54+81$
   (h) $72+99$
   (i) $124+38$
   (j) $561+84$
   (k) $387+44$
   (l) $988+33$
   (m) $821+218$
   (n) $431+314$
   (o) $772+277$
   (p) $181+432$
   (q) $8+19+27$
   (r) $81+37+63$
   (s) $921+37+822$
   (t) $3.71+83.20$
   (u) $161.5+37.2$
   (v) $22.1+83.7+15.6$
   (w) $0.711+0.012$
   (x) $0.135+0.911+0.223$
   (y) $0.00713+0.00037$

4. Perform the following subtractions:

(a) $11-4$

(b) $5-3$

(c) $8-8$

(d) $\pi-2$

(e) $93-87$

(f) $57-13$

(g) $76-23$

(h) $181-26$

(i) $837-378$

(j) $21.83-11.76$

(k) $1.75-0.77$

(l) $832.11-483.54$

(m) $0.0391-0.0139$

(n) $1.001-0.091$

(o) $99.99-9.99$

(p) $31.01-\pi$

5. Perform the following combined operations:

(a) $3+7-5$

(b) $11+9-5$

(c) $21-6-6$

(d) $1.33+7.11-8.32$

(e) $7.012-0.071+0.08$

(f) $118-56-11$

(g) $\pi-1.13+2.6$

(h) $78.30-\pi-31.63$

6. Perform the following multiplications:

(a) 3×4

(b) 7×8

(c) 9×7

(d) $(4)(6)$

(e) 11×36

(f) 65×21

(g) 18×83

(h) 58×55

(i) 116×74

(j) 832×27

(k) 6×234

(l) 697×769

(m) $(1.33)(2.91)$

(n) $(3.4)(7.32)$

(o) $(0.307)(1.28)$

(p) $(18.3)(21.4)$

(q) $(4)(\pi)$

(r) $(1.73)(\pi)$

(s) $(0.711)(\pi)$

(t) $(3)(7)(5)$

(u) $(7.4)(2.3)(5.5)$

(v) $(0.72)(0.27)(0.81)$

(w) $(63)(0.51)(\pi)$

(x) $(7.1)(3.4)(1.2)$

(y) $7(213.1)\pi$

(z) $(116)(0.0012)$

(aa) $5,800,000(0.92)$

(bb) $(0.00071)(0.0013)$

(cc) $(7)(16)(3)(8)$

(dd) $(21)(3)(5)(12)$

(ee) $(11.1)(6)(3.2)(0.7)$

(ff) $(6.3)(2.8)(1.7)(0.4)$

(gg) $(0.13)(0.71)(1.22)(0.9)$

(hh) $(3.1)(\pi)(0.87)(23)$

(ii) $(3)(5)(4)(6)(7)$

(jj) 116.3π

7. Perform the following divisions:

(a) $11/6$

(b) $3/9$

(c) $7/20$

(d) $16/43$

(e) $\pi/4$

(f) $16/\pi$

(g) $1.7/3.8$

(h) $7.33/1.85$

(i) $\pi\div6$

(j) $432\div6.7$

(k) $217\div11.6$

(l) $51.3\div84.2$

(m) $0.77\div1.73$

(n) $0.017\div0.011$

(o) $0.81\div1.08$

(p) $4.033\div17.2$

(q) $216.3\div841$

(r) $0.199\div432$

(s) $723\div0.08$

(t) $66\div\pi$

(u) $\dfrac{11}{31.2}$

(y) $\dfrac{0.0073}{0.162}$

(v) $\dfrac{29}{1.78}$

(z) $\dfrac{0.113}{1.003}$

(w) $\dfrac{0.34}{21.6}$

(aa) $\dfrac{91.3}{42.5}$

(x) $\dfrac{121.2}{231.5}$

(bb) $\dfrac{0.00091}{0.00012}$

8. Perform the following combined operations:

(a) $(3+7)(9-4)$

(g) $\dfrac{(723)(0.72)}{(421)(23.8)}$

(b) $(1.6-0.7)\pi$

(h) $\dfrac{(0.87)\pi}{(1.12)(0.06)}$

(c) $\dfrac{(21.3+1.77)}{\pi}$

(i) $\dfrac{(38.7-21.1)}{(4.11)(3.16)}$

(d) $\dfrac{3\pi}{4}$

(j) $\dfrac{(21.6-3.1)(6.72)}{(18.6+2.84)}$

(e) $\dfrac{(16.1-11.2)}{(91.3+1.22)}$

(k) $\dfrac{3\pi(2.88)}{(17.1-4.8)(0.22)}$

(f) $\dfrac{(21.2)(13)}{(7.1)(8.4)}$

(l) $\dfrac{(1.31)(6.72)(1.87)}{(9.66)(3.12)(1.16)}$

2.3 SOME APPLICATIONS

The purpose of this section is twofold. First of all, a number of different electric circuit quantities are introduced for the benefit of all who have not yet encountered these quantities elsewhere. Second, a variety of electric circuit problems are presented in order to provide the opportunity to use a hand calculator in obtaining solutions.

Electric Circuit Quantities

The letter I is the symbol used for a *current* in an electric circuit, and current is measured in units of *amperes*, abbreviated A. Thus the equation $I=5$A means, "The electric current (I) is equal to 5 amperes."

V is the symbol used to represent *potential difference* or *voltage* in an electric circuit, and voltage is measured in units of *volts*, abbreviated V. Thus the equation $V=23$ V means, "The voltage (V) is equal to 23 volts."

An electric *resistance* in a circuit is designated with the letter R, and resistance is measured in units of *ohms*, abbreviated Ω (the Greek letter omega). The equation $R=4700$ Ω means, "The resistance (R) is equal to 4700 ohms."

The letter P is used to designate electrical *power*, and power is measured in units of *watts*, abbreviated W. Thus, $P = 1.5$ W means, "The power (P) is equal to 1.5 watts."

It will always be assumed that the electric circuit quantities are sufficiently accurate to justify solutions containing 4 significant figures.

Electric Circuit Formulas

Ohm's Law is an expression that defines the proper relationship between current, voltage, and resistance. In equation form, Ohm's Law can be written as

OHM'S LAW

$$I = V/R \qquad\qquad (2.1)$$

where I is current in amperes

V is voltage in volts

R is resistance in ohms

Equation (2.1) is useful in calculating circuit current when voltage and resistance are known.

Two other forms of Ohm's Law are

$$V = IR \qquad\qquad (2.2a)$$

$$R = V/I \qquad\qquad (2.2b)$$

Equation (2.2a) is useful in determining voltage from known values of current and resistance, while Eq. (2.2b) is used to calculate resistance, using given values of voltage and current.

EXAMPLE 2.3   Determine the current in a 110 V lamp if the lamp resistance is known to be 121 Ω.

SOLUTION.   Equation (2.1) will be used to calculate the current using given values of voltage and resistance. Thus, given $V = 110$ V and $R = 121$ Ω, Eq. (2.1) yields

$$I = \frac{V}{R} = \frac{110}{121} = 0.9091 \text{ A}$$

Notice that the calculated value of current contains four significant figures.

An equation that can be used to calculate electric power is

POWER

$$P = VI \qquad (2.3)$$

where P = power in watts

V = voltage in volts

I = current in amperes

This expression is useful in calculating power, using given values of current and voltage.

EXAMPLE 2.4 Calculate the power dissipated in the lamp in Example 2.3.

SOLUTION. Equation (2.3) will be used to calculate power using the given voltage and the calculated current. Thus, given $V = 110$ V and calculated $I = 0.9091$ A, Eq. (2.3) yields

$$P = VI = 110 \times 0.9091 = 100 \text{ W}$$

Note the round off to four significant figures.

Exercises 2.3

The following problems can be solved by using Eq. (2.1) to calculate current, using given values of voltage and resistance. (Remember, use 4-digit accuracy.)

1. What is the current in a 6.3 V lamp if it has a resistance of 52.5 Ω?

2. Calculate the current in a light-emitting-diode (LED) if the voltage across the diode is 1.8 V and the diode resistance is equivalent to 100 Ω.

3. A 3300 Ω carbon-composition resistor is connected across the terminals of a 12 V battery. What is the current in the resistor?

4. The heating element of a 1500 W electric heater has an operating resistance of 9.14 Ω. What current is required for the heater if it is connected to a 117 V source?

5. What is the current required for a 1500 W heater designed for a 230 V source if its operating resistance is 35.3 Ω?

6. What must be the current in a 680 Ω resistor if a voltmeter connected across the resistor reads 3.88 V?

7. The voltage measured across a 220 Ω bias resistor in a transistor amplifier circuit is 0.36 V. What must be the current in the resistor?

8. The hot resistance of a certain soldering iron is 288 Ω. What is the current in the heating element if it is used on a 117 V line?

9. Each of the following resistors is connected across the terminals of a 28 V power supply. Calculate each resistor current.

   (a) 270 Ω   (b) 680 Ω   (c) 5600 Ω   (d) 33,000 Ω   (e) 680,000 Ω

10. A 1800 Ω resistor is connected across the terminals of a variable power supply. Calculate the current in the resistor for each of the following power-supply terminal voltages:

   (a) 0 V   (b) 10 V   (c) 45 V   (d) 85 V   (e) 250 V

   The following problems can be solved by using Eq. (2.2a) to calculate voltage, using given values of current and resistance.

11. What is the voltage across the terminals of a 250 Ω resistor if the resistor is conducting an electric current equal to 0.433 A?

12. An electric heater has an operating resistance of 9.14 Ω. If it conducts a current equal to 12.8 A, what must be its terminal voltage?

13. It is desired to adjust the terminal voltage of a power supply such that a current of 0.175 A will pass through a 750 Ω resistor connected across its terminals. Determine the proper supply voltage.

14. A 2200 Ω carbon-composition resistor is connected across the terminals of a power supply. Determine the proper supply voltage to provide a current through the resistor equal to:

   (a) 0.0011 A   (b) 0.0035 A   (c) 0.0412 A   (d) 0.0731 A
   (e) 0.375 A

15. Each of the following resistors is to be connected across the terminals of a power supply. Determine the necessary terminal voltage to provide a current of 0.111 A in each resistor.

   (a) 110 Ω   (b) 330 Ω   (c) 2700 Ω   (d) 5600 Ω   (e) 10,000 Ω

   The following problems can be solved by using Eq. (2.2b) to calculate the resistance, using given values of voltage and current.

16. What must be the operating resistance of a 6.3 V pilot lamp that requires a current of 0.095 A?

17. A 117 V soldering iron requires a current of 0.211 A. What must be the resistance of its heating element?

18. A temperature sensing element has 0.027 V across its terminals when conducting a current equal to 0.000048 A. What must be the resistance of the element?

19. What resistance must be connected across the terminals of a 12 V source to provide a current in the resistance equal to:
    (a) 0.001 A?   (b) 0.033 A?   (c) 0.415 A?   (d) 1.35 A?
    (e) 7.82 A?

20. What resistance must be connected across the terminals of a 28 V source to provide a source current equal to:
    (a) 0.007 A?   (b) 0.038 A?   (c) 0.095 A?   (d) 0.472 A?
    (e) 3.88 A?

The following problems can be solved by using Eq. (2.3) to calculate power, using given values of current and voltage.

21. Calculate the power dissipated in a 6.3 V lamp conducting a current equal to 0.12 A.

22. What is the power dissipated in a 3300 Ω resistor if the resistor current is 0.0036 A and the voltage across the resistor is 12 V?

23. The nameplate data on an electric wall heater states 117 V and 12.8 A. What is the power rating?

24. What is the power rating of an electric heater if it is designed to operate with 230 V at a current of 6.52 A?

25. What is the power rating of a 117 V soldering iron if it operates with a current of 0.406 A?

26. The cathode bias resistor in a vacuum tube circuit conducts a current equal to 0.012 A. What is the power dissipated in the resistor if its terminal voltage is 11.4 V?

27. Is it safe to use a 100 Ω resistor rated at 1/2 W maximum dissipation if the resistor voltage is 50 V and the resistor current is 0.5 A? Explain.

28. Calculate the power dissipated in each of the resistors in Problem 19.

29. Calculate the power dissipated in each of the resistors in Problem 20.

In general, the following problems will require the use of two of the formulas given in this section.

30. Calculate the power dissipated in a 330 Ω resistor having a voltage equal to 65 V across its terminals. (*Hint:* Use Eq. (2.1) to calculate the current, and then use Eq. (2.3) to calculate the power.)

31. Determine whether or not it is safe to connect a 4700 Ω resistor rated at 1/2 W across the terminals of a power supply having a terminal voltage of 50 V.

32. What is the power dissipated in a 560 Ω emitter bias resistor in a transistor amplifier if the resistor current is 0.043 A?

33. What must be the resistance of the heating element in a 47 1/2 W soldering iron designed to operate on 117 V?

34. What must be the hot resistance of a 100 W, 117 volt lamp?
35. What must be the resistance of the heating element in an electric heater rated at 230 V and 750 W?
36. Calculate the power dissipated in each of the resistors in Problem 9.
37. Calculate the power dissipated in each of the resistors in Problem 10.
38. Claculate the power dissipated in each of the resistors in Problem 14.
39. Calculate the power dissipated in each of the resistors in Problem 15.

UNIT 3
NEGATIVE NUMBERS

3.1   ADDITION AND SUBTRACTION

In a previous section the *number line* was used to introduce the concept of a negative number. Negative numbers play an important role in electric circuit problems, and it is essential to know how to deal with them. The number line can be used to develop the rules for adding or subtracting signed numbers.

The Number Line

Every number has a *sign* associated with it. A *positive* number is expressed by placing a + sign directly in front of the number, although it is assumed to be a positive number if there is no sign. For instance, +5 and 5 are both positive numbers. A *negative* number is expressed by placing a − sign directly in front of the number. Clearly there are only two types of signed numbers to deal with—positive numbers and negative numbers.

On the number line shown in Figure 3.1, a positive number is represented by moving in the positive direction (to the right) a distance equal to the magnitude of the given number. Likewise, a negative number is represented by moving in the negative direction (to the left) a distance equal to the magnitude of the negative number.

Addition

The addition of two numbers is accomplished by moving in the proper direction a distance equal to the magnitude of each number. Then the sum is simply the *net distance* from the reference point, and the sign of this sum is positive if the final position is to the right of reference, or negative if it is to the left.

There are really only two cases to consider: first, the addition of two numbers having the same signs; and second, the addition of two numbers having opposite signs.

First Case: Addition of two numbers having the same sign.

If both numbers are positive, addition is accomplished on the number line by moving *to the right* a distance equal to the first number and then an additional distance equal to the second number. The sum is the *total distance* moved, and it is clearly a positive number. For instance, the addition of +3 and +4, that is, $(3+4=7)$, is shown in Figure 3.2(a).

If both numbers are negative, addition is accomplished on the number line by moving *to the left* a distance equal to the first number and then an additional distance equal to the second number. The sum again is the *total distance* moved, and it is clearly a negative number. For instance, the addition of −3 and −4, that is, $[(-3)+(-4)=-7]$, is shown in Figure 3.2(b).

30

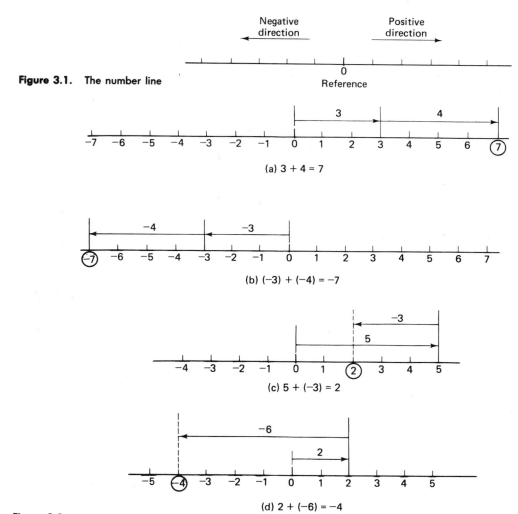

Figure 3.1. The number line

(a) 3 + 4 = 7

(b) (−3) + (−4) = −7

(c) 5 + (−3) = 2

(d) 2 + (−6) = −4

Figure 3.2

The basic rule for adding numbers having like signs can be stated as follows: *Find the sum of the magnitudes and attach the common sign.*

Second Case: Addition of two numbers having unlike signs.

The addition of two numbers having opposite signs is accomplished on the number line by moving to the right a distance equal to the positive number, and then moving to the left a distance equal to the negative number. The sum is then the total distance between the starting point and the finishing point. The sum is positive if the finishing point is to the right of reference and negative if it is to the left.

For instance, the sum of 5 and (-3) is shown in Figure 3.2(c) as $[5+(-3)=2]$. As a second example, the sum of 2 and (-6) is shown in Figure 3.2(d) as $[2+(-6)=-4]$.

The basic rule for adding two numbers having opposite signs can be stated as follows: *Find the difference of the magnitudes and attach the sign of the number having the largest magnitude.*

The number line is simply a graphical presentation used as an aid in understanding operations with signed numbers. It is neither practical nor advisable to use the number line in all calculations, and it is therefore desirable to learn the basic rules for adding signed numbers. The two rules are summarized here for convenience. When adding signed numbers, the sum obtained is referred to as the *algebraic sum*, and thus

---

ADDITION

To find the algebraic sum of two numbers:

(1) Like signs—Find the sum of the magnitudes and attach the common sign.

(2) Unlike signs—Find the difference of magnitudes and attach the sign of the number having the largest magnitude.

Notice that when finding the algebraic sum of two numbers, the order in which the numbers are written is immaterial. For example, $2+3=3+2=5$, or $5+(-3)=(-3)+5=2$.

Subtraction

Subtraction of signed numbers is accomplished by using the rules for addition. Here too there are only two cases to consider—the subtraction of a positive number, and the subtraction of a negative number.

Subtracting a positive number can be thought of as moving back a distance moved to the right on the number line. But this is equivalent to moving to the left the same distance on the number line. In other words, subtracting a positive number is the same as adding the negative of that number. For instance, subtracting $(+2)$ is the same as adding (-2); in equation form, $-(2)=+(-2)$. Therefore, to subtract a positive number, simply change its sign to a negative and then add, using the rules for addition. For example, to subtract 3 from 7, write

$$7-3=$$

Change the sign of the number being subtracted and add

$$7+(-3)=$$

Use the rules for addition

$$7+(-3)=4$$

As another example, to subtract 6 from 4, write

$$4-6=$$

Change the sign of the number being subtracted and add

$$4+(-6)=$$

Use the rules for addition

$$4+(-6)=-2$$

In a similar fashion, subtracting a negative number can be thought of as moving back a distance moved to the left on the number line. But this is equivalent to moving to the right the same distance on the number line. In other words, subtracting a negative number is the same as adding the positive of that number. For instance, subtracting (-3) is the same as adding $(+3)$; in equation form, $-(-3)=+(3)$. Thus, to subtract a negative number, simply change its sign to positive and add, using the rules for addition. For example, to subtract (-3) from 7, write

$$7-(-3)=$$

Change the sign of the number being subtracted and add

$$7+(3)=$$

Use the rules for addition

$$7+(3)=10$$

As another example, to subtract (-3) from (-7), write

$$(-7)-(-3)=$$

Change the sign of the number being subtracted and add

$$(-7)+(3)=$$

Use the rules for addition

$$(-7)+(3)=-4$$

To summarize, the subtraction of one signed number from another is accomplished according to the following rule:

SUBTRACTION

To find the difference between two signed numbers, change the sign of the number being subtracted and follow the rules for addition. The result is referred to as the *algebraic difference*.

Parentheses

Parentheses were used in some of the examples above to clarify intent. Clearly the other signs of grouping could also be used in a similar way, e.g., [ ] or { }. The following statements show how to *add* or how to *remove* parentheses and are useful when finding algebraic sums or differences.

"Plus a plus equals a plus."	$+(+3) = +3 = 3$
"Minus a minus equals a plus."	$-(-3) = +3 = 3$
"Plus a minus equals a minus."	$+(-3) = -3$
"Minus a plus equals a minus."	$-(+3) = -3$

Exercises 3.1

1. Find the algebraic sums of the following:

   (a) $21 + 7$

   (b) $(-13) + (-9)$

   (c) $(-1.7) + (-7.9)$

   (d) $15 + (-7)$

   (e) $(-15) + 7$
   (f) $(21.8) + (-85)$
   (g) $(-3.88) + (-1.72)$
   (h) $(0.07) + (-0.11)$

   (i) $\dfrac{0.0072}{0.0013}$

   (j) $\dfrac{1,462}{-3,917}$

   (k) $\dfrac{-16.7}{-11.8}$

   (l) $\dfrac{-2.86}{1.32}$

   (m) $3 + (-8) + (-5)$
   (n) $(-1.7) + (-7.2) + (8.6)$
   (o) $1.5 + (-3.1) + 2.2$
   (p) $0.0013 + (-0.0023) + (-0.0081)$

2. Find the algebraic differences of each of the following:

   (a) $15 - 6$
   (b) $7 - (-5)$
   (c) $(-5) - (3)$
   (d) $(-51) - (-62)$
   (e) $25 - 9$
   (f) $9 - 15$

   (g) $(-6) - (5)$
   (h) $(-21) - (-18)$
   (i) $(-63.7) - (81.2)$
   (j) $0.062 - (-0.8)$
   (k) $21 - (-9)$
   (l) $(-0.117) - (0.021)$

3. Perform the indicated operations:

   (a) $21 - (-9) + (6) + (-15)$
   (b) $-21.2 + (-6.8) - (-18.3)$
   (c) $1.12 - 3.61 + 3.07$

   (d) $-4.7 - 11.3 - 6.8 + 3.8$
   (e) $1821 - 828 - 241$
   (f) $-72.3 + 17.6 + 43.6 - 22.8$

4. If the outside air temperature increases from $-10°C$ to $+4.5°C$, what is the total temperature change?

5. What is the total change in temperature if an initial reading is recorded as $+21.0°C$ and the final reading is:

   (a) $+45.5°C$?   (b) $+4.5°C$?   (c) $-1.5°C$?   (d) $-11.0°C$?

6. In a certain electric circuit, the potential (voltage) at point A is measured at $+15.0$ V. What is the *potential difference* between points A and B if point B has a measured potential (voltage) of:

   (a) $+5.0$ V?   (e) -15.0 V?
   (b) $+85.5$ V?   (f) -45.0 V?
   (c) 0.0 V?   (g) $+15.0$ V?
   (d) -11.0 V?

3.2   SIGNED ALGEBRAIC TERMS

Remember that an algebraic term is an algebraic expression having no + or − signs (it may have × or ÷ signs). For example, x, IV, V^2/R and πr^2 are terms. But, $x+3$, $V^2/R-P$, and $3ax-21y$ are not terms; they are simply algebraic expressions containing more than one term.

Like algebraic terms, or *similar* terms, are those terms having exactly the same literal part, although their numerical coefficients may be different. For example, $3i^2R$ and $7i^2R$ are like terms, as are $7.3VI$ and $0.7VI$. However, $3IR$, $4V/R$, and $2VI$ are all *unlike* or *dissimilar* terms.

An algebraic expression containing only a single term is called a *monomial*, while an expression having two or more terms is called a *polynomial*. Polynomials can be simplified by adding algebraically all like terms. For example:

$$2xy-3z-7xy+5z=(2xy-7xy)+(5z-3z)=-5xy+2z$$

or

$$0.25IR-6V-3IR+13V=7V-2.75IR$$

Furthermore, polynomials can be added to or subtracted from polynomials by algebraically adding all like terms.

EXAMPLE 3.1. Perform the indicated operations:	SOLUTION. Simplify by collecting all like terms.
(a) $3ab-6ab+7a$	(a) $(3ab-6ab)+7a=-3ab+7a$
(b) $3x,\ -7y,\ 6x,\ 6y$ (add)	(b) $3x+6x-7y+6y=9x-y$
(c) $2a^2+3b^2+a^2+2b^2$	(c) $2a^2+a^2+3b^2+2b^2=3a^2+5b^2$
(d) $2a^2+3b^2-a^2-2b^2$	(d) $2a^2-a^2+3b^2-2b^2=a^2+b^2$
(e) $7I^2R+4V+3I^2R+3$	(e) $7I^2R+3I^2R+4V+3=10I^2R+4V+3$

It has been stated previously that signs of grouping, ( ), [ ], { }, are used to clarify operations on an expression. Parentheses can be inserted in a polynomial to group certain terms, or parentheses can be removed from an expression.

If parentheses are inserted around a number of terms in an expression, the sign of each term remains unchanged provided a plus sign $(+)$ precedes the added parentheses. For example,

$$3ab + 2a + 5b = 3ab + (2a + 5b)$$

or

$$3ab + 2a - 5b = 3ab + (2a - 5b)$$

However, in order to place a minus sign $(-)$ before the parentheses, it is necessary to change the sign of every term inside the parentheses. For example,

$$3ab + 2a + 5b = 3ab - (-2a - 5b)$$

or

$$3ab + 2a - 5b = 3ab - (-2a + 5b)$$

Notice that the *removal* of parentheses must conform to similar rules. For example,

$$3ab + (2a + 5b) = 3ab + 2a + 5b$$

or

$$3ab - (-2a + 5b) = 3ab + 2a - 5b$$

The following rules can be used when removing parentheses or any other signs of grouping:

PARENTHESES

To remove parentheses:
(1) If the parentheses are preceded by a plus sign $(+)$, leave the signs of the contained terms unchanged.
(2) If the parentheses are preceded by a minus sign $(-)$, change the sign of every enclosed term.

The simplification of algebraic expressions can usually be accomplished by removing parentheses and then combining like terms.

EXAMPLE 3.2. Simplify by removing parentheses and combining like terms:
(a) $(2a^2 + 3b^2) + (a^2 + 2b^2)$
(b) $(2a^2 + 3b^2) - (a^2 + 2b^2)$
(c) $(7I^2R + 4V) - (-3I^2R + 3)$

SOLUTION.
(a) $2a^2 + 3b^2 + a^2 + 2b^2 = 2a^2 + a^2 + 3b^2 + 2b^2$
$\qquad\qquad\qquad\qquad = 3a^2 + 5b^2$
(b) $2a^2 + 3b^2 - a^2 - 2b^2 = a^2 + b^2$
(c) $7I^2R + 4V + 3I^2R - 3 = 10I^2R + 4V - 3$

Exercises 3.2

1. Find the sums:
   (a) $5x$, $3x$, $-7x$
   (b) $3i^2R$, $-4i^2R$
   (c) $9iv$, $11iv$
   (d) $3\pi r^2$, $-4\pi r^2$
   (e) $2a^2b$, $4a^2b$, $-3a^2b$
   (f) $(-8p + vi)$, $(4p - 3vi)$

2. Find the differences:

   (a) $3hw$ less $5hw$ (b) $3hw$ less $-5hw$ (c) $7.8vi$ less $-5.1vi$
   (d) $-\pi r^2$ less $0.3\pi r^2$ (e) $(6v^2/R)+(3iv)$ less $4v^2/R$

3. Simplify by combining like terms:

   (a) $3IR+6V-7X-4IR-2X$ (h) $(4XY+2)+(-5-6XY)$
   (b) $31ab-\pi r^2+y-0.5\pi r^2-11ab$ (i) $1-(3x^2y+7a)$
   (c) $4.3x^2-0.2y-0.7x^2+1.2y$ (j) $0-(4V/R)+6$
   (d) $0.08W-2V^2/R+W/7-0.1V^2/R$ (k) $(x^2+3x-1)-(2x^2-2x-7)$
   (e) $abc-ac+ab-3ac-2ab-4abc$ (l) $(6I^2R+7P)+(-6I^2R-7P)$
   (f) $(0.4a^2-6b^2)-(4b^2-a^2)$ (m) $21a-2ab+3ac-7bc-3bc$
   (g) $(3v^2/R)-(2v^2/R+7P)$ (n) $0.2y^3-0.5y^2-0.1y^2+0.1y^3$

3.3   MULTIPLICATION

Signed Numbers

The multiplication of two signed numbers is accomplished by first determining the product of the two magnitudes, and then affixing the appropriate sign to this product. There are only four possible combinations of signs to consider:

   1. (plus)·(plus)

   2. (plus)·(minus)

   3. (minus)·(plus)

   4. (minus)·(minus)

However, multiplication can be considered a process of "repeated additions," and the appropriate sign is therefore not difficult to determine.

Consider first the product of two positive numbers. For instance, $(2)\cdot(4)$ simply means "two is to be added four times." On the number line, such a sum provides 8 for an answer, as illustrated in Figure 3.3(a). Clearly the sign to be attached to the product of two positive numbers is a plus sign. That is,

$$(plus)\cdot(plus)=(plus)$$

The second and third cases above are really one and the same, namely, the product of a positive number and a negative number. For instance, $(-2)\cdot(4)$ can be thought of as "minus two taken four times." On the number line shown in Figure 3.3(b), this product yields

$$(-2)\cdot(4)=-8$$

Alternately, using parentheses and brackets, this expression can be written as

$$(-2)\cdot(4)=-(2)\cdot(4)$$
$$=-[(2)\cdot(4)]$$

This can be thought of as "the negative of two added four times," and the result is

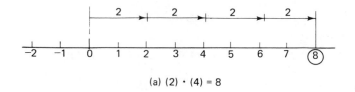

(a) (2) · (4) = 8

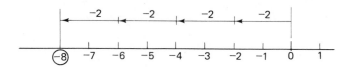

(b) (−2) · (4) = −8   **Figure 3.3**

seen to be -8; since $-[(2)\cdot(4)]=-[8]=-8$. Clearly, $(-2)\cdot(4)=(4)\cdot(-2)$, and thus the appropriate sign for the product of a positive and a negative number is negative. That is,

$$(\text{plus})\cdot(\text{minus})=(\text{minus})\cdot(\text{plus})=(\text{minus})$$

The fourth case above concerns the sign attached to the product of two negative numbers. For instance, $(-2)\cdot(-4)$. Using signs of grouping, this expression can be written as

$$(-2)\cdot(-4) = -[(2)\cdot(-4)] \qquad \text{(removing the first minus sign)}$$
$$= -\{-[(2)\cdot(4)]\} \qquad \text{(removing the second minus sign)}$$
$$= [(2)\cdot(4)] \qquad \text{(since "minus a minus equals plus")}$$
$$= (2)\cdot(4)$$
$$= 8$$

Clearly, the appropriate sign to attach to the product of two negative numbers is a plus sign. That is,

$$(\text{minus})\cdot(\text{minus})=(\text{plus})$$

These results can now be summarized in the form of the following rules for multiplication of two signed numbers:

MULTIPLICATION

To find the product of two signed numbers:
(1) Find the product of two magnitudes.
(2) The sign of this product is *positive* if the two numbers have *like* signs; it is *negative* if they have *unlike* signs.

EXAMPLE 3.3. Find the products:	SOLUTION.
(a) 3×7	(a) $3 \times 7 = 21$
(b) $(-8) \cdot (4)$	Like signs. Product is positive.
(c) $(8) \cdot (-4)$	(b) Magnitude $= 8 \cdot 4 = 32$
(d) $(-5) \cdot (-9)$	Unlike signs yield a negative product. Thus,
	$$(-8) \cdot (4) = -32$$
	(c) Same as (b). Thus,
	$$(8) \cdot (-4) = -32$$
	(d) Magnitude $= 5 \cdot 9 = 45$
	Like signs yield a positive product. Thus, $(-5) \cdot (-9) = 45$

Law of Exponents for Multiplication

The definition of an *exponent* was given in Section 1.3, and the concept was applied to both literal numbers such as h^2 or v^4, and whole numbers such as 2^3 or 10^5. In a term having an exponent (or power), the number being operated on by the exponent is called the *base*. That is, the base of the term h^2 is h, the base of v^4 is v, the base of 2^3 is 2, and so on.

The exponent is used to indicate how many times the base is multiplied by itself. For example,

$$h^2 = h \times h \qquad v^4 = v \times v \times v \times v \qquad 2^3 = 2 \times 2 \times 2 \qquad 10^5 = 10 \times 10 \times 10 \times 10 \times 10$$

Notice that if the exponent is 1, it is usually not written. For instance, $v^1 = v$, or $2^1 = 2$.

The product of two numbers having the same base can be determined by making use of exponents. To illustrate,

$$v^2 = v \cdot v \qquad \text{and} \qquad v^3 = v \cdot v \cdot v$$

Thus,
$$v^2 \cdot v^3 = v \cdot v \cdot v \cdot v \cdot v = v^5$$

The product can be determined by simply *adding* the components. The rule for multiplication using exponents is thus,

---

EXPONENTS

The product of two or more numbers having the *same base* can be found by *adding* algebraically their exponents. The base may have any value except zero.

In equation form, the *Law of Exponents for multiplication* is

$$a^m \cdot a^n = a^{m+n} \qquad \text{(provided } a \neq 0) \tag{3.1}$$

EXAMPLE 3.4. Find the products:	SOLUTION. Using the Law of Exponents, the products can be determined by simply adding the exponents.
(a) b^4 and b^3	(a) $(b^4)(b^3) = b^{4+3} = b^7$
(b) v^2 and $-v^5$	(b) $(v^2)(-v^5) = -v^{2+5} = -v^7$
(c) 2^6 and 2^4	(c) $(2^6)(2^4) = 2^{6+4} = 2^{10}$
(d) 3^2 and 3^{-5} and 3^6	(d) $(3^2)(3^{-5})(3^6) = 3^{2-5+6} = 3^3$

Algebraic Terms

Determining the product of two algebraic terms includes exactly the same steps used for signed numbers, but must also include finding the product of the literal numbers. The Law of Exponents for multiplication is used to find the product of the literals. The final product then consists of a coefficient, the literal numbers, and the appropriate sign. The multiplication of two algebraic terms is therefore accomplished by determining the three components in the final product.

MULTIPLICATION

The product of two algebraic terms includes:
(1) A coefficient equal to the product of the coefficients of two given terms.
(2) The literal numbers, found by taking the product of the literals in each of the two given terms. Use the Law of Exponents. It is customary to write the literals in algebraic order.
(3) The sign is *positive* if the two given terms have *like signs*; it is *negative* if they have *unlike signs*.

EXAMPLE 3.5. Find the product of $(-4ab)$ and $(-5ax)$.	SOLUTION.
	$$(-4ab)\cdot(-5ax) = 20a^2bx$$
	Note: (1) The magnitude of the coefficient is $(4)(5) = 20$.
	(2) The literals are $abax = a\cdot a\cdot b\cdot x = a^2bx$.
	(3) The sign is positive (like signs). Remember,
	$$a\cdot a = a^1\cdot a^1 = a^2$$

Exercises 3.3

1. Find the products of each of the following:

   (a) 4 and 7
   (b) -5 and 9
   (c) -6 and -3
   (d) 15 and -11
   (e) 0.1 and -0.7
   (f) -0.50 and -0.67
   (g) 3.2 and -7.5
   (h) -0.21 and 0.12
   (i) 7.44 and 4.77
   (j) -11.1 and 0.86
   (k) 7120 and -63
   (l) -0.002 and 5
   (m) a^2 and a^4
   (n) x^4 and $-x$

   (o) 10^4 and 10^7
   (p) 2^3, 2^4, and 2^5
   (q) $-x$ and x^3
   (r) $-I^2$ and R
   (s) $-v$ and $-i$
   (t) V^2 and $-V^2$
   (u) $-2R$ and $0.33R$
   (v) -9 and z
   (w) $-3x$ and $-7y$
   (x) $2a^2$ and $-7b$
   (y) I and I and R
   (z) z^a and z^{-b}
   (aa) $-Q^2$, Q^{-5}, and Q^6
   (bb) a^m and a^3

2. Use Ohm's Law, $V = IR$, to determine V if:

   (a) $I = 7$, $R = 28$ (b) $I = -0.065$, $R = 1500$ (c) $I = -0.018$, $R = 4700$

3. Use $P = VI$ to calculate power P if:

   (a) $V = 10$ V, $I = 4$ A
   (b) $V = 65$ V, $I = -0.18$ A
   (c) $V = -28$ V, $I = 0.075$ A
   (d) $V = -18$ V, $I = -0.77$ A

4. Given $I_1 = -0.172$ A, $I_2 = 1.38$ A, $V_1 = 10$ V, $V_2 = -15$ V, $R_1 = 470$ Ω, $R_2 = 680$ Ω, and $R_3 = 2200$ Ω, evaluate the following:

   (a) $I_1 V_1$
   (b) $I_1 V_2$
   (c) $I_2 V_1$
   (d) $I_2 V_2$
   (e) $I_1 R_1$
   (f) $I_1 R_2$
   (g) $I_1 R_3$
   (h) $I_2 R_1$
   (i) $I_2 R_2$
   (j) $I_2 R_3$

   (k) $I_1^2 R_1$
   (l) $I_2^2 R_2$
   (m) $I_1^2 R_3$
   (n) $I_1^2 R_2$
   (o) $I_2^2 R_1$
   (p) $I_2^2 R_3$
   (q) $R_1 R_2$
   (r) $R_1 R_3$
   (s) $R_2 R_3$
   (t) $R_1 R_2 R_3$

5. Find the products of the following:

   (a) $3v \cdot 5$
   (b) $(-3x)(-5x)$
   (c) $(5m)(-7m^2 n^3)$
   (d) $(-3a^2)(-4b^3)$
   (e) $8x^a \cdot 3x^b y$

   (f) $(13c^2 d)(-5acd)$
   (g) $(-3P)(-3P)(-3P)$
   (h) $x^4(-9xyz)$
   (i) $(-6i^2 R)(-2iv)$
   (j) $3e^3 t \cdot 1/3tx$

(k) $9xy(-5yt)t^2$

(l) $3x^a y^b \cdot 4xy^c$

(m) $(0.7zw^{-2})(-2w)$

(n) $a^2 \cdot 3a^3 \cdot (-7a^{-1})$

(o) $2z^{-3}(-5z^4 Y)(-Y^{-2})$

(p) $ab^n \cdot b^m c^p \cdot a^x c$

(q) $2 \times 10^{-7} \times 4y \times 10^{-3}$

(r) $(32gt^2)(-1.8v)$

(s) $(-10^7)(-10^{-6})(7x^3)$

(t) $(-7 \times 10^3)(-7 \times 10^{-3})(-7 \times 10^{-3})$

3.4   DIVISION

Signed Numbers

Since the division of two numbers is simply the *inverse* of multiplication, the rules for finding the quotient of two signed numbers are similar to those used for multiplication. Specifically, the magnitudes of the two given numbers are used to determine the magnitude of the quotient, and the proper sign is then determined exactly as in multiplication. Thus,

---

DIVISION

To find the quotient of two signed numbers:
(1) Find the quotient of the two magnitudes.
(2) The sign of this quotient is *positive* if the numbers have *like* signs; it is *negative* if they have *unlike* signs.

EXAMPLE 3.6.   Find the following quotients:

(a) $8 \div 6$

(b) $(-9) \div (3)$

(c) $(15)/(-5)$

(d) $(-64)/(-4)$

SOLUTION.

(a) Magnitude: $8/6 = 1.333\ldots$
     The sign is positive (like signs). Thus,
$$8 \div 6 = 1.333\ldots$$

(b) Magnitude: $9/3 = 3$
     Sign is negative (unlike signs). Thus,
$$(-9) \div (3) = -3$$

(c) Magnitude: $15/3 = 5$
     Sign is negative (unlike signs). Thus,
$$\frac{15}{-3} = -5$$

(d) Magnitude: $64/4 = 16$
     Sign is positive (like signs). Thus,
$$\frac{-64}{-4} = 16$$

Law of Exponents for Division

Recall that the exponent is used to indicate how many times the base is multiplied by itself. That is,

$$v^1 = v, \qquad v^2 = v \cdot v, \qquad \text{and} \qquad v^5 = v \cdot v \cdot v \cdot v \cdot v$$

The division of any number by itself is simply unity (1). That is, $v/v = 1$, or $x/x = 1$.

Now, examine the division of v^5 by v^2; the result is

$$\frac{v^5}{v^2} = \frac{v \cdot v \cdot v \cdot v \cdot v}{v \cdot v} = v \cdot v \cdot v \left(\frac{v}{v} \right) \left(\frac{v}{v} \right) = v \cdot v \cdot v (1)(1) = v \cdot v \cdot v = v^3$$

This could be rewritten as

$$\frac{v^5}{v^2} = \frac{v \cdot v \cdot v \cdot \cancel{v} \cdot \cancel{v}}{\cancel{v} \cdot \cancel{v}} = v \cdot v \cdot v = v^3$$

Notice that two of the v's in the numerator can be thought of as canceling two of the denominator v's. In any case, the correct result can be obtained by *subtracting* the denominator exponent from the numerator exponent. The rule for division using exponents is thus,

---

EXPONENTS DIVISION

The quotient of two numbers having the *same base* can be found by subtracting the denominator exponent from the numerator exponent. The base may have any value except zero.

In equation form, the *Law of Exponents for division* is

$$a^m \div a^n = a^{m-n} \qquad \text{(provided } a \neq 0) \tag{3.2}$$

EXAMPLE 3.7.   Divide the first term by the second:

(a) b^4 and b^3
(b) v^2 and $-v^5$
(c) 2^6 and 2^4
(d) x^7 and x^{-5}
(e) $-i^{-2}$ and $-i^3$

SOLUTION.
(a) $b^4 \div b^3 = b^{4-3} = b$
(b) $(v^2)/(-v^5) = -v^{2-5} = -v^{-3}$
(c) $2^6/2^4 = 2^{6-4} = 2^2$
(d) $x^7/x^{-5} = x^{7+5} = x^{12}$
(e) $(-i^{-2})/(-i^3) = i^{-2-3} = i^{-5}$

An examination of Eq. (3.2) will show how it is possible to have the value of the exponent equal to zero. That is, $a^{m-n} = a^0$ if $m = n$. What is the meaning of a^0?

Notice that if $m = n$, the numerator term is *identical* to the denominator term, and the value of the expression must be equal to one. That is, if $m = n$, then,

$$a^m \div a^n = \frac{a^m}{a^m} = a^{m-m} = a^0 = 1$$

For instance, $2^3 \div 2^3 = 2^{3-3} = 2^0 = 1$, since

$$2^3 \div 2^3 = \frac{2 \cdot 2 \cdot 2}{2 \cdot 2 \cdot 2} = \left(\frac{2}{2}\right)\left(\frac{2}{2}\right)\left(\frac{2}{2}\right) = 1 \cdot 1 \cdot 1 = 1$$

From this it should be clear that *any base having zero as its exponent has a numerical value of one.** For example, a^0, z^0, N^0, 10^0, 2^0, and $(3 - Y)^0$ are all equal to 1.

If $m = 0$ in Eq. (3.2), then

$$a^m \div a^n = a^0 \div a^n = \frac{1}{a^n} \qquad \text{or} \qquad a^m \div a^n = a^{0-n} = a^{-n}$$

Thus,

$$\frac{1}{a^n} = a^{-n} \tag{3.3}$$

This important relationship shows that a term can be moved from the denominator of a fraction up to the numerator by simply changing the sign of the exponent. A term can also be moved from the numerator down to the denominator if the sign of the exponent is changed.

| EXAMPLE 3.8. Write as a single term by moving the denominator into the numerator: (a) x^3/y^2 (b) $2^3/2^{-2}$ | SOLUTION. (a) Change the sign of the exponent of y and write $$x^3/y^2 = x^3 y^{-2}$$ (b) $2^3/2^{-2} = 2^3 \cdot 2^{-(-2)} = 2^3 \cdot 2^2 = 2^5$ |

Algebraic Terms

Determining the quotient of two algebraic terms includes exactly the same steps used for signed numbers, but must also include finding the quotient of the literal numbers. The Law of Exponents for division is used to find the quotient of the literals. The final quotient then consists of a coefficient, the literal number, and the appropriate sign. The division of two algebraic terms is therefore accomplished by determining the three components in the final quotient. Thus,

---

*Using zero as a base is of course excluded.

DIVISION

The quotient of two algebraic terms includes:
(1) A coefficient equal to the quotient of the coefficients of the two given terms.
(2) The literal numbers, found by taking the quotient of the literals in each of the two given terms. Use the Law of Exponents. It is customary to write the literals in alphabetic order.
(3) The sign is *positive* if the two given terms have *like* signs; it is *negative* if they have *unlike* signs.

EXAMPLE 3.9. Find the quotient of $7a^2b^3$ divided by $(-3ab)$

SOLUTION.

$$\frac{7a^2b^3}{-3ab} = -\left(\frac{7}{3}\right)\left(\frac{a^2b^3}{ab}\right)$$

$$= -2.3a^{2-1}b^{3-1}$$

$$= -2.3ab^2$$

Note: (1) The magnitude is $7/3 = 2.3$.
(2) The literals are

$$\frac{a^2b^3}{ab} = a^2a^{-1}b^3b^{-1}$$

$$= a^{2-1}b^{3-1} = ab^2$$

(3) The sign is negative (unlike signs).

Exercises 3.4

1. Find the quotients of the following:
    (a) $28/6$
    (b) $87/(-5)$
    (c) $(-95)/(16)$
    (d) $(-11)/(20)$
    (e) $(-11.5) \div (-6.4)$
    (f) $(-1.72)/(-7.21)$
    (g) $(-16.3)/(7.88)$
    (h) $(0.0788)/(-0.101)$
    (i) $(1)/(-8.33)$
    (j) $(-1)/(0.714)$

2. Use the Law of Exponents for division to simplify:
   (a) v^2/v (g) $2^3/2^2$
   (b) a^7/a^4 (h) $(-x)/(x^3)$
   (c) i/i^2 (i) $1/v^2$
   (d) i^2R/i (j) $(i^2) \div (-i^2)$
   (e) $(x^3)/(-x^2)$ (k) $(-10^3)/(10^3)$
   (f) $(10^4)/(-10^7)$ (l) $(-V^3)/(-V^2)$

3. Use Ohm's Law, $I = V/R$, to determine I if:
   (a) $V=7$, $R=56$ (b) $V=-10$, $R=680$ (c) $V=38$, $R=2700$

4. Use $V = P/I$ to determine V if:
   (a) $P=8.6$, $I=1.12$ (c) $P=0.013$, $I=0.0087$
   (b) $P=1.15$, $I=-0.872$ (d) $P=3.61$, $I=-0.711$

5. Given $I_1 = -1.18$ A, $I_2 = 0.722$ A, $V_1 = 2.88$ V, $V_2 = -17.6$ V, $R_1 = 680$ Ω, and $R_2 = 2700$ Ω, evaluate the following:
   (a) V_1/I_1 (g) V_2/R_1
   (b) V_1/I_2 (h) V_2/R_2
   (c) V_2/I_1 (i) V_1^2/R_1
   (d) V_2/I_2 (j) V_1^2/R_2
   (e) V_1/R_1 (k) V_2^2/R_1
   (f) V_1/R_2 (l) V_2^2/R_2

6. Simplify the following:
   (a) $3x^2 \div 5x^2$ (p) $(-0.67)^{-3}$
   (b) $(-15a) \div (6a)$ (q) $(-2)^{-6}$
   (c) $32v \div 8v^2$ (r) $(3y+7)^0$
   (d) $17y^4/(-5y^{-3})$ (s) $3^2 - 2^{-3}$
   (e) $(-0.5a) \div (0.75ab)$ (t) $(1.5)^{-1}$
   (f) $2^6/2^{-5}$ (u) $2^3/3^{-2}$
   (g) $(-3^2)/(3^5)$ (v) $z^6 \div 3z^2$
   (h) $(1)/(4a)^2$ (w) $4i^3 \div i^2$
   (i) $3x^0$ (x) $(-15v^2)/(-6v^3)$
   (j) $(-5)(4)^0$ (y) x^5/y^2
   (k) 4^{-3} (z) $3v^2/5i^2$
   (l) $(0.25)^{-2}$ (aa) $(-4y^4)/(-3y^{-2})$
   (m) $2^0 + 2^6$ (bb) $(2v \times 10^6)/(10^{-2})$
   (n) $(-5)^{-2}$ (cc) $(-10^7)/(3 \times 10^{-2})$
   (o) $4^7(4^{-7})$ (dd) $2/x^{-3}$

7. Divide the following:
   (a) $25x^3$ by $5x^2$ (f) $-e^2i^2$ by $4e^2i$
   (b) ab by $3b^2$ (g) $64a^2bx^3$ by $16a^2bx$
   (c) $-3v^2i$ by $-2iv$ (h) $14x^2y^2z^2$ by $7x^2y^2z^2$
   (d) $2.7z$ by $-9xgz$ (i) 1 by $5x^2y$
   (e) $21az$ by $10a^2z^3$

8. The area of a circle is given by $A = \pi r^2$, where r is the radius. If the radius of a second circle has a radius twice that of the first circle, how many times greater is its area?

9. The power P dissipated in a resistor R is given by $P = I^2 R$, where I is the electric current in the resistor. If the current I is doubled, by what factor does the power increase?

10. If the current in Problem 9 above is halved, by what factor does the power decrease?

11. The power P dissipated in a resistance R is given by $P = V^2/R$, where V is the voltage across the resistor terminals. What happens to the power if the voltage is: (a) doubled? (b) halved?

UNIT 4
NUMBER REPRESENTATION

4.1 POWERS OF TEN

Representation

At this point it should be well understood that an *exponent* specifies the number of times a *base* is to be multiplied by itself. For instance,

$$2^3 = (2)(2)(2) \qquad i^2 = (i)(i) \qquad 10^4 = (10)(10)(10)(10)$$

Since the decimal number system is based on *ten* digits, it will be most valuable to have a clear understanding of "the power of ten"—that is, the values and uses of 10^n.

It is easy to evaluate 10^n for different values of n. Simply assign different values for n, calculate 10^n, and tabulate the results.

(a) Let $n = 0$, then $10^n = 10^0 = 1$ (Remember any number with a zero exponent is equal to 1.)

(b) Let $n = 1$, then $10^n = 10^1 = 10$

(c) Let $n = 2$, then $10^n = 10^2 = (10)(10) = 100$

(d) Let $n = 3$, then $10^n = 10^3 = (10)(10)(10) = 1000$ and so on.

Now, for completeness, negative values of n must be included. Thus,

(e) Let $n = -1$, then $10^n = 10^{-1} = \dfrac{1}{10} = 0.1$

(f) Let $n = -2$, then $10^n = 10^{-2} = \dfrac{1}{10^2} = \dfrac{1}{(10)(10)} = \dfrac{1}{100} = 0.01$

(g) Let $n = -3$, then $10^n = 10^{-3} = \dfrac{1}{10^3} = \dfrac{1}{1000} = 0.001$ and so on.

The values of 10^n for a number of different values of n are tabulated in Table 4.1.

The first entry states "one hundred thousand" is equal to *ten to the fifth power* (10^5). Similarly, "ten thousand" is equal to *ten to the fourth power* (10^4), and "one thousand" is equal to *ten to the third power* (10^3), and so on. *Ten to the first power* (10^1) is simply 10.

For numbers less than one, the exponent is negative. "One-tenth" is equal to *ten to the minus one* (10^{-1}), "one one-hundredth" is equal to *ten to the minus two* (10^{-2}), and so on.

Note carefully that the value of n in each entry in Table 4.1 is exactly equal to the number of places the decimal point must be moved in order to make the

Table 4.1
Powers of Ten

Quantity	10^n	n
100,000	10^5	5
10,000	10^4	4
1000	10^3	3
100	10^2	2
10	10^1	1
1	10^0	0
0.1	10^{-1}	-1
0.01	10^{-2}	-2
0.001	10^{-3}	-3
0.0001	10^{-4}	-4
0.00001	10^{-5}	-5
		$\vdots$

number in the left-hand column exactly equal to one (1). The exponent is *positive* if the number is greater than 1, and *negative* for a number less than 1.

Powers of ten can be used as a "mathematical shorthand" for expressing the location of a decimal point in a number. For instance, it is much easier to write 10^7 instead of 10,000,000. Count the required number of decimal places moved.

$$1\,0\,0\,0\,0\,0\,0\,0$$

Similarly, it is much easier to write 10^{-12} in place of 0.000 000 000 001. Again, count the required number of 12 decimal places moved. Note that the exponent is *negative* for numbers less than 1.

$$0\,0\,0\,0\,0\,0\,0\,0\,0\,0\,0\,1$$

EXAMPLE 4.1.   Write the power of ten for each of the following numbers:
(a) 1,000,000
(b) 0.000 001

SOLUTION.
(a) Count the number of decimal places moved to make the given number equal 1.

$$1\,0\,0\,0\,0\,0\,0$$

(6 places)

The power is positive for a number greater than 1. Thus,

$$1{,}000{,}000 = 10^6$$

(b) Count the number of decimal places moved. The power is negative for a number less than 1. Thus,

$$0.000001 = 10^{-6}$$

(6 places)

Prefixes

Powers of ten are so widely used in science and engineering that a number of special symbols called *prefixes* have been developed. The following table gives some common prefixes along with their abbreviations and corresponding powers of ten. Note carefully the difference between M and m!

PREFIX	ABBREVIATION	POWER OF TEN
tera	T	10^{12}
giga	G	10^{9}
mega	M	10^{6}
kilo	k	10^{3}
hecto	h	10^{2}
deka	da	10^{1}
deci	d	10^{-1}
centi	c	10^{-2}
milli	m	10^{-3}
micro	μ	10^{-6}
nano	n	10^{-9}
pico	p	10^{-12}

The usage of these prefixes is quite straightforward as seen in the following examples:

1000 volts $= 10^3$ volts $= 1$ kilovolt $= 1$ kV

1000 ohms $= 10^3$ ohms $= 1$ kilohm $= 1$ kΩ

0.001 amperes $= 10^{-3}$ amperes $= 1$ milliampere $= 1$ mA

1,000,000 watts $= 10^6$ watts $= 1$ megawatt $= 1$ MW

Exercises 4.1

1. Write the following numbers as powers of ten:

   (a) 100   (b) 1   (c) 100,000   (d) 10,000,000
   (e) 1,000,000,000 (f) 0.1   (g) 0.001   (h) 0.000 0001
   (i)   0.000 000 001 (j) 0.000 000 000 001

2. Write the following numbers in decimal point form:

   (a) 10^0 (b) 10^1 (c) 10^3 (d) 10^5 (e) 10^8
   (f) 10^{-1} (g) 10^{-3} (h) 10^{-5} (i) 10^{-8} (j) 10^{-12}

3. What is the prefix and the prefix abbreviation for the following powers of ten?

   (a) 10^1 (b) 10^3 (c) 10^{-3} (d) 10^6
   (e) 10^{-6} (f) 10^{-9} (g) 10^{-12} (h) 10^9

4. Use prefix abbreviations and unit abbreviations to express the following:

   (a) One millivolt   (e) One thousand watts
   (b) One thousand volts   (f) One million ohms
   (c) One thousand ohms   (g) One million watts
   (d) One one-thousandth of a watt

5. Write in words, and write the correct decimal number for:

   (a) 1 mV   (b) 1 mA   (c) 1 mW   (d) 1 kΩ
   (e) 1 MΩ   (f) 1 kV   (g) 1 MV   (h) 1 kW

4.2 SCIENTIFIC NOTATION

Numbers such as 5,600,000 or 0.000 0013 are often encountered in electrical/electronics problems. Such numbers are tedious and cumbersome, and a technique for writing them in a much more concise form has been developed. The technique is called *scientific notation*, and it is really quite simple.

Consider the number 2,510,000. There are essentially two types of digits in this number. The digits 2, 5, and 1 express the significant data contained in the number, while the four trailing zeros serve to locate the decimal point. Similarly, the significant data in the number 0.000 0036 are contained in the digits 3 and 6, while the five zeros between the decimal point and the 3 are used to indicate the proper decimal point position. All the necessary information in any number can be expressed in the following way:

(a) Use *significant figures* to express the significant data.

(b) Use *powers of ten* to locate the decimal point.

Notice that any decimal number can be expressed as the product of significant digits and an appropriate power of ten. For instance,

$$2000 = (2)(1000) = (2)(10^3) = 2 \times 10^3$$

The significant figure is (2), and the appropriate power of ten is (1000) or 10^3.

As another example,

$$0.008 = (8)(10^{-3}) = 8 \times 10^{-3}$$

where (8) is the significant figure and (10^{-3}) is the power of ten.

Now, here is how to write a number in *scientific notation*:

SCIENTIFIC NOTATION

To write a number in *scientific notation*:
(1) Write the significant figures as a mixed number having a value between 1 and 10.
(2) Locate the decimal point using the appropriate power of ten.
(3) The desired result is the product of the significant figures and the appropriate power of ten.

EXAMPLE 4.2   Write the following numbers in scientific notation:
(a) 2,510,000
(b) 0.00036

SOLUTION.
(a) The significant figures are 2, 5, and 1; the mixed number between 1 and 10 is thus 2.51. The decimal point must be moved as follows:

$$2. \underset{\text{(6 places)}}{510\ 000}$$

Positive since the original number is greater than 1. The appropriate power of ten is thus 10^6. Therefore,

$$2,510,000 = 2.51 \times 10^6$$

(b) The significant figures are 3 and 6; the proper mixed number is 3.6. The decimal point is moved as follows:

$$\underset{\text{(4 places)}}{0.00036}$$

Negative since the original number is less than 1. The appropriate power of ten is 10^{-4}. Thus,

$$0.00036 = 3.6 \times 10^{-4}$$

A certain amount of care must be taken when determining the significant figures in a given number. For instance, only the 3 and the 8 are significant in the number 0.0$\underline{38}$. But 3, 8, and the final 0 are all three significant in the number 0.0$\underline{380}$. The final zero is written to indicate a known value, and is not written to help locate the decimal point.

EXAMPLE 4.3. Write the following numbers in scientific notation: (a) 21.0 (b) 0.0071 (c) 0.00710	SOLUTION. (a) The significant figures are 2, 1, and 0. The final zero is written to establish a known value. Thus, $$21.0 = 2.10 \times 10$$ (1 place; positive) (b) The significant figures are 7 and 1. Thus, $$0.0071 = 7.1 \times 10^{-3}$$ (3 places: negative) (c) The significant figures are 7, 1, and 0. Note, the final zero expresses a known value. Thus, $$0.00710 = 7.10 \times 10^{-3}$$ (3 places; negative)

Exercises 4.2

1. Write the *significant figures* in the following numbers:
   (a) 4610   (b) 85.6   (c) 21.070   (d) 0.0107   (e) 0.01070

2. Write the following numbers in power of ten form:
   (a) 1,000,000   (b) 100   (c) 0.0001   (d) 0.000 001
   (e) 10,000,000

3. Write the following numbers in scientific notation:
   (a) 438   (b) 9800   (c) 0.1006   (d) 27.10   (e) 0.000 082
   (f) 7,610,000   (g) 0.0002   (h) 6.72

4. Write the following numbers in decimal form:
   (a) 7.31×10^5 (b) 9.60×10^{-2} (c) 6.183×10^2 (d) 3.101×10^9

5. Write the following numbers in scientific notation:
   (a) 77×10^4 (b) 38.6×10^{-5} (c) 381×10^2 (d) 11.7×10^{-5}

4.3   SOME APPLICATIONS

Using scientific notation as a mathematical shorthand to express otherwise cumbersome numbers is a very useful technique. It is equally valuable to learn how to manipulate numbers written in scientific notation. Specifically, it is necessary to learn how to perform the four basic operations $(+, -, \times, \div)$.

Addition and Subtraction

In order to find a *sum* or a *difference* directly using numbers written in scientific notation, it is first necessary to express them with the same power of ten. For instance, 2.1×10^3 and 3.7×10^3 both have the same power of ten (10^3). These two expressions can be added directly by simply adding their significant figures and multiplying this sum by the common power of ten. That is,

$$2.1 \times 10^3 + 3.7 \times 10^3 = (2.1 + 3.7) \times 10^3 = 5.8 \times 10^3$$

It is also possible to find the difference between these two expressions directly since they have the same power of ten. For instance,

$$3.7 \times 10^3 - 2.1 \times 10^3 = (3.7 - 2.1) \times 10^3 = 1.6 \times 10^3$$

Note that two expressions such as 2×10^3 and 3×10^4 cannot be added (or subtracted) *directly*—because they do not have the same power of ten! They could of course be rewritten with the same power of ten and then added (or subtracted). For instance, choose 10^3 as the desired power of ten and find the sum of 2×10^3 and 3×10^4.

Leave unchanged $\qquad\qquad\qquad 2 \times 10^3$

Rewrite as $\qquad\qquad\qquad\qquad 3 \times 10^4 = 30 \times 10^3$

Now add $\qquad\qquad 2 \times 10^3 + 30 \times 10^3 = (2 + 30) \times 10^3 = 32 \times 10^3$

Of course, any other power of ten could also be used. For instance, choose 10^4 as the desired power of ten and find the sum.

Leave unchanged $\qquad\qquad\qquad 3 \times 10^4$

Rewrite as $\qquad\qquad\qquad\qquad 2 \times 10^3 = 0.2 \times 10^4$

Now add $\qquad\qquad 3 \times 10^4 + 0.2 \times 10^4 = (3 + 0.2) \times 10^4 = 3.2 \times 10^4$

You might like to try the same problem using 10^2 as the desired power of ten.

To summarize, two expressions can be added or subtracted directly in scientific notation only if they are expressed with the same power of ten.

To find the *sum or difference* of two mathematical expressions written in scientific notation:
(1) Express the numbers with a common power of ten.
(2) Find the sum (or difference) of the significant figures and multiply by the common power of ten.

EXAMPLE 4.4.   Perform the following:
(a) $2 \times 10^2 + 5 \times 10^2$
(b) $7.2 \times 10^{-3} - 5.1 \times 10^{-3}$
(c) $9.10 \times 10^4 + 3.3 \times 10^3$

SOLUTION.
(a) Both expressions have the same power of ten. Thus, add the significant figures and multiply by the common power of ten.

$$2 \times 10^2 + 5 \times 10^2 = (2+5) \times 10^2 = 7 \times 10^2$$

(b) The power of ten is the same. Thus, find the difference directly.

$$7.2 \times 10^{-3} - 5.1 \times 10^{-3} = (7.2 - 5.1) \times 10^{-3}$$
$$= 2.1 \times 10^{-3}$$

(c) Choose 10^3 as the desired power of ten. Rewrite $9.10 \times 10^4 = 91.0 \times 10^3$ Retain unchanged 3.3×10^3 Add directly.

$$91.0 \times 10^3 + 3.3 \times 10^3 = (91.0 + 3.3) \times 10^3$$
$$= 94.3 \times 10^3$$

Multiplication

Finding the product of two or more numbers written in scientific notation is straightforward and is best illustrated with an example. Consider the multiplication of 2×10^3 and 4×10^5.

Write $\qquad 2 \times 10^3 \times 4 \times 10^5$

Since the *order* of multiplication is unimportant

$$(a \times b = b \times a)$$

Rewrite as $\qquad 2 \times 10^3 \times 4 \times 10^5 = 2 \times 3 \times 10^3 \times 10^5$

Now, group as $\qquad = (2 \times 3) \times (10^3 \times 10^5)$

And simplify as $\qquad = (6) \times (10^{3+5})$

$$= 6 \times 10^8$$

Now it is easy to see that the product is found by multiplying the significant figures, and combining the power of ten according to the Law of Exponents. Thus,

---

To find the *product* of two numbers written in scientific notation:
(1) Find the product of the significant figures.
(2) Multiply this product by the power of ten obtained by combining the two given powers of ten.

EXAMPLE 4.5. Find the products:
(a) 2×10^2 and 4×10^3
(b) 3×10^5 and 2×10^{-3}

SOLUTION.

(a) $2 \times 10^2 \times 4 \times 10^3 =$

Rewrite and group $= (2 \times 4) \times (10^2 \times 10^3)$

Simplify $= (8) \times (10^{2+3})$

$= 8 \times 10^5$

(b) $3 \times 10^5 \times 2 \times 10^{-3} =$

Rewrite and group $= (3 \times 2) \times (10^5 \times 10^{-3})$

Simplify $= (6) \times (10^{5-3})$

$= 6 \times 10^2$

Finding the product of more than two expressions is quite easily extended, as shown by the following example.

$$2 \times 10^2 \times 3 \times 10^3 \times 4 \times 10^{-4} = (2 \times 3 \times 4) \times (10^2 \times 10^3 \times 10^{-4})$$
$$= (24)(10^{2+3-4})$$
$$= 24 \times 10^1 = 240$$

Division

Finding the quotient of two numbers written in scientific notation is also straightforward and is easily illustrated with an example. For instance, consider the quotient of

$$(6 \times 10^4) \div (3 \times 10^2)$$

Rewrite as $\dfrac{6 \times 10^4}{3 \times 10^2} =$

Group as $= (6/3) \times (10^4 / 10^2)$

And simplify as $= (2) \times (10^4 \times 10^{-2})$

$= (2) \times (10^{4-2}) = 2 \times 10^2$

Notice that the quotient is found by finding the quotient of the significant figures and determining the correct power of ten by using the Law of Exponents. Thus,

> To find the *quotient* of two numbers written in scientific notation:
> (1) Find the quotient of the significant figures.
> (2) Multiply this quotient by the power of ten obtained by combining the two given powers of ten.

EXAMPLE 4.6.   Find the quotient:

$$(8 \times 10^4)/(4 \times 10^3)$$

SOLUTION.

$$\frac{8 \times 10^4}{4 \times 10^3} =$$

Rewrite and group $= (8/4) \times (10^4/10^3)$

Simplify $\qquad = (2) \times (10^{4-3})$

$$= 2 \times 10 = 20$$

Most electronic hand calculators are capable of handling numbers written in scientific notation, and it is important to learn how to use your own particular calculator accordingly.

Exercises 4.3

1. Use the following numbers to practice making entries of numbers in scientific notation on your own hand calculator:

   (a) 1.5×10^2 (b) 2.7×10^3 (c) 3.88×10^4 (d) 7.89×10^5
   (e) 6.31×10^{-1} (f) 5.22×10^{-2} (g) 4.78×10^{-3} (h) 10^7
   (i) 10^{-6} (j) 8.91×10^{-4} (k) 9.03×10^{-5} (l) 1.02×10^{-9}
   (m) 3×10^6 (n) 8.11×10^{-6} (o) 0.2×10^3 (p) 0.71×10^{-3}

2. Perform the following operations:

   (a) $2 \times 10^2 + 3 \times 10^2$
   (b) $4.1 \times 10^3 + 1.8 \times 10^3$
   (c) $5.1 \times 10^{-6} + 1.8 \times 10^{-6}$
   (d) $1.8 \times 10^3 + 3.1 \times 10^2$
   (e) $3.7 \times 10 + 1.9 \times 10^2$
   (f) $8.6 \times 10^2 + 1.1 \times 10^3$
   (g) $3 \times 10^{-3} + 4.1 \times 10^{-2}$
   (h) $1.78 \times 10^4 + 7.11 \times 10^3$
   (i) $6.1 \times 10^{-9} + 7.11 \times 10^{-8}$
   (j) $5 \times 10^2 - 3 \times 10^2$
   (k) $7 \times 10^{-2} - 9 \times 10^{-2}$
   (l) $6.8 \times 10^{-6} - 1.7 \times 10^{-6}$
   (m) $6.3 \times 10^3 - 7.4 \times 10^2$
   (n) $8.1 \times 10^4 - 9.2 \times 10^3$
   (o) $4 \times 10^3 + 3 \times 10^4 \times 2 \times 10^2$
   (p) $5.1 \times 10^{-3} + 7.8 \times 10^{-2} - 3.1 \times 10^{-3}$
   (q) $10^3 + 10^4 - 10^2$
   (r) $10^{-4} + 10^{-5} - 10^{-3}$

   (s) $(2 \times 10^2)(3 \times 10^2)$
   (t) $(1.1 \times 10^2)(7.8 \times 10^3)$
   (u) $(6.7 \times 10^{-1})(3.8 \times 10^2)$
   (v) $(7.7 \times 10^{-3})(6.8 \times 10^3)$
   (w) $(3.3 \times 10^3)(1.7 \times 10^{-4})$
   (x) $(5.2 \times 10^{-6})(6.3 \times 10^3)$
   (y) $(4.7 \times 10^3)(3.8 \times 10^{-6})$
   (z) $(1.1 \times 10^{-3})(10^{-3})$
   (aa) $(10^6)(8.7 \times 10^{-4})$
   (bb) $(10^6)(10^{-3})(10^{-1})$
   (cc) $(1.8 \times 10^3)(4)(6.2 \times 10^{-6})$
   (dd) $(1.5 \times 10^3)(3.3 \times 10^3)(4.7 \times 10^{-6})$
   (ee) $(3 \times 10^6)/(2 \times 10^3)$
   (ff) $(1.8 \times 10^3)/(9 \times 10^2)$
   (gg) $(4.7 \times 10^3)/(2.8 \times 10^6)$
   (hh) $(4.3 \times 10^3) \div (10^6)$
   (ii) $(7.1) \div (4.7 \times 10^3)$
   (jj) $(87.2) \div (3.3 \times 10^3)$

(kk) $10^2 \div 10^3$

(ll) $(0.71) \div (2.2 \times 10^3)$

(mm) $(3.72) \div (1.77 \times 10^{-3})$

(nn) $\dfrac{(2.7 \times 10^2)(3 \times 10^3)}{4.1 \times 10^4}$

(oo) $\dfrac{1.63 \times 10^{-2}}{(2.1 \times 10^{-3})(7.1 \times 10^{-3})}$

(pp) $\dfrac{(1.52 \times 10^2)(3.7 \times 10^3)}{(7.4 \times 10^3)(6.1 \times 10^4)}$

3. Use Ohm's Law, $V = IR$, to determine V if:

   (a) $I = 3$ mA, $R = 4.7$ kΩ
   (b) $I = 15$ μA, $R = 220$ kΩ
   (c) $I = 0.17$ A, $R = 680$ Ω
   (d) $I = 24.3$ mA, $R = 3.3$ kΩ

4. Use Ohm's Law, $I = V/R$, to determine I if:

   (a) $V = 10$ V, $R = 2.2$ kΩ
   (b) $V = 28$ V, $R = 680$ Ω
   (c) $V = 110$ V, $R = 5.6$ kΩ
   (d) $V = 270$ mV, $R = 3.3$ kΩ

5. Given $I_1 = 3.51$ mA, $I_2 = 74.3$ mA, $R_1 = 4.7$ kΩ, $R_2 = 680$ Ω, $V_1 = 12$ V, and $V_2 = 5$ V, evaluate the following:

   (a) $I_1 R_1$ (b) $I_1 R_2$ (c) $I_2 R_1$ (d) $I_2 R_2$
   (e) V_1/R_1 (f) V_1/R_2 (g) V_2/R_1 (h) V_2/R_2
   (i) V_1/I_1 (j) V_1/I_2 (k) V_2/I_1 (l) V_2/I_2
   (m) $I_1^2 R_1$ (n) $I_1^2 R_2$ (o) $I_2^2 R_1$ (p) $I_2^2 R_2$
   (q) V_1^2/R_1 (r) V_1^2/R_2 (s) V_2^2/R_1 (t) V_2^2/R_2
   (u) $I_1 V_1$ (v) $I_1 V_2$ (w) $I_2 V_1$ (x) $I_2 V_2$

UNIT 5
SQUARE ROOTS AND CUBE ROOTS

5.1 SQUARE ROOTS

Take another look at the equation for the Law of Exponents, $a^m \cdot a^n = a^{m+n}$. Consider the special case where $m+n=1$; then, $a^m \cdot a^n = a^1 = a$. There are a number of possible values for m and n that will satisfy $m+n=1$; but look at the case when $m=n$. In this event, the only possible values are $m=n=1/2$. This means $a^m \cdot a^n = a^{1/2} \cdot a^{1/2} = a^{1/2+1/2} = a^1 = a$. But what exactly is the meaning of $a^{1/2}$?

By definition, the *square root* of some number Q is a number q such that $q \cdot q = Q$. For example, the square root of 9 is 3, since $3 \cdot 3 = 9$. The square root of 25 is 5, since $5 \cdot 5 = 25$; the square root of 16 is 4, since $4 \cdot 4 = 16$; and the square root of x^2 is x, since $x \cdot x = x^2$.

Now, since $a^{1/2} \cdot a^{1/2} = a$, it is clear that $a^{1/2}$ is simply the square root of a. Thus,

$$\text{The square root of } a = a^{1/2}$$

Another way to express the square root of a number is to use the *radical sign* $\sqrt{}$. Using this sign, the square root of the number a is written as $\sqrt{a}$.

Therefore,
$$\text{The square root of } a = a^{1/2} = \sqrt{a}$$

Notice that $(-2)(-2)=4$. Therefore the number -2 satisfies the definition of the square root of 4 (so does the number $+2$, since $2 \cdot 2 = 4$). In fact, every number has *two* square roots, equal in magnitude but opposite in sign. The *positive root* is known as the *principal root*, and this is generally the number sought when one is asked to find the square root. However, when asked to determine the square roots of some number Q, it is correct to say that they are $+q$ and $-q$. This result is generally written as $\pm q$, or,

$$\text{The square roots of } Q = Q^{1/2} = \sqrt{Q} = \pm q$$

The square root function is usually one of the functions available on an electronic hand calculator. At this point you should learn how to determine the square root of a number on your own particular hand calculator, and then use it when solving the problems in this section.

EXAMPLE 5.1. Find the *principal root* of the following: (a) 64 (b) 2.25	SOLUTION. (a) $64 = (64)^{1/2} = 8$ As a check, $8 \cdot 8 = 64$ (b) $\sqrt{2.25} = 1.5$ As a check, $1.5 \times 1.5 = 2.25$

EXAMPLE 5.2. Find the square roots of the following: (a) 64 (b) 6.25	SOLUTION. (a) $\sqrt{64} = \pm 8$ As a check, $8 \cdot 8 = (-8)(-8) = 64$ (b) $\sqrt{6.25} = \pm 2.5$ As a check, $(2.5)(2.5) = (-2.5)(-2.5) = 6.25$

Notice that $\sqrt{x^4} = \pm x^2$ since $(x^2)(x^2) = x^4$, and $(-x^2)(-x^2) = x^4$ also. Similarly, $\sqrt{a^6 b^8} = \pm a^3 b^4$, since

$$(a^3 b^4)(a^3 b^4) = (-a^3 b^4)(-a^3 b^4) = a^6 b^8$$

From the foregoing, it is easy to see that the square root of a literal number is simply the same literal with its exponent divided by two. Thus the square roots of a monomial can be determined as follows:

SQUARE ROOT/MONOMIAL

To find the square roots of a monomial:
(1) Find the square roots of the numerical coefficient.
(2) Divide the exponents of the literals by two.

EXAMPLE 5.3. Evaluate: (a) $\sqrt{z^4}$ (b) $(4a^6 d^{10})^{1/2}$ (c) $(25 V^2)^{1/2}$	SOLUTION. (a) $\sqrt{z^4} = \pm z^{4/2} = \pm z^2$ As a check, $(z^2)(z^2) = (-z^2)(-z^2) = z^4$ (b) $(4a^6 d^{10})^{1/2} = \pm \sqrt{4}\, a^{6/2} d^{10/2} = \pm 2a^3 d^5$ As a check, $(2a^3 d^5)(2a^3 d^5) = (-2a^3 d^5)(-2a^3 d^5) = 4a^6 d^{10}$ (c) $(25 V^2)^{1/2} = \pm \sqrt{25}\, V^{2/2} = \pm 5 V$ As a check, $(5V)(5V) = (-5V)(-5V) = 25 V^2$

Exercises 5.1

1. Find the *principal roots* of the following:

    (a) 49   (b) 12.25   (c) 225   (d) 30.25
    (e) 169   (f) a^2 (g) a (h) 4356

2. Evaluate the following:

(a) $(36)^{1/2}$ (b) $(20.25)^{1/2}$ (c) $\sqrt{16b^4}$ (d) $\sqrt{121z^2}$

(e) $(144e^2p^4)^{1/2}$ (f) $\sqrt{1000t^8}$ (g) $\sqrt{1/4}$ (h) $(81x^4y^6z^{12})^{1/2}$

(i) $\sqrt{36 \cdot 4}$ (j) $\sqrt{3^2a^4}$ (k) $(196P^{16})^{1/2}$ (l) $\sqrt{a^0}$

(m) $1/\sqrt{16}$ (n) $5/(25)^{1/2}$ (o) $3 - \sqrt{9}$ (p) $\sqrt{16}/\sqrt{1/4}$

5.2   CUBE ROOTS

Look again at the equation for the Law of Exponents for the case where $m + n = 1$; $a^m \cdot a^n = a^{m+n} = a^1 = a$. This time, divide the exponent 1 into three equal parts. That is, let

$$1 = \frac{1}{3} + \frac{1}{3} + \frac{1}{3}$$

Then, write $a^1 = a^{1/3} \cdot a^{1/3} \cdot a^{1/3}$. The term $a^{1/3}$ is called the *cube root* of a. The cube root can also be expressed by using the radical sign in conjunction with a small 3 as $\sqrt[3]{}$. Thus,

$$\text{The cube root of } a = a^{1/3} = \sqrt[3]{a}$$

By definition, the cube root of some number Q is a number q such that $q \cdot q \cdot q = Q$. For example, the cube root of 8 is 2, since $2 \cdot 2 \cdot 2 = 8$. The cube root of 27 is 3, since $3 \cdot 3 \cdot 3 = 27$; the cube root of v^3 is v, since $v \cdot v \cdot v = v^3$; the cube root of -8 is -2, since $(-2)(-2)(-2) = -8$.

Notice that the cube root of a positive number is a positive number, while the cube root of a negative number is a negative number. There are always *three* possible cube roots of a number, but two of them involve *imaginary numbers*—a topic that will be discussed in a following section. For now, be content to find the number q such that $q \cdot q \cdot q = Q$, and the sign of q will be the same as that of Q.

EXAMPLE 5.4. Find the cube root:	SOLUTION.
(a) 64	(a) $\sqrt[3]{64} = 4$
(b) -27	As a check, $4 \cdot 4 \cdot 4 = 64$
(c) -125	(b) $\sqrt[3]{-27} = -3$
(d) 216	As a check, $(-3)(-3)(-3) = -27$
	(c) $\sqrt[3]{-125} = -5$
	As a check, $(-5)(-5)(-5) = -125$
	(d) $\sqrt[3]{216} = 6$
	As a check, $(6)(6)(6) = 216$

Notice that $\sqrt[3]{A^3} = (A^3)^{1/3} = A^{3/3} = A$, since $A \cdot A \cdot A = A^3$. It is quite easy to see how to find the cube root of a monomial. Here are the rules:

CUBE ROOT/MONOMIAL

To find the cube root of a monomial:
(1) Find the cube root of the numerical coefficient and affix the same sign.
(2) Divide the exponents of the literals by three.

EXAMPLE 5.5.   Evaluate:

(a) $\sqrt[3]{P^6}$
(b) $(27x^3y^9)^{1/3}$
(c) $(-64y^{12})^{1/3}$
(d) $1/\sqrt[3]{-125}$

SOLUTION.

(a) $\sqrt[3]{P^6} = P^{6/3} = P^2$
   As a check, $P^2 \cdot P^2 \cdot P^2 = P^6$

(b) $(27x^3y^9)^{1/3} = \sqrt[3]{27} \; x^{3/3}y^{9/3} = 3xy^3$
   As a check, $(3xy^3)(3xy^3)(3xy^3) = 27x^3y^9$

(c) $(-64y^{12})^{1/3} = \sqrt[3]{-64} \; y^{12/3} = -4y^4$
   As a check, $(-4y^4)(-4y^4)(-4y^4) = -64y^{12}$

(d) $\dfrac{1}{\sqrt[3]{-125}} = \dfrac{1}{-5} = -0.2$
   As a check,

$$\left(\frac{1}{-5}\right)\left(\frac{1}{-5}\right)\left(\frac{1}{-5}\right) = \frac{1}{-125}$$

Exercises 5.2

1. Find the cube roots of the following:

   (a) 27   (b) -216 (c) 343   (d) 1331
   (e) 1000   (f) 10^6 (g) 10^{-9} (h) $-R^6$

2. Evaluate the following:

   (a) $\sqrt[3]{8^3}$ (b) $\left(\frac{1}{8}Q^{15}\right)^{1/3}$ (c) $\sqrt[3]{1/27}$ (d) $\sqrt[3]{-27a^6b^9c^{12}}$

   (e) $\sqrt[3]{-8/Q^6}$ (f) $(-10^{-3})^{1/3}$ (g) $1/\sqrt[3]{-64}$ (h) $1/(27^3)^{1/3}$

   (i) $\sqrt[3]{t^0}$ (j) $(-8000)^{1/3}$ (k) $\sqrt[3]{-1/64}$ (l) $\sqrt[3]{27}/(-64)^{1/3}$

5.3   EVALUATION

There are two rules concerning radicals that can often be used to simplify their evaluation. They are

$$\sqrt{a \cdot b} = \sqrt{a} \cdot \sqrt{b} \tag{5.1}$$

$$\sqrt{\frac{a}{b}} = \frac{\sqrt{a}}{\sqrt{b}} \tag{5.2}$$

To see how they can be used to simplify calculations, consider the following examples:

$$\sqrt{900} = \sqrt{9 \times 10^2} = \sqrt{9} \cdot \sqrt{10^2} = 3 \cdot 10 = 30$$
$$\sqrt{324} = \sqrt{81 \cdot 4} = \sqrt{81} \cdot \sqrt{4} = 9 \cdot 2 = 18$$
$$\sqrt{4/9} = \sqrt{4} / \sqrt{9} = 2/3$$
$$\sqrt{\frac{9}{25} x^4} = \frac{\sqrt{9x^4}}{\sqrt{25}} = \frac{\sqrt{9} \sqrt{x^4}}{\sqrt{25}} = \frac{3}{5} x^2$$

Almost all of the previous examples and problems have intentionally been selected to simplify the numerical calculations. However, things do not always work out so nicely in the *real world*, and you should learn to use your hand calculator to evaluate square and cube roots. The solutions can always be checked by multiplying out the roots—and this should always be done!

EXAMPLE 5.6.   Evaluate:

(a) $\sqrt[3]{-63}$

(b) $(5.3x^2)^{1/2}$

(c) $\sqrt{50}$

(d) $\sqrt{48}$

SOLUTION.

(a) $\sqrt[3]{-63} = -3.979$
    Check: $(-3.979)(-3.979)(-3.979) = -63$

(b) $(5.3x^2)^{1/2} = 2.3x$
    Check: $(2.3x)(2.3x) = 5.3x^2$

(c) $\sqrt{50} = \sqrt{25} \cdot \sqrt{2} = 5 \times 1.414 = 7.071$
    Check: $(7.071)(7.071) = 50$

(d) $\sqrt{48} = \sqrt{16 \cdot 3} = \sqrt{16} \cdot \sqrt{3} = 4 \times 1.732 = 6.928$
    Check: $(6.928)(6.928) = 48$

There are two roots that occur frequently in electrical problems, and it will be worthwhile to memorize them. They are

---

ROOTS

$$\sqrt{2} = 1.414 \qquad \sqrt{3} = 1.732$$

Exercises 5.3

1. Evaluate:

   (a) $\sqrt{32}$

   (b) $(75)^{1/2}$

   (c) $\sqrt{6x^2}$

   (d) $\sqrt{9/16}$

   (e) $\sqrt{4/18}$

   (f) $(18x^2/25y^4)^{1/2}$

   (g) $\sqrt[3]{9/y^3}$

   (h) $(-17/z^6)^{1/3}$

   (i) $\sqrt{23p^6}$

   (j) $\sqrt[3]{9.6v^3}$

   (k) $\sqrt{0.79}$

   (l) $\sqrt{0.196}$

   (m) $\sqrt[3]{0.66}$

   (n) $(0.495)^{1/3}$

   (o) $(6\times10^4)^{1/2}$

   (p) $(-9\times10^6)^{1/3}$

   (q) $\sqrt{3\times10^{-6}}$

   (r) $\sqrt{8\times10^{-12}}$

   (s) $\sqrt{3x^4}/\sqrt[3]{7x^6}$

   (t) $(a^3b^6c^9)^{1/3}/(-10^{-9})^{1/3}$

2. The voltage V across a resistance R in which the power dissipation is P watts is given by $V=\sqrt{PR}$. Find V if $P=1500$ W and $R=10$ Ω.

3. What voltage V must be applied to a 30 W soldering iron if the resistance of the iron is 440 Ω? The proper equation is $V=\sqrt{PR}$.

4. A 3000 W heating element has a resistance of 19 Ω. What voltage V must be applied to its terminals? $V=\sqrt{PR}$.

5. The heating current I in a resistance R in which the power dissipation is P watts is given by $I=\sqrt{P/R}$. Find I if:

   (a) $P=1000$ W, $R=10$ Ω   (b) $P=20$ W, $R=10$ Ω

   (c) $P=30$ mW, $R=1$ kΩ   (d) $P=250$ mW, $R=470$ Ω

6. Calculate the current I to heat an electric iron if its element resistance is 14 Ω and it dissipates 900 W. $I=\sqrt{P/R}$.

7. Calculate the heater current in a 1.5 kW, 240 V heating element if the element resistance is 38.4 Ω. $I=\sqrt{P/R}$.

8. Calculate the heater current necessary to deliver 30 mW to the filament of a vacuum tube if the heater resistance is 130 Ω. $I=\sqrt{P/R}$.

9. The radius r of a circle having an area A is given by $r=\sqrt{A/\pi}$. Find r if:

   (a) $A=100$ in.^{2} (b) $A=6$ ft^2 (c) $A=9\pi$ in.^{2} (d) $A=50\pi$ cm^2

10. Calculate the length of the sides of a square having an area of:

    (a) 6 ft^2 (b) 3×10^{-3} cm^2 (c) 15 in.^{2}

    (*Hint:* Remember $A=h\cdot w$ for a rectangle.)

11. The well-known Pythagorean Theorem allows one to determine one side of a right triangle provided the other two sides are known (see Figure 5.1). The relationships are: $a = \sqrt{c^2 - b^2}$ and $b = \sqrt{c^2 - a^2}$. Calculate:

(a) a if $c = 6$ and $b = 4$
(b) a if $c = 5$ and $b = 3$
(c) b if $c = 5$ and $a = 4$
(d) b if $c = 16$ and $a = 10$

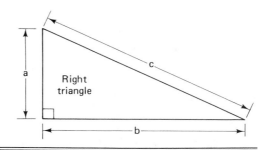

Figure 5.1

UNIT 6
PERCENTAGES

6.1   PERCENT

Sometimes it is convenient to speak of a *fractional portion* of a given quantity. For example, it can be said that a 20-gal tank is *half full*. What is really meant is, it contains 10 gal, or 10 gal/20 gal=1/2 of its given capacity. It could also be said that it is 1/4 full (5 gal), or 1/8 full (2.5 gal), but in order to gain precision it is necessary to resort to fractions such as 115/256. Such fractions are not very convenient, and it is generally preferable to use the decimal equivalent, which would be 115/256=0.4492.

In order to express a fractional quantity as a *percent* (%), it is only necessary to multiply the decimal fraction by 100. Thus, half full means 1/2, which is 0.5, and this becomes $0.5 \times 100 = 50\%$. Similarly, 115/256=0.4492, which is $0.4492 \times 100 = 44.92\%$.

EXAMPLE 6.1. Express as a percent (%):	SOLUTION.
(a) 1/4	(a) $1/4 = 0.25$ $0.25 \times 100 = 25\%$
(b) 3/8	(b) $3/8 = 0.375$ $0.375 \times 100 = 37.5\%$
(c) 6 parts out of 25	(c) $6/25 = 0.24$ $0.24 \times 100 = 24\%$

EXAMPLE 6.2. Express the following percentages as decimals:	SOLUTION. Divide each percentage by 100.
(a) 75%	(a) $75\%/100 = 0.75$
(b) 10%	(b) $10\%/100 = 0.1$
(c) $33\frac{1}{3}\%$	(c) $33\frac{1}{3}\%/100 = 0.333\ldots$

Notice that if the tank in question is full, the appropriate fraction is $1/1 = 1.0$, which becomes 100%. Thus 100% means the *entire quantity*, or the *maximum value* possible. On the other hand, if the tank is empty, the appropriate fraction is $0/1 = 0.0$, which becomes 0%. Thus 0% corresponds to the *minimum* value possible.

In general, the fractional portion of a given quantity expressed as a percent will fall within the limits $0 \leqslant \% \leqslant 100$. Frequently percent is simply taken to be the number of parts in 100. Thus, 2% means 2 parts per hundred, 15% means 15 parts in 100, and so on.

Exercises 6.1

1.   Express each of the following as a percent:

    (a)   0.38   (b)   0.716   (c)   0.1129   (d)   0.9999
    (e)   5/16   (f)   0   (g)   4/11   (h)   13/100

(i) 75/1000   (j) $\sqrt{2}/64$ (k) 1.0   (l) 9/16

(m) 7/8   (n) 25/32   (o) 5 parts in 50   (p) 1 part in 15

2. Express in decimal form:

(a) 50%   (b) $66\frac{2}{3}\%$ (c) 99.9%   (d) 12.5%

(e) 1%   (f) 2.45%   (g) 100%   (h) 6 parts in 15

6.2 EFFICIENCY

One of the most basic concepts in science can be expressed by saying, "You can't get something for nothing." In practice, if you want to drive your car, it must be supplied with gasoline (energy). If you want to operate an electric motor, it must be connected to a source of electric energy. In short, if you expect work as an *output* from any sort of system, it must be supplied with some form of energy as an *input*. Naturally, the output (W_{out}) must be less than the input (W_{in}). Otherwise, it would be possible to construct a *perpetual motion machine*—a machine that would run forever without the need for input energy. Since this is not possible,

$$W_{out} < W_{in}$$

A measure of how well a system performs is found by finding the ratio of W_{out}/W_{in}. A similar measure is given by the ratio of output power to input power (P_{out}/P_{in}). Since the output is always less than the input, this quantity is a fraction, and it is called the *efficiency* (η) of the system (η is the Greek letter eta, pronounced *ētáh*). Thus,

EFFICIENCY

$$\eta = \frac{W_{out}}{W_{in}} = \frac{P_{out}}{P_{in}} \tag{6.1}$$

W_{out} and W_{in} must have the same units; the same of course is true for P_{out} and P_{in}.

EXAMPLE 6.3. The input power of an electric motor is 800 W, and its output is equivalent to 746 W. What is the efficiency?

SOLUTION.

$$\eta = \frac{P_{out}}{P_{in}} = \frac{746 \text{ W}}{800 \text{ W}} = 0.932$$

Since the efficiency η of a useful system is always less than 1.0, but is also always greater than zero, the limits on η are given by

$$0 < \eta < 1.0$$

It seems natural then to express efficiency as a percent, and all that need be done is multiply η by 100. Thus

---

PERCENT EFFICIENCY

Efficiency expressed as a percent is equal to

$$\eta(\%) = 100\eta$$

The limits on percent efficiency are

$$0\% < \eta(\%) < 100\%$$

In Example 6.3, the motor efficiency is 0.932. The percent efficiency is then given as

$$\eta(\%) = 100\eta = 100 \times 0.932 = 93.2\%$$

Notice that in a *perfect*, *ideal* system, $W_{out} = W_{in}$ and $\eta = 1.0$. The percent efficiency in a *perfect*, *ideal* system is therefore 100%. This is not realizable, but serves as an ideal goal. Thus, the closer $\eta(\%)$ approaches 100%, the better.

Exercises 6.2

1. Calculate the efficiency of each of the following:
   (a) $W_{out} = 80$, $W_{in} = 100$ (b) $W_{out} = 1500$ ft-lb, $W_{in} = 2000$ ft-lb
   (c) $P_{out} = 65$ W, $P_{in} = 70$ W   (d) $P_{out} = 35$ mW, $P_{in} = 45$ mW

2. What is the percent efficiency of a 240 V electric motor whose output is equivalent to 1492 watts when the motor current is 7.0 amperes? The input power is $P = VI$.

3. An electric motor rated at 120 V and 10 A does work at the rate of 1100 W. What is its percent efficiency? Recall, $P = VI$.

4. An electronic hi-fi amplifier will deliver 50 W output power to a speaker. What is its percent efficiency if it requires 500 mA from a 120 V source? $P = VI$.

5. If the output energy W_{out} is 75 for a system having a percent efficiency of 75%, what must be its input energy (W_{in})?

6. An electronic power amplifier has a percent efficiency of 80%. What is its output power if the input power is 100 W?

7. Which motor is the most efficient: (a) 117 V @ 7.5 A or (b) 240 V @ 3.2 A? They both provide the same output power. Recall, $P = VI$.

6.3   TOLERANCE

It is quite difficult to produce a product having a particular dimension or value that is *exact*. That is to say, it might be desirable to obtain a bolt having a diameter of 0.500 in., but most likely the bolt would have an actual diameter somewhere between 0.490 in. and 0.510 in. Such a bolt is said to have a *tolerance* of ±0.010 in., and its diameter would be given as 0.500 in. ±0.010 in. Notice that the *minimum* diameter is found as the desired diameter minus the tolerance (0.500 in. − 0.010 in. = 0.490 in.), and the *maximum* diameter is the desired value plus the tolerance (0.500 in. + 0.010 in. = 0.510 in.).

EXAMPLE 6.4. Express the limits of a 3/8 in. diameter hole to be drilled in a piece of steel if the tolerance is ±0.005 in.	**SOLUTION.** $$d = 3/8 \text{ in.} = 0.375 \text{ in.}$$ The upper limit is 0.375 in. + 0.005 in. = 0.380 in. The lower limit is 0.375 in. − 0.005 in. = 0.370 in. Thus, $$0.370 \text{ in.} \leqslant d \leqslant 0.380 \text{ in.}$$
EXAMPLE 6.5. A wire-wound resistor having a desired value of 200 Ω is to be constructed within a tolerance of ±20 Ω. What are the limits of resistance?	**SOLUTION.** $$\text{Max } R = 200 + 20 = 220 \ \Omega$$ $$\text{Min } R = 200 - 20 = 180 \ \Omega$$ Thus, $$180 \ \Omega \leqslant R \leqslant 220 \ \Omega$$

Sometimes it is more convenient to express the tolerance as a *percent of the desired value*. Thus a 200 Ω resistor having a tolerance of ±20 Ω is said to have a percent tolerance of

$$\text{Tol}(\%) = \frac{\pm 20 \ \Omega}{200 \ \Omega} \times 100 = \pm 10\%$$

An equation for the tolerance expressed as a percent is

PERCENT TOLERANCE

$$\text{Tol}(\%) = \frac{\pm \text{tolerance}}{\text{desired value}} \times 100$$

EXAMPLE 6.6. What is the Tol(%) of a resistor (R) constructed within the limits $4230 \ \Omega \leqslant R \leqslant 5170 \ \Omega$?

SOLUTION.   This is the same as

$$R = 4700 \pm 470 \ \Omega$$

Thus,

$$\text{Tol}(\%) = \frac{\pm 470}{4700} \times 100 = \pm 10\%$$

EXAMPLE 6.7. What are the maximum and minimum values of a 680 $\Omega \pm 10\%$ resistor?

SOLUTION.

$$10\% \text{ of } 680 \ \Omega = 0.1 \times 680 = 68 \ \Omega$$

$$\therefore R_{\text{max}} = 680 + 68 = 748 \ \Omega$$

$$\therefore R_{\text{min}} = 680 - 68 = 612 \ \Omega$$

Carbon resistors are commonly constructed with tolerances of $\pm 1\%$, $\pm 5\%$, $\pm 10\%$, and $\pm 20\%$. A pictorial representation of a 1000 Ω resistor and the limits specified by the various tolerances are shown in Figure 6.1. It is easy to see why resistors constructed to a precision of $\pm 1\%$ are more difficult to achieve and are thus more expensive.

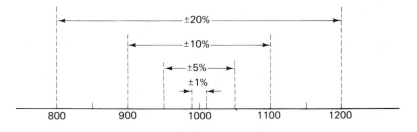

Figure 6.1

Exercises 6.3

1. Express the limits of a 1/4 in. bolt having a tolerance of ± 0.01 in.
2. Express the limits of a steel ball having a diameter of 0.070 in. ± 0.003 in.
3. What are the upper and lower limits on a $1\frac{3}{4}$ in. hole drilled to within ± 0.02 in.?
4. What is the tolerance on a nominal 0.200 in. rivet if the limits are 0.210 in. maximum and 0.190 in. minimum?
5. What are the limits on a 500 Ω resistor having a tolerance of $\pm 25 \ \Omega$?
6. What is the Tol(%) for the resistor in Problem #5 above?

7. What is the Tol(%) of a resistor R having limits 95 $\Omega \leqslant R \leqslant$ 105 Ω?

8. What are the upper and lower values of a 3300 Ω resistor having tolerances of:
   (a) $\pm 20\%$?   (b) $\pm 10\%$?   (c) $\pm 5\%$?   (d) $\pm 1\%$?

9. What are the upper and lower values of a 47 kΩ resistor having tolerances of:
   (a) $\pm 20\%$?   (b) $\pm 10\%$?   (c) $\pm 5\%$?   (d) $\pm 1\%$?

10. What are the limits on a 33 kΩ resistor having a tolerance of:
    (a) $\pm 20\%$?   (b) $\pm 10\%$?   (c) $\pm 5\%$?   (d) $\pm 1\%$?

11. It is desired to have a resistor whose value is 5.6 kΩ, but its value must not exceed 6000 Ω. What tolerance must one use:
    (a) 20%,   (b) 10%,   (c) 5%, or   (d) 1%?

12. Is a 2.7 k$\Omega \pm 10\%$ resistor acceptable to obtain a resistance whose value must not exceed 3000 Ω? Show why or why not.

13. A 1.5 kΩ resistor is found to have an actual value of 1300 Ω. Does it have a tolerance within:
    (a) $\pm 1\%$,   (b) $\pm 5\%$,   (c) $\pm 10\%$, or   (d) $\pm 20\%$?

14. What must be the tolerance on a 560 Ω resistor if its actual value must never exceed 600 Ω:
    (a) $\pm 1\%$,   (b) $\pm 5\%$,   (c) $\pm 10\%$, or   (d) $\pm 20\%$?

6.4   PERCENT ERROR

Suppose a certain voltmeter having a *full-scale* reading of 50 V is used to measure a known *voltage standard* of 50 volts to determine the accuracy of the voltmeter. If the reading taken is 49 volts, there is an error of $50 - 49 = 1$ V. This error is found by taking the difference between the known *ideal* value and the *actual* value. Similarly, a reading of 51 V would lead to an error of $51 - 50 = 1$ V.

An indication of the accuracy of the measurement can be found in terms of the *percent error* (% ERR). In equation form:

$$\text{PERCENT ERROR}$$

$$\% \text{ ERR} = \frac{|\text{Ideal} - \text{Actual}|}{\text{Ideal}} \times 100$$

Notice that in the numerator the *magnitude* of the error is required (e.g., $50 - 49 = 1$ V, or $51 - 50 = 1$ V).

EXAMPLE 6.8.   Calculate the % ERR for the voltmeter reading taken in the preceding discussion.

SOLUTION.

$$\%ERR = \frac{|50-49|}{50} \times 100 = \frac{1}{50} \times 100 = 2\%$$

or

$$\%ERR = \frac{|50-51|}{50} \times 100 = \frac{1}{50} \times 100 = 2\%$$

Exercises 6.4

1. Calculate the % ERR of the following:

   (a) Ideal = 100 V, actual = 97 V
   (b) Ideal = 58 A, actual = 59 A
   (c) Ideal = 60 W, actual = 56 W
   (d) Ideal = 5.6 kΩ, actual = 5000 Ω
   (e) Ideal = 67 mV, actual = 71 mV
   (f) Ideal = 650 mW, actual = 0.5 W
   (g) Ideal = 80 mA, actual = 70 mA
   (h) Ideal = 33 cm, actual = 33.5 cm
   (i) Ideal = 0.750 in., actual = 0.788 in.

2. If the resistance of a 550 Ω precision resistor is reduced by 8%, what is the value of the resistance?

UNIT 7
ALGEBRAIC EQUATIONS
AND INEQUALITIES

7.1 FORMATION

An algebraic equation can be formed by connecting two algebraic expressions with an equality sign ($=$). The equation $x=y$ says, "x is equal to y." An algebraic inequality is formed by connecting two expressions with connectives such as $<$, $\leqslant$, $>$, or $\geqslant$. For example, the inequality $a \leqslant b$ says, "a is less than or equal to b," whereas the inequality $a \geqslant b$ says, "a is greater than or equal to b." Thus, an algebraic equation or inequality is simply a mathematical formulation of a sentence, and mathematical symbols are used to replace words. For example:

$+$ plus, the sum of, added to

$-$ minus, the difference between, subtracted from

$\times$ times, the product of, multiplied by

$\div$ per, the quotient of, divided by

$=$ is equal to, is equivalent to, is the same as

$\leqslant$ is less than or equal to

$<$ is less than

$\geqslant$ is greater than or equal to

$>$ is greater than

EXAMPLE 7.1. Translate the following sentences into algebraic expressions: (a) The sum of a and b (b) 5 subtracted from z (c) The product of V and I (d) The quotient of P over R	SOLUTION. (a) $a+b$ (b) $z-5$ (c) $V \cdot I$ or VI (d) P/R or $P \div R$
EXAMPLE 7.2. Translate the following sentences into algebraic equations: (a) The sum S of the numbers a, b, and c. (b) The sum S of x and y diminished by 7. (c) Four times the number z added to 16 is equal to Y. (d) The resistance R is equal to the product of R_1 and R_2 divided by their sum.	SOLUTION. (a) $S=a+b+c$ (b) $S=x+y-7$ (c) $Y=4z+16$ (d) $R=\dfrac{R_1 R_2}{R_1+R_2}$

EXAMPLE 7.3. Write an inequality that states that the resistance R of a 510 $\Omega \pm 10\%$ resistor is greater than 459 Ω, but less than 561 Ω.	SOLUTION. $$459\ \Omega < R < 561\ \Omega$$
EXAMPLE 7.4. Write in words the meaning of the inequality $$950\ \Omega \leqslant R \leqslant 1050\ \Omega$$	SOLUTION. R is less than or equal to 1050 Ω, but greater than or equal to 950 Ω.

Exercises 7.1

1. Find the algebraic expressions for each of the following:
   (a) The sum of 16 and P.
   (b) x plus 9.
   (c) Q added to q and r.
   (d) The sum of z and Y diminished by 4.
   (e) The difference between R_1 and R_2.
   (f) The difference between $17x$ and $11y$.
   (g) The difference between three times x and five times y.
   (h) The product of F and d.
   (i) Six times P subtracted from four times the quantity $(x - 7)$.
   (j) The product of 80(kWh) and 4(¢/kWh).
   (k) The product divided by the sum of the two numbers R_a and R_b.
   (l) The quotient specified by 15 miles per hour.

2. Find an algebraic equation or inequality for each of the following. It may be necessary to define appropriate symbols—see part (a) below.
   (a) The sum of twice a number added to six times another number. (*Hint*: Let $S =$ the sum, $a =$ the first number, and $b =$ the second number.)
   (b) The total resistance is the sum of three resistances.
   (c) The voltage is equal to $3V$ minus IR.
   (d) The current is the total current minus the branch current.
   (e) The total power is the product of the voltage and the current.
   (f) The total power is the product of the square of the current and the resistance.
   (g) The total power is the quotient of the square of the voltage divided by the resistance.
   (h) The current is the voltage divided by the resistance.
   (i) The voltage is the product of the current and the resistance.
   (j) The resistance is the voltage divided by the current.
   (k) Work is equal to force times distance.

(l)   Distance is equal to velocity times the time.

(m)   The area of a circle is the product of π and the square of the radius.

(n)   The area of a circle is the product of π and the square of the diameter divided by 4.

(o)   The area of a rectangle is equal to the product of its length and its width.

(p)   The volume of a rectangular box is the product of the width, the length, and the depth.

(q)   The sum of the three angles in a triangle is equal to 180°.

(r)   The cost is equal to the total number of hours multiplied by the cost per kilowatthour.

(s)   The output power is always less than the input power.

(t)   The output power is less than or equal to the input power.

(u)   The height is greater than the width.

(v)   The resistance is greater than or equal to R.

(w)   The % error is less than 5% but greater than 1%.

(x)   The resistance is between 180 and 260 ohms.

(y)   The voltage is 50 volts or less, and at the same time, 49 volts or more.

(z)   The product of 4 times the resistance squared is less than or equal to 1/5 the product of the voltage divided by the current.

(aa) The square root of the voltage squared and divided by the resistance is equal to 1/8 the resistance diminished by 6.

(bb) Velocity is equal to miles per hour.

7.2   MANIPULATION OF EQUATIONS

The equality sign ($=$) in an algebraic equation can be thought of as the *fulcrum* of a balance, as shown in Figure 7.1. The left-hand expression ($2x$) in Figure 7.1 is said to be equal to the right-hand expression ($y+3$), and the system is *in balance*.

It is evident that any number N could be added to the left-hand side and still maintain balance as long as the *same* number is added to the right-hand side. Thus the equation can be altered by adding N to each side to obtain

$$2x + N = y + 3 + N$$

and the equation is still valid.

Similarly each side can be diminished by the same quantity (for example, Q) and still maintain balance. The equation would be altered as

$$2x - Q = y + 3 - Q$$

It is also possible to multiply or divide each side by the same number—but remember, *whatever is done on one side must be done on the other side* in order to preserve the equality! For example, multiply both sides by 3 to get

$$3(2x) = 3(y + 3)$$

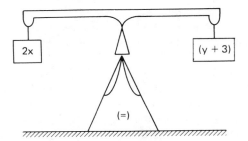

Figure 7.1

or divide both sides by R to get

$$\frac{2x}{R} = \frac{y+3}{R}$$

These operations can be summarized by the following rules:

ALGEBRAIC OPERATIONS

The equality in an algebraic equation is maintained under the following operations:
  (1) Add the same number to both sides.
  (2) Subtract the same number from both sides.
  (3) Multiply both sides by the same number.
  (4) Divide both sides by the same number.

These four fundamental rules can be used to manipulate algebraic equations. It is possible to express an equation in terms of any one of its variables (*solve* the equation for a variable), or determine the numerical value of a variable that satisfies the equation. Such a numerical value is called a *root* or *solution* of the equation. This amounts to solving the equation for its root, or finding a solution for the equation.

Ohm's Law expresses in equation form how the voltage across a resistance is related to the current through the resistance. Namely,

$$I = \frac{V}{R} \tag{7.1}$$

This equation is in a form convenient for solving for the current if the voltage and resistance are known. For example, if $V = 100$ V and $R = 50$ Ω, then,

$$I = \frac{V}{R} = \frac{100 \text{ V}}{50 \text{ } \Omega} = 2 \text{ A}$$

In this case, $I = 2$ A is the solution (root) of the equation.

If the current and resistance are known, the equation can be rearranged to solve for the voltage as follows. *Multiply both sides* by R. Thus,

$$R \times I = \frac{V}{\cancel{R}} \times \cancel{R}$$

The two R's on the right-hand side cancel, and the result is

$$RI = V$$

This is equivalent to

$$V = IR \tag{7.2}$$

Note that $R \times I = V$ is the same thing as $V = R \times I = IR$; that is, $3 \times 2 = 2 \times 3$. Now it is possible to find the solution for V given any value of I and R. Suppose for instance $I = 3$ mA and $R = 5$ kΩ. Then,

$$V = IR = 3 \text{ mA} \times 5 \text{ k}\Omega$$
$$= (3)(10^{-3})(5)(10^3) = 15 \text{ V}$$

If the current and voltage are known, the equation can be solved for R by *dividing both sides* of Eq. (7.2) by I. Thus,

$$\frac{V}{I} = \frac{\cancel{I}R}{\cancel{I}}$$

The I's on the right side cancel, and the result is

$$R = \frac{V}{I} \tag{7.3}$$

Again, note that $V/I = R$ is the same as $R = V/I$. Now, suppose $V = 25$ V and $I = 10$ mA; the solution for R is then,

$$R = \frac{V}{I} = \frac{25}{10 \text{ mA}} = \frac{25}{10(10^{-3})} = \frac{25}{10} \times 10^3 = 2.5 \text{ k}\Omega$$

The previous discussion used rules (3) and (4); here is a problem using rules (1) and (2). Find the solution for x, given

$$5x - 4 = 5 + 2x$$

First add 4 to each side to obtain

$$5x - \cancel{4} + \cancel{4} = 4 + 5 + 2x$$
$$5x = 9 + 2x$$

Then subtract $2x$ from each side to obtain

$$5x - 2x = 9 + \cancel{2x} - \cancel{2x}$$
$$3x = 9$$

Finally divide both sides by 3 to obtain

$$\frac{\cancel{3}x}{\cancel{3}} = \frac{9}{3}$$
$$x = 3$$

There is almost always more than one way to solve a problem, and you should develop your skills by working many different problems. This section provides a brief introduction to the manipulation of algebraic equations, and you will have ample opportunity to apply these ideas and develop your skills in the sections that follow.

| EXAMPLE 7.5. Solve for °C in the following equation:

$$°F = \frac{5}{9}(°C - 32)$$ | SOLUTION.

$$°F = \frac{5}{9}(°C - 32)$$

Multiply both sides by 9/5
$$\frac{9}{5}°F = \frac{\cancel{9}}{\cancel{5}} \times \frac{\cancel{5}}{\cancel{9}}(°C - 32)$$
$$\frac{9}{5}°F = °C - 32$$

Add 32 to both sides
$$\frac{9}{5}°F + 32 = °C - \cancel{32} + \cancel{32}$$
$$°C = \frac{9}{5}°F + 32$$ |

Notice that the first two operations (adding or subtracting from each side of the equation) can be achieved by an operation known as *transposition*. That is, a term may be *transposed* (moved) from one side of the equation to the other if its sign is changed. In general it is desirable to get the variable in question all by itself on the left side of the equation. For example,

$$5x - 4 = 5 + 2x$$

First transpose $(2x)$ $5x - 2x - 4 = 5$

$$3x - 4 = 5$$

Then transpose (-4) $3x = 5 + 4$

$$3x = 9$$

And divide both sides by (3) $\dfrac{\cancel{3}x}{\cancel{3}} = \dfrac{9}{3}$

$$x = 3$$

Here are two more examples, one containing an exponent and one containing a square root.

EXAMPLE 7.6.   Solve for I if $P = 10$ W and $R = 20$ Ω.

$$I^2 = \frac{P}{R}$$

SOLUTION.   It is first necessary to take the square root of each side. Determine the principal root.

$$I = \sqrt{\frac{P}{R}} = \sqrt{\frac{10}{20}} = \sqrt{0.50} = 0.707 \text{ A}$$

EXAMPLE 7.7.   Solve for A if $r = 3$ in.

$$r = \sqrt{\frac{A}{\pi}}$$

SOLUTION.   First, square both sides of the equation to obtain

$$r^2 = \frac{A}{\pi}$$

Then multiply both sides by π to obtain

$$\pi \cdot r^2 = \frac{A}{\pi} \cdot \pi$$

Thus,

$$A = \pi r^2 = \pi \cdot 3 \cdot 3 = 28.27 \text{ in.}^2$$

Exercises 7.2

1. Given $I = V/R$, solve for (a) V; (b) R.
2. Given $P = I^2R$, solve for (a) R; (b) I.
3. Given $P = IV$, solve for (a) I; (b) V.
4. Given $P = V^2/R$, solve for (a) R; (b) V.
5. Given $F = k\dfrac{Q_1 Q_2}{d^2}$, solve for (a) Q_1; (b) k.
6. Given $I = Q/t$, solve for (a) Q; (b) t.
7. Given $V = W/Q$, solve for (a) W; (b) Q.
8. Given $W = F \times d$, solve for (a) F; (b) d.
9. Given $W = Pt$, solve for (a) P; (b) t.
10. Given $W = VQ$, solve for (a) V; (b) Q.
11. Given $Pt = VQ$, solve for (a) P; (b) V.
12. Given $P = W/t$, solve for (a) W; (b) t.
13. Given $I = \sqrt{P/R}$, solve for (a) P (b) R.
14. Given $V = \sqrt{PR}$, solve for (a) P; (b) R.
15. Given $\eta = P_{out}/P_{in}$, solve for P_{in}.
16. Given $R_s = (V - V_t)/I_L$, solve for (a) V_t; (b) V.

17. Given $R = \rho L/A$, solve for (a) ρ; (b) L; (c) A.

18. Given $\rho = \dfrac{1}{q\mu N}$, solve for (a) q; (b) μ; (c) N.

19. Given $\rho = \dfrac{1}{q(\mu_n + \mu_p)N}$, solve for (a) N; (b) μ_n.

20. Given $R = R_{20}[1 + \alpha(T - 20)]$, solve for T.

21. Given $R = \dfrac{R_1 R_2}{R_1 + R_2}$, solve for R_1.

22. Given $V_x = VR_x/R_T$, solve for (a) V; (b) R_x; (c) R_T.

23. Given $I_x = I_T \dfrac{R_{\text{opp}}}{R_x + R_{\text{opp}}}$, solve for (a) I_T; (b) R_{opp}; (c) R_x.

24. Given $R_T = R_1 + R_2 + R_3$, solve for R_1.

25. Given $X_L = 2\pi f L$, solve for (a) f; (b) L.

26. Given $X_c = 1/2\pi f C$, solve for (a) f; (b) C.

27. Given $Q = \omega L/R$, solve for (a) ω; (b) R.

28. Given $A = \pi r^2$, solve for r.

29. Given $\text{Vol} = hld$, solve for d.

30. Given $c^2 = a^2 + b^2$, solve for (a) a; (b) b; (c) c.

31. Given $z^2 = R^2 + x^2$, solve for (a) z; (b) R; (c) x.

32. Given $R_V = \dfrac{V_T}{I_{FS}} - R_n$, solve for (a) I_{FS}; (b) R_n; (c) V_T.

33. Given $T = BlINd$, solve for (a) I; (b) N.

34. Given $R = 1/\mu A$, solve for (a) μ; (b) l.

35. Given $L_m = k\sqrt{L_1 L_2}$, solve for (a) k; (b) L_1.

36. Given $L_m = \dfrac{L_A - L_B}{4}$, solve for (a) L_A; (b) L_B.

37. Given $L = \dfrac{r^2 N^2}{8r + 11t}$, solve for t.

38. Given $C = \varepsilon \dfrac{A}{d}$, solve for (a) A; (b) d.

39. Given $W = \dfrac{1}{2}CV_0^2$, solve for V_0.

40. Given $C = \sqrt{a^2 + b^2}$, solve for a.

41. Given $f = \dfrac{1}{2\pi\sqrt{LC}}$, solve for (a) L; (b) C.

42. Given $P + VI = E^2/R$, solve for (a) V; (b) P; (c) R; (d) E.

43. Solve the following equations and check all solutions:

(a) $x + 2 = 5$ (b) $V - 3 = 7$ (c) $P + 1.5 = 7$

(d) $7 = P + 2$

(e) $3x = 9$

(f) $2V - 3 = 15$

(g) $P/2 = P + 7$

(h) $\dfrac{1}{4} i = 6$

(i) $2\dfrac{1}{2} x = \dfrac{3}{4}$

(j) $v + 7 = 3v - 9$

(k) $\dfrac{v + 2}{6} = 7$

(l) $z - 5 = (3z + 2)/2$

(m) $3.8x - 7 = 0.7x + 1$

(n) $2y + 7 = y - 19$

(o) $\dfrac{120x - 16}{4} = 3x$

44.   Find an equation, solve, and check each of the following:

(a)   The sum of two resistances is 1.5 kΩ. If one is 700 Ω, what is the other?

(b)   The sum of two voltages is 36 V. If one is twice the other, what are the voltages?

(c)   Two resistors are chosen such that their sum is 800 Ω and one resistor has a value 1/4 the other. What are the resistor sizes?

(d)   Two currents are chosen such that one is 3/5 the other and their sum is 50 mA. What are the currents?

(e)   The smaller of two resistors is 100 Ω. What is the size of the larger one if their difference is 250 Ω?

(f)   A rectangle has a perimeter (the sum of the length of its sides) of 85 cm. What are the lengths of the sides if:

(1) It is square?

(2) The length is twice the width?

(3) The length is 5 cm longer than the width?

(4) The length is 5 cm shorter than the width?

(g)   If 39 is added to a certain number, the result is the same as 5 times the number diminished by 9. What is the number?

UNIT 8
SERIES CIRCUITS

8.1 CIRCUIT SYMBOLS

An *electric circuit* consists of a closed path around which an electric current can exist. There must be a *source* of electric energy, such as an electric generator, or an electronic power supply, or a battery. There must also be a device that uses the electric energy supplied (*load*), such as an electric heater, a toaster, a television set, an electric motor, etc. In general, an electric circuit can be formed by connecting a *source* to a *load*.

For example, the electric circuit in the transistor radio in Figure 8.1 consists of a 9 V battery (source) connected to the transistor radio circuit board (load). The electric current exists around the closed path formed by connecting the battery and the circuit board.

It is common practice to use symbols for a battery and a resistive load, as shown in Figure 8.2. Using symbols, it is possible to draw a circuit diagram (schematic) for the transistor radio, as shown in Figure 8.3. Note that the source V is the 9 V battery, and the resistance R represents the circuit board load. The closed path around which the electric current I exists is shown by the dotted line. A number of common electric circuit symbols are shown in the Appendix at the end of this text.

The two components (the source V and the load R) in Figure 8.3 are said to be connected in *series* since the current is the same in both components, and this is called a *series circuit*.

EXAMPLE 8.1. Use symbols to draw the series circuit formed by connecting the following:	SOLUTION.

EXAMPLE 8.1.   Use symbols to draw the series circuit formed by connecting the following:
(a) A 12 V auto battery and a 12 V lamp.
(b) Two 1.5 V flashlight cells connected in series, and a 3 V lamp in series with them.
(c) A 117 V ac source and a 1.5 kW resistance heater.
(d) A 9 V battery and two 10 kΩ resistors.

SOLUTION.

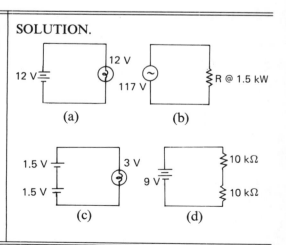

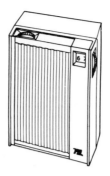

(a) Transistor radio

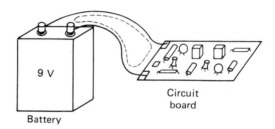

9 V

Battery

Circuit board

Figure 8.1

(b) Source and load

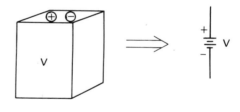

V

V

(a) Battery

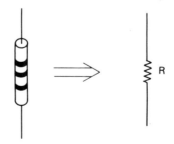

R

Figure 8.2

(b) Resistance

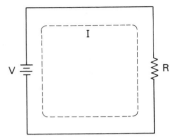

Figure 8.3

Exercises 8.1

Use circuit symbols to draw an electric circuit diagram for the following components connected in series:

1. A 12 V battery and a 2 W lamp
2. A 1.5 V battery and a 3.3 kΩ resistor
3. A 100 V dc source and a 100 Ω, 10 W resistor
4. A 230 V ac source and a 3 kW heater
5. A 117 V ac source and a 1 hp motor
6. A 30 V dc source, a 1 kΩ resistor, and a 2.2 kΩ resistor

8.2 OHM'S LAW

There are three variable quantities in the series circuit in Figure 8.3: These are the voltage V, the current I, and the resistance R. Ohm's Law is the formula that includes these three variables, and as has been seen previously, Ohm's Law can be solved for each variable. That is,

$$I = \frac{V}{R} \qquad V = IR \qquad R = \frac{V}{I}$$

These three formulas can be easily displayed using the diagram in Figure 8.4. It is

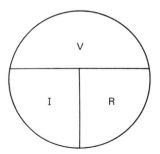

Figure 8.4

possible to find a numerical value for one variable, provided the numerical values for the other two variables are known. In Figure 8.4, if one variable is covered, the relationship describing it in terms of the other two variables can be seen.

EXAMPLE 8.2.   In Figure 8.1, the battery is 9 V and the circuit current is 30 mA. Find the resistive load R of the circuit board.

SOLUTION. This is a simple series circuit, as shown in Figure 8.3. Using Ohm's Law,

$$R = \frac{V}{I} = \frac{9\text{ V}}{30\text{ mA}} = 300\ \Omega$$

EXAMPLE 8.3.   If the 12 V lamp used in Example 8.1(a) has a resistance of 120 Ω, what is the circuit current?

SOLUTION.   Using Ohm's Law,

$$I = V/R = 12\text{ V}/120\ \Omega = 0.1\text{ A}$$

EXAMPLE 8.4.   What must be the value of the voltage V in Figure 8.5?

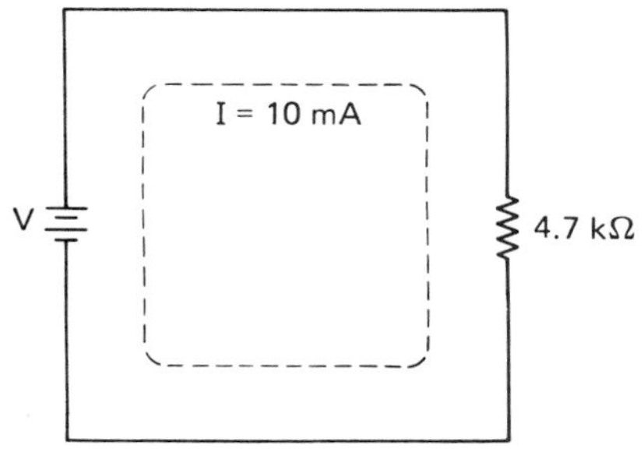

SOLUTION.

$$V = IR = 10\text{ mA} \times 4.7\text{ k}\Omega = 47\text{ V}$$

I = 10 mA

V

4.7 kΩ

Figure 8.5

Exercises 8.2

1. The voltage measured across the 330 Ω resistor in Figure 8.6 is 5.8 V. What must be the current through the resistor?

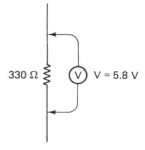

330 Ω   Ⓥ   V = 5.8 V

Figure 8.6

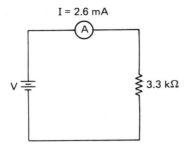

I = 2.6 mA

Figure 8.7

2. The measured current in the 3.3 kΩ resistor in Figure 8.7 is 2.6 mA. What must be the applied voltage V?

3. What must be the resistance of a 6.3 V lamp if it draws a current of 100 mA?

4. The resistance of a heating element is 15 Ω. What current will it require if it is connected to a 230 V source?

5. What must be the current through the 470 Ω cathode bias resistor R_k in Figure 8.8 if the voltage across it is 3 V?

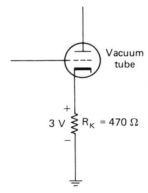

Vacuum tube

3 V R_K = 470 Ω

Figure 8.8

6. What should be the reading of a voltmeter connected across a 2.7 kΩ resistor if the resistor current is known to be 20 mA?

7. The picture tube voltage in a TV is 27 kV, and it draws a current of 2.5 mA. What is the equivalent load resistance?

8. What is the equivalent load resistance of a transistor radio that draws 11.5 mA from a 8.9 V battery?

9. What must be the voltage V_{R_E} across the 10 kΩ emitter resistor R_E in Figure 8.9 if the current is known to be 0.9 mA?

10. What must be the current in the 39 kΩ base resistor R_B in Figure 8.9 if the voltage across it is known to be 0.2 V?

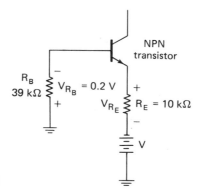

Figure 8.9

11. The emitter resistor R_E in Figure 8.9 must be changed such that it will conduct a current of 0.3 mA when the voltage across it (V_{R_E}) is 8.8 V. Find R_E.

8.3   POWER

The energy delivered by the source is consumed by the load in Figure 8.3, and the *rate* at which this energy is delivered (or consumed) is defined as the power P. Equation (8.1) provides a formula for calculating the power in this series circuit. The power is simply the product of the voltage V and the current I; that is,

$$P = VI \qquad (8.1)$$

Recall the basic unit for power is the watt, provided the voltage is given in volts and the current is given in amperes.

EXAMPLE 8.5 What is the power consumed by the transistor radio in Example 8.2?	SOLUTION.
	$P = VI = 9 \text{ V} \times 0.03 \text{ A} = 0.27 \text{ W}$
	since
	$30 \text{ mA} = 0.03 \text{ A}$

Two other formulas for calculating the power in the simple series circuit are shown below:

$$P = \frac{V^2}{R} \qquad (8.2)$$

$$P = I^2 R \qquad (8.3)$$

EXAMPLE 8.6. Use Eq. (8.2) to calculate the power in a 12 V, 120 Ω lamp.	SOLUTION. $$P = V^2/R = \frac{12^2}{120} = 1.2 \text{ W}$$
EXAMPLE 8.7. Use Ohm's Law and Eq. (8.1) to derive Eq. (8.3).	SOLUTION. $$P = VI \qquad V = IR$$ Since $V = IR$, substitute IR for V in the first equation to obtain $$P = VI = (IR)I = I^2R$$

Since power is the rate per unit time at which energy is delivered or consumed,

$$P = \frac{W}{t} \tag{8.4}$$

where P is the power in watts, W is the energy in joules, and t is the time in seconds. This formula can be used to calculate the power, provided the energy and the time are known. On the other hand, it is possible to solve for the energy by multiplying both sides by t as follows:

$$P \times t = \frac{W}{\cancel{t}} \times \cancel{t}$$

or
$$W = Pt \tag{8.5}$$

EXAMPLE 8.8. How much energy is consumed by a 100 W lamp in a period of 24 hours?	SOLUTION. Using Eq. (8.5), $$W = Pt = 100 \times 24 \times 60 \times 60 = 8.64 \times 10^6$$ joules Note that the time is $24 \times 60 \times 60$ seconds.

Exercises 8.3

1. A load draws 10 A when connected to a 117 V source. What is the power delivered to the load?
2. What is the power delivered by a 9 V battery supplying a current of 8 mA?
3. What is the current in a 6.3 V, 1 W lamp?
4. What power is dissipated in the resistor in Figure 8.6?
5. What is the power in the resistor in Figure 8.7?

6. Calculate the power in the cathode resistor R_k in Figure 8.8.

7. What is the maximum voltage that can be safely applied to a 5.6 kΩ, 2 W resistor?

8. What is the maximum current allowed in a 470 Ω, 1/2 W resistor?

9. Show whether or not it is safe to connect a 50 Ω, 10 W resistor to a 117 V source.

10. What is the current drawn by a $47\frac{1}{2}$ W soldering iron connected to a 117 V source?

11. One horsepower (hp) is equivalent to 746 W. Solve the following:
    (a) What current is needed for a 1 hp motor connected to a 230 V source?
    (b) What is the hp of a 117 V motor that draws 12.75 A?
    (c) A motor draws 7 A when connected to a 230 V source. If it operates with an efficiency of 90%, what is its output horsepower?
    (d) What is the efficiency of a 117 V motor rated at 1/2 hp if it draws a current of 3.5 A?

12. What is the electric power in watts if electric energy is consumed at the rate of:
    (a) 1 joule per sec?   (b) 10 joules per sec?
    (c) 500 joules per min?   (d) 10^5 joules per hour?

13. Electric energy (W) measured in kilowatthours (kWh) is found by taking the product of the power (P) in kW and the time (t) in hours (h). Thus,

$$W(\text{kWh}) = P(\text{kW}) \times t(\text{h})$$

    (a) What is the total energy supplied to a 1.5 kW heater during a 10 hour period?
    (b) A TV set rated at 400 W is operated for 15 hours. What is the cost for energy if the base price is 4¢ per kWh?
    (c) What is the cost to operate a 1 1/2 hp motor for 360 hours if energy costs 5.7¢ per kWh?

14. How many joules of electric energy are consumed by:
    (a) A 47 1/2 W soldering iron in one hour?
    (b) A 1 hp motor in one hour (1 hp=746 W)?
    (c) A 1.2 W pilot lamp in one hour?

15. If energy costs 5¢ per kWh, what is the cost of 1 joule of energy?

ALGEBRA

UNIT 1
GRAPHING

1.1 LINEAR SCALE GRAPHS

In the field of electronics a great deal of information is conveniently displayed in *graphical* form. There are a number of different methods for displaying information graphically, but any graph is simply a pictorial representation of a relationship between two or more quantities.

The information in Table 1.1 shows the variation in resistance for a diffused resistor in a silicon integrated circuit (IC) as the temperature is varied.

Table 1.1

R (kΩ)	1.03	1.00	1.03	1.06	1.10	1.13	1.16	1.20
Temperature (°C)	−40	−20	0	20	40	60	80	100

The data were obtained by placing the resistor in an enclosure in which the temperature can be controlled, and then measuring the value of R at a number of different temperatures. These are *experimental* data obtained by actual laboratory measurement, as contrasted with *theoretical* data that would be obtained by finding solutions to an equation or formula.

The data in Table 1.1 can be presented in graphical form, as shown in Figure 1.1. The pictorial presentation offered by the graph makes it quite easy to see just how the resistance varies as the temperature changes. Each measured value in the data table corresponds to a single point on the graph. For example, at −20°C the value of R is 1 kΩ; this is the point directly above −20°C on the horizontal line (scale), and to the left of 1.0 on the vertical scale. All the other data points are located (plotted) in a similar manner.

Notice that the graph in Figure 1.1 really consists of a number of straight-line segments drawn between adjacent data points. This graph gives a good indication of the manner in which R varies with temperature changes, but the line appears segmented (angular) because not enough data points were taken. If the resistance were measured at intervals of 5°C, instead of the 20°C intervals in Table 1.1, there would be four times as many data points, and the graph would appear as in Figure 1.2(a). Notice that here the segments are much smaller, and the graph is much *smoother*. The number of readings taken could of course be increased, and eventually the graph would be a smooth, continuous curve, as shown in Figure 1.2(b).

Generally speaking, it is desirable to include as many data points as necessary in order to allow us to draw a smooth (not segmented) curve that passes through all of the data points. One must take care in choosing how many data points; clearly the two data points plotted in Figure 1.3 are insufficient to show the true variation in R.

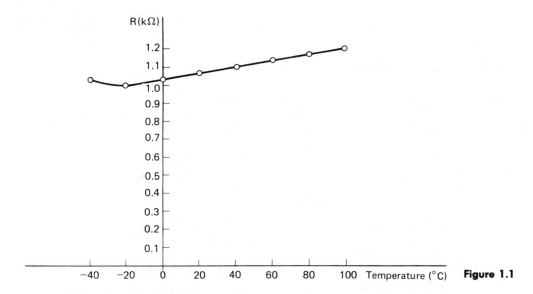

Figure 1.1

One must also choose the horizontal and vertical scales such that the resulting graph displays the desired data in a useful form. For example, the data in Table 1.1 are plotted in Figures 1.4(a) and (b), but the graphs are not satisfactory due to a poor choice of scales. In Figure 1.4(a), the interval on the horizontal scale is much too large, and the graph is "squeezed" together (shortened). In Figure 1.4(b), the interval on the vertical scale is too large, and the resistance variation is not revealed —the graph is flattened out. On the other hand, if the scale intervals are chosen too small, it will not be possible to plot all the data.

All of the graphs discussed so far have *linear* scales; that is, the scales are divided into equal intervals, and each interval has the same value. For example, the linear scale below has ten equal intervals, and each interval is equivalent to 5°C.

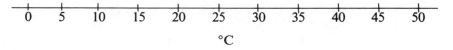

°C

Exercises 1.1

Plot graphs for each of the following sets of data. Choose the horizontal and vertical scales carefully and label each graph clearly.

1. A passivating layer of silicon dioxide (SiO_2) is deposited on the surface of a silicon wafer during the processing of integrated circuits (ICs). The thickness of this grown layer depends on time, as shown by the following data. Plot oxide

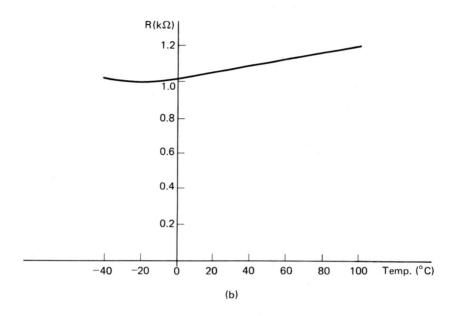

(b)

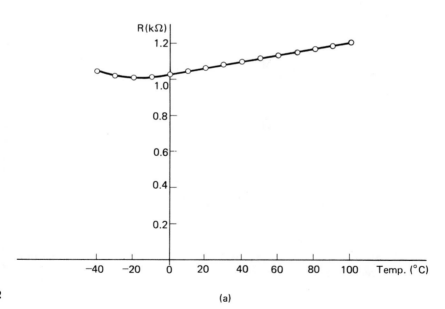

Figure 1.2

(a)

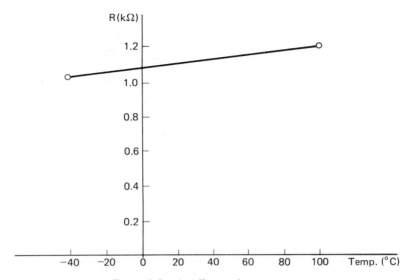

Figure 1.3.   Insufficient data points.

thickness in microns on the vertical scale and time in sec on the horizontal scale.

Thickness (microns)	1.0	2.5	3.0	3.7	4.1	4.6	5.0	5.3	5.8	6.0
Time (sec)	10	20	30	40	50	60	70	80	90	100

2. The values of diode voltage V_D and diode current I_D were recorded for a semiconductor diode. Plot I_D on the vertical scale and V_D on the horizontal scale.

V_D (V)	0	0.1	0.2	0.3	0.33	0.35	0.37	0.4
I_D (mA)	0	0.055 μA	2.98 mA	0.163	0.54	1.20	2.68	8.89

3. The values of plate voltage and plate current for a 6CL6 vacuum tube triode were recorded. Plot with plate current on the vertical scale and plate voltage on the horizontal scale.

V_p (V)	110	130	150	170	190	210	230	250
I_p (mA)	0.5	3.0	6.5	13.0	21.5	32.0	44.0	58.0

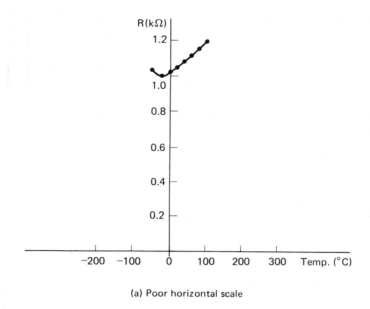

(a) Poor horizontal scale

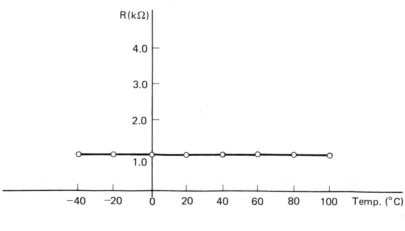

Figure 1.4

(b) Poor vertical scale

4. The variations in the resistance of a carbon composition resistor as the temperature varied were recored. Plot with resistance on the vertical scale and temperature on the horizontal scale.

R (kΩ)	1.010	1.005	1.000	0.995	0.990	0.985	0.980	0.975
Temperature (°C)	0	10	20	30	40	50	60	70

5. The terminal voltage V_T of a power supply was recorded as its load current I_L was varied. Plot V_T versus I_L with V_T on the vertical scale.

V_T (V)	10.0	9.98	9.96	9.94	9.92	9.90	9.88	9.86	9.84
I_L (mA)	0	20	40	60	80	100	120	140	160

1.2 LOGARITHMIC SCALE GRAPHS

Sometimes the data to be plotted vary over so great a range that a linear scale is not satisfactory, and it will be necessary to use a *logarithmic scale*. The C and D scales on a slide rule are logarithmic scales. A five-interval log scale is shown in Figure 1.5(a). Note that each interval has a value of two, but the *length* of each interval is different. Also, note carefully that the scale begins at 1 (not zero) and ends at 10. A ten-interval log scale is shown in Figure 1.5(b).

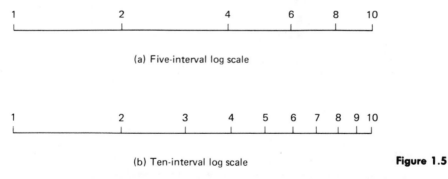

(a) Five-interval log scale

(b) Ten-interval log scale
Figure 1.5

Each of the scales in Figure 1.5 is called a *cycle*, and a number of cycles could be used to form a scale for a graph. The key to using a log scale is realizing that the end points of each cycle are separated by a power of ten. For example, here is a three-cycle log scale beginning at 1 (10^0) and ending at 1000 (10^3).

The same three-cycle could be labeled to cover a range from 0.01 to 10 as follows:

$$10^{-2} \qquad 10^{-1} \qquad 1 \qquad 10$$

Graphs having one log scale and one linear scale as in Figure 1.6(a) are called *semilog* graphs. It is also possible to have graphs in which both scales are

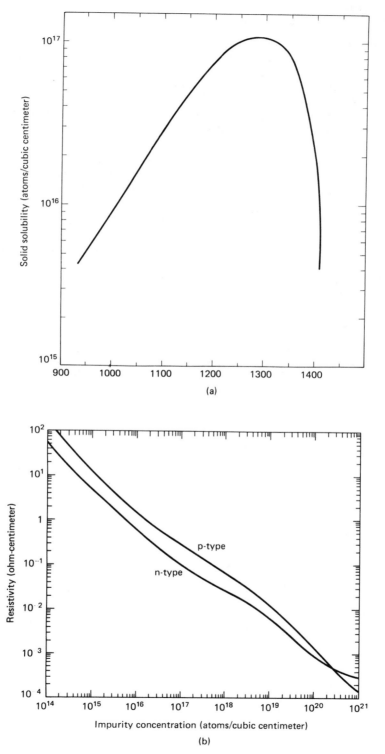

Figure 1.6. (a) Solid solubility of gold (Au) in silicon (Si); (b) resistivity of silicon (Si) at 300°K.

logarithmic as in Figure 1.6(b), and these are called *log-log* graphs. Special graph paper printed on $8\frac{1}{2}$ in. by 11 in. sheets in either semilog or log-log format is readily available. It is always necessary to specify how many cycles are desired when purchasing such paper.

Exercises 1.2

1. The solid solubility of aluminum (Al) in silicon (Si), added as an impurity during the manufacture of semiconductor devices, is recorded below. Plot the data as follows: horizontal scale—temperature on a linear scale beginning at 600°C and ending at 1500°C; vertical scale—solid solubility on a 3-cycle log-scale beginning at 10^{17} and ending at 10^{20} atoms/cm.

Solid solubility (Atoms/cm³)	6×10^{18}	9×10^{18}	0.5×10^{19}	1×10^{19}	1.9×10^{19}	2×10^{19}	1.9×10^{19}	1×10^{19}	2×10^{18}
Temperature (°C)	600	700	800	900	1000	1100	1200	1300	1400

2. The voltage gain A_V of a hi-fi amplifier has been measured at various frequencies (a frequency response). Plot as follows: horizontal scale—a 4-cycle log-scale from 10 to 10^5 Hz; vertical scale—a linear scale from 0 to 40 dB.

A_V (dB)	11	23	35	38	38	38	38	38	38	38	38	35	23	11
freq (Hz)	10	20	40	60	100	200	500	10^3	2×10^3	5×10^3	10^4	2×10^4	4×10^4	8×10^4

3. Use a log-log graph to plot resistivity (ρ) of n-type silicon on the vertical scale versus impurity concentration on the horizontal scale.

$\rho(\Omega\text{-cm})$	50	4.5	0.6	0.1	2.8×10^{-2}	6.2×10^{-3}	9×10^{-4}
Impurity concentration (cm⁻³)	10^{14}	10^{15}	10^{16}	10^{17}	10^{18}	10^{19}	10^{20}

4. Plot the following data on a log-log graph.

Oxide thickness (microns)	8×10^{-2}	1.5×10^{-1}	2.6×10^{-1}	4.5×10^{-1}	8.5×10^{-1}	1.5	2.8
Time (min)	10	30	100	300	1000	3000	10^4

5. Read values for the resistivity of silicon from the graph in Figure 1.6(b) for the following values of impurity concentration:
   (a) 10^{15} (b) 10^{18} (c) 3×10^{17} (d) 2×10^{15} (e) 8×10^{19}

UNIT 2
STRAIGHT-LINE GRAPHS

2.1 LINEAR FUNCTIONS

In the previous section it was shown how to make graphs by plotting data that have been obtained experimentally. It is also possible to construct graphs that give us a pictorial presentation of a mathematical equation or a formula.

Recall that an equation can be solved for one variable in terms of another variable. For example, Ohm's Law can be expressed as $V = RI$. V is called the *dependent variable* since its value depends on I; I is called the *independent variable*; and R is a *constant*. When Ohm's Law is written in this form, we say, "V is a function of I." A *function* is defined as:

FUNCTION

If a correspondence exists between any two variables x and y such that there exists exactly one value of y for each value of x, then y is a function of x.

In order to graph an equation, construct a *data table* as shown in Figure 2.1. Then assign values for the independent variable (x) as desired, and use the given equation to calculate the *theoretical* values for the dependent variable y. Then plot these data on a graph having suitable scales, in the same way the experimental data in the previous section were graphed.

EXAMPLE 2.1. Given Ohm's Law in the form $V = RI$, complete the data table by calculating the values for V for each given value of I. $R = 25 \ \Omega$.

I(A)	0	1	2	3	4	5	6	7
V(V)								

SOLUTION. Use $V = RI = 25I$ to calculate each value of V.
For $I = 0$, $V = 25 \times 0 = 0$.
For $I = 1$, $V = 25 \times 1 = 25$, and so on.

I(A)	0	1	2	3	4	5	6	7
V(V)	0	25	50	75	100	125	150	175

EXAMPLE 2.2.   Graph the data calculated in Example 2.1.

SOLUTION   Locate each point (I, V) in the data table and draw a straight line connecting the points.

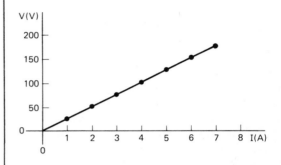

x							
y							

Figure 2.1.   Data table.

Notice that the graph of Ohm's Law, $V = RI$, in Example 2.2 is a *straight line*. In fact, the graph of any equation of the form $y = mx + b$ will be a straight line. In this equation, y is the dependent variable, x is the independent variable, and m and b are constants. Ohm's Law is in this form if we set $b = 0$, $m = R$, $y = V$, and $x = I$ (try it!).

Since the graph of any equation in this form is a straight line, this is called a *linear equation*, and we can say, "y is a linear function of x." Note that both y and x are raised to the first power (that is, $y^1 = y$ and $x^1 = x$), and this is said to be an equation of the *first degree*. Thus, Ohm's Law (and any equation of the form $y = mx + b$) is a linear equation; it is the equation of a straight line, and it is a first-degree equation.

Exercises 2.1

Complete a data table and plot a graph for each of the following linear equations:

1. $V = RI$; $R = 1$ kΩ $0 \leqslant I \leqslant 10$ mA;   in 1 mA steps

2. $I = \dfrac{1}{R} V$: $R = 2.2$ kΩ $0 \leqslant V \leqslant 100$ V;   in 10 V steps

3. $X_L = \omega L$: $L = 5$ H $0 \leqslant \omega \leqslant 1000$ Hz;   in 100 Hz steps

4. $q = CV$: $C = 2$ F $0 \leqslant V \leqslant 50$ V;   in 5 V steps

5. $Q = \omega L / R$: $\dfrac{L}{R} = 10^{-3}$ $200 \leqslant \omega \leqslant 1000$ Hz;   in 100 Hz steps

6. $P = VI$: $V = 230$ V $0 \leqslant I \leqslant 25$ A;   in 5 A steps

7. $C = \pi d$: $\pi = 3.14159$ $0 \leqslant d \leqslant 36$ in.;   in 3 in. steps

8. $A = h \times W$: $h = 7$ cm $0 \leqslant W \leqslant 45$ cm;   in 5 cm steps

9. $y = 7x$: – $0 \leqslant x \leqslant 6$;   in 0.5 steps

10. $y = 2x + 1$: – $0 \leqslant x \leqslant 5$;   in 1.0 steps

2.2   SLOPE

When constructing the graph of a linear function, it is possible to choose both positive and negative values for the independent variable. A standard *coordinate system* as shown in Figure 2.2 is widely used to plot any function. It accounts for both positive and negative values.

The horizontal line is called the *x-axis*, the vertical line is the *y-axis*, and their intersection (O) is called the *origin*. Horizontal distances measured to the *right* of the origin are *positive*, while distances to the *left* of the origin are *negative*. Similarly, vertical distances *above* the origin are *positive*, while those *below* the origin are *negative*. The two axes divide the plane into four *quadrants*, labeled I, II, III, and IV in Figure 2.2.

Any point on the coordinate system in Figure 2.2 can be uniquely determined by specifying a value for x and a value for y. For example, the point P_1 is the only place where $x = 2$ and $y = 3$. For any unique point, the value of x is said to be the *x-coordinate* (*abscissa*), the value of y is said to be the *y-coordinate* (*ordinate*), and the two values together are called the *coordinates* of the point P.

The coordinates of any point are written in the form $P(x,y)$. For example, the point P_1 in Figure 2.2 is specified by writing $P_1(2,3)$.

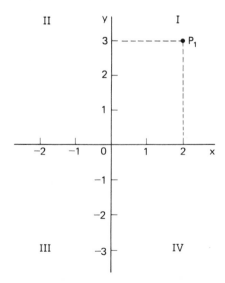

Figure 2.2

EXAMPLE 2.3. Plot the following points on an $x - y$ coordinate system:
(a) $A(3,3)$
(b) $B(-1,5)$
(c) $C(-4,-3)$
(d) $D(2,-4)$

SOLUTION.
(a) $A(3,3)$
    $x = 3$ and $y = 3$
(b) $B(-1,5)$
    $x = -1$
    $y = 5$
(c) $C(-4,-3)$
    $x = -4$
    $y = -3$
(d) $D(2,-4)$
    $x = 2$
    $y = -4$

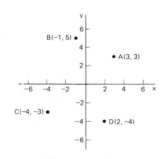

EXAMPLE 2.4. Complete the data table and plot on an x-y coordinate system for the function $y = -4x$.

x	-3	-2	-1	0	1	2	3
y							

SOLUTION. Calculate each point as follows: $x = -3$, $y = -4x = -4(-3) = 12$, and so on.

x	-3	-2	-1	0	1	2	3
y	12	8	4	0	-4	-8	-12

Plot the points and connect them with a straight line.

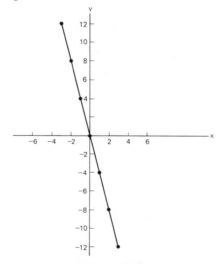

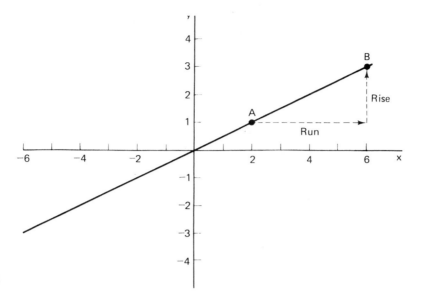

Figure 2.3

Consider now the straight line graphed in Figure 2.3. Choose any two points on the graph, for example, points *A* and *B* as shown. To move from point *A* to point *B*, travel a horizontal distance to the right (*run*) and then a vertical distance up (*rise*) as shown by the dotted lines. For any straight line, the *slope* of that line is defined as the quotient of the rise divided by the run. Note carefully that run is movement to the right and rise is movement up.

SLOPE

The slope of any straight line is the number obtained by taking the quotient of rise over run between any two points on the line.

In equation form,

$$\text{Slope} = \frac{\text{rise}}{\text{run}} \tag{2.1}$$

EXAMPLE 2.5. Use points *A* and *B* to calculate the slope of the line in Figure 2.3.	SOLUTION. Run $= 6 - 2 = 4$, and rise $= 3 - 1 = 2$. $$\text{Slope} = \frac{\text{rise}}{\text{run}} = \frac{2}{4} = \frac{1}{2}$$

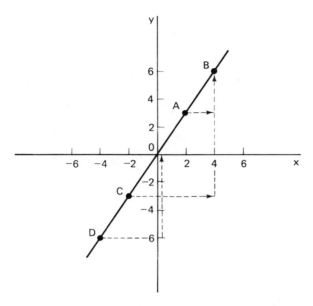

Figure 2.4

Although it may not seem true at first glance, it is a fact that the slope of a straight line is constant. Thus we can choose *any* two points desired on the line, and use the run and the rise between these two points to calculate the slope. No matter which two points are chosen, the slope will always be the same. This is illustrated in Figure 2.4.

| EXAMPLE 2.6. Calculate the slope of the line in Figure 2.4, using points:
(a) *A* and *B*
(b) *B* and *C*
(c) *D* and *0* | SOLUTION.
(a) Moving from point *A* to *B*, the run is 2 and the rise is 3.

Slope = rise/run = 3/2

(b) Moving from *C* to *B*, the run is 6 and the rise is 9.

Slope = 9/6 = 3/2

(c) Moving from *D* to 0, the run is 4 and the rise is 6.

Slope = 6/4 = 3/2 |

Examine the straight line graphed in Figure 2.5. Choose the origin *O* and point *A* as two points on the line in order to determine the slope. In moving from *O* to *A*, the run is to the left—this is the *negative* direction on the *x*-axis; and thus the run is a negative number, that is, run = −1. The rise is still upward and is +2.

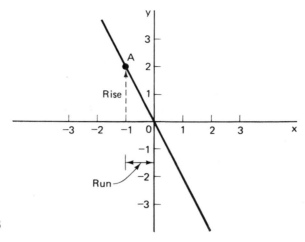

Figure 2.5

Thus, the slope is found to be

$$\text{Slope} = \frac{\text{rise}}{\text{run}} = \frac{+2}{-1} = -2$$

Notice that the same result would be obtained by starting at A and moving to O. In this case, the run would be $+1$ (moving from A to the y-axis), and the rise would be -2 (moving *down* from A to the x-axis). Thus the slope would be

$$\text{Slope} = \frac{\text{rise}}{\text{run}} = \frac{-2}{+1} = -2$$

Since the slope of a straight line is determined by moving between any two points on the line, the rise can be considered as a *change in y*, written Δy, as shown in Figure 2.6. Similarly, the run can be considered as a *change in x*, written Δx.

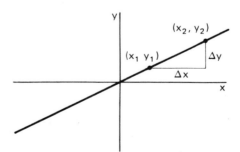

Figure 2.6

Notice that $\Delta y = y_2 - y_1$ and $\Delta x = x_2 - x_1$. Thus,

$$\text{Slope} = \frac{\text{rise}}{\text{run}} = \frac{\Delta y}{\Delta x} = \frac{y_2 - y_1}{x_2 - x_1} \tag{2.2}$$

EXAMPLE 2.7. Find the slope of each line in Figure 2.7.

SOLUTION. Using Eq. (2.2):
Line AA', between 0 and a,

$$\text{Slope} = \frac{3}{3} = 1$$

Line BB', the rise is zero,

$$\text{Slope} = \frac{0}{x} = 0$$

Line CC', between $y = -5$ and $y = +1$,

$$\text{Slope} = \frac{6}{-2} = -3$$

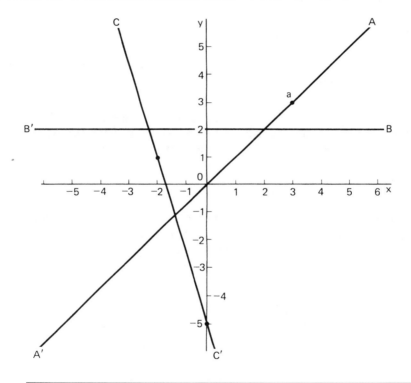

Figure 2.7

Exercises 2.2

1. Determine the slope for each equation in Exercises 2.1.
2. A straight line passes through the origin and the point (5,9). Determine its slope.

3. Determine the slope of the straight line passing through the points:
   (a) $(0, 1)$ and $(3, 5)$ (b) $(-2, -2)$ and $(1, 1)$ (c) $(1, 1)$ and $(-3, 2)$
   (d) $(-3, -3)$ and $(-2, 5)$ (e) $(-1, 2)$ and $(4, 2)$ (f) $(1, -2)$ and $(1, 3)$
   Note that division by zero $(y/0)$ is undefined, but we can say it approaches infinity (∞)—that is, $y/0 = ? \rightarrow \infty$.
4. Plot the following lines and determine their slopes. What can you say about the slopes of parallel lines?
   (a) $y = 2x + 2$ (b) $y = 2x - 3$ (c) $y = 2x$
5. It is desired to draw a straight line with a slope of -3 that passes through the point $(1, 4)$. Find the point where the line crosses:
   (a) The x-axis   (b) The y-axis

2.3   THE SLOPE-INTERCEPT EQUATION

In Section 2.1 it is stated that any equation of the form

$$y = \underset{\text{(slope)}}{mx} + \underset{(y\text{-intercept})}{b} \qquad (2.3)$$

is an equation of a straight line. It is also stated that y and x are the dependent and independent variables, and that m and b are constants. In fact, m is the slope of the line and b is the y-intercept (the point where the line crosses the y-axis). Thus, Eq. (2.3) is known as the slope-intercept-equation of a straight line.

As an example, the equation $y = 2x + 1$ has a slope of $+2$, and it crosses the y-axis at the point $(0, 1)$. Note that if $x = 0$, $y = 1$—thus the point $(0, 1)$. This equation is plotted in Figure 2.8.

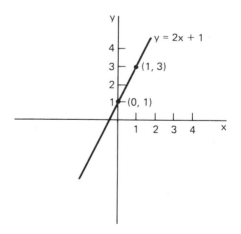

Figure 2.8

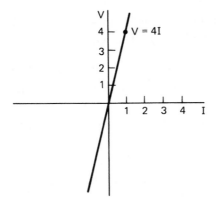

Figure 2.9

Ohm's Law can be written as $V = RI$. If the value of R is chosen as 4, then $V = 4I$. This is the equation of a straight line having a slope of 4 and crossing the V-axis at the point $(0,0)$. The variables have been changed by letting $x = V$ and $x = I$, but the equation is still in the form $y = mx + b$, where $b = 0$ and $m = 4$. The graph of this equation is shown in Figure 2.9.

Notice that in order to plot any straight line, it is necessary to determine only two points. It is not necessary to have an entire data table! Given any equation in the standard slope-intercept form of Eq. (2.3), the value of one point is immediately available—the y-intercept; it is equal to b. Then use the slope $m = \text{rise}/\text{run}$ to move from b to the second point. Then draw the straight line through these two points. For example, the equation $y = 2x + 1$ is plotted in Figure 2.8. It is easy to draw this graph by: (1) plotting the y-intercept $(0,1)$; (2) using the slope $m = 2 = \text{rise}/\text{run} = 2/1$ to move from $(0,1)$ to $(1,3)$—that is, run 1 and then rise 2; and (3) drawing a straight line through these two points.

To summarize:

GRAPHING

To graph the equation of a straight line in slope-intercept form, $y = mx + b$:
(1) Plot the y-intercept b.
(2) Use the slope m to locate a second point.
(3) Draw the straight line through the two points.

EXAMPLE 2.8. Graph the equation

$$y = -x - 2$$

SOLUTION.
(1) The y-intercept is at -2; thus the point $(0, -2)$.

(2) The slope $= m = -1 = 1/-1 = \text{rise}/\text{run}$. Thus rise $+1$ and run (left) -1 to the point $(-1, -1)$.

(3) Draw the line as in Figure 2.10.

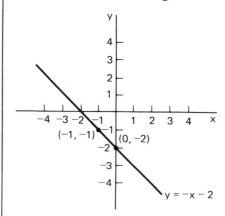

Figure 2.10

The slope-intercept form of an equation can be used to determine a suitable equation for experimental data. It is easy to look at a graph of a straight line and determine (a) the y-intercept b, and (b) the slope m.

Consider the data graphed in Example 2.1. Since the line passes through the origin, the y-intercept is clearly $0(b=0)$. The slope is seen to be $m = 100/4 = 25$. And thus the slope-intercept equation can be written as

$$y = mx + b$$
$$y = 25x + 0$$

or
$$V = 25I$$

This is the equation used in Example 2.1.

To summarize:

SLOPE-INTERCEPT EQUATION

To determine the equation of a straight line from graphical data:
(1) Determine the y-intercept b.
(2) Determine the slope m.
(3) Write in the form $y = mx + b$. Change variables if necessary.

EXAMPLE 2.9. Determine an equation for each line in Figure 2.7.

SOLUTION.

Line AA':

$$y\text{-intercept} = b = 0$$

$$\text{Slope} = 1$$

using $y = mx + b$, $y = 1x + 0$. Thus, $y = x$.

Line BB':

$$b = 2$$

$$\text{Slope} = m = 0$$

Thus,

$$y = mx + b = 0 \times x + 2$$

$$y = 2$$

Line CC':

$$b = -5$$

$$\text{Slope} = m = -3$$

Thus,

$$y = mx + b$$

$$y = -3x - 5$$

Exercises 2.3

1. Draw the graph of each of the following equations by finding two points from the slope and the y-intercept. Plot a couple of extra points to serve as a check. You may have to arrange the equation to get it into standard slope-intercept form.

   (a) $y = 2x$
   (b) $y = 3x + 1$
   (c) $y = x - 2$
   (d) $y = -2x + 1$
   (e) $y = -x - 3$
   (f) $y + x = 2$
   (g) $2y - x = 3$
   (h) $3y + 6x - 9 = 0$

   (i) $3y = x - 1$
   (j) $V = 9I$
   (k) $P = 20I + 3$
   (l) $P = 10V + 10$
   (m) $I = \dfrac{1}{10} V$
   (n) $R = 0.01T + R_0$
   (o) $V_T = -0.1I + 6$

2. Plot the following three graphs on the same coordinate system: $x - y = 2$, $x - y = 0$, $x - y = -3$. What do the three lines have in common?

3. Draw the graphs of the equations $y = 3x$, $y = -7x$, and $y = x$ on the same coordinate system. What do they have in common?

4. Graph the following equations:

   (a) $y = 3$ (b) $y = -1$ (c) $x = 2$ (d) $x = -3$

5. A graph of resistance versus temperature for a piece of iron wire is given in Figure 2.11. Write an equation giving R as a function of T in the form of Eq. (2.3).

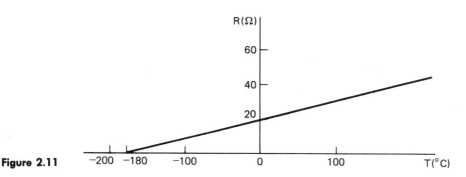

Figure 2.11

6. A voltage source and its $V - I$ curve are shown in Figure 2.12. Write an equation in the form of Eq. (2.3), giving V_T as a function of I_L.

7. The voltage source in Figure 2.12 would be a *perfect*, *ideal* source if $R_S = 0\ \Omega$. Draw the $V - I$ curve (V_T vs. I_L) and write an equation with V_T as a function of I_L for such an ideal source.

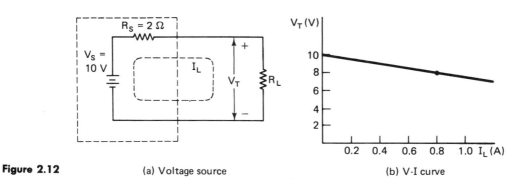

Figure 2.12   (a) Voltage source   (b) V-I curve

8. The voltage source in Figure 2.13 was set up in a laboratory and the following measurements were taken: (i) V_T was found to be 28 V dc with R_L disconnected; (ii) V_T was found to be 27.8 V dc with an R_L of 100 Ω connected. Draw a graph of V_T versus I_L (V_T on the vertical axis), and write an equation giving V_T as a function of I_L.

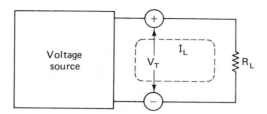

Figure 2.13

9. Repeat Problem 8 for a 100 V dc source whose terminal voltage drops to 97 V dc with a 150 Ω load connected to its terminals.

10. A plot of diode current I_Z versus diode voltage V_Z for a zener diode operating in the reverse breakdown region is shown in Figure 2.14. Write an equation giving I_Z as a function of V_Z in the form of Eq. (2.3).

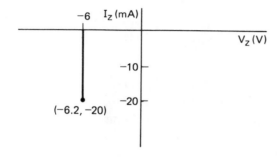

Figure 2.14

UNIT 3
ALGEBRAIC FRACTIONS—I

3.1 REDUCTION

If Ohm's Law is written in the form $I = V/R$, the term V/R is called an *algebraic fraction*; V is the *numerator* and R is the *denominator*. Another algebraic fraction is encountered when the equation for conductance G is written in terms of resistance R; that is, $G = 1/R$. In the special case where the numerator is unity (1), use the term *reciprocal*, and say, "G is the reciprocal of R," or, "G is one over R," or, "G is one divided by R." Algebraic fractions are frequently encountered, and it is necessary to know how to handle them.

Fortunately, the rules for operating on algebraic fractions are precisely the same as the rules for arithmetic fractions. If you are not sure of your ability to handle arithmetic fractions, now is a good time for review.

The *reduction to lowest terms* can be accomplished by making use of the Laws of Exponents, and the following fundamental principle for algebraic fractions:

FRACTIONS

The value of a fraction is not changed if both the numerator and the denominator are multiplied or divided by the same quantity (zero excluded).

For example, divide both the numerator and denominator of the fraction $10/15$ by 5 to show that

$$\frac{10}{15} = \frac{\dfrac{10}{5}}{\dfrac{15}{5}} = \frac{2}{3}$$

Similarly, division of numerator and denominator by some constant a, yields

$$\frac{ax}{ay} = \frac{\dfrac{ax}{a}}{\dfrac{ay}{a}} = \frac{x}{y}$$

The reduction of $21x^2y/28xy^2$ to lowest terms yields

$$\frac{21x^2y}{28xy^2} = \frac{\dfrac{21x^2y}{7xy}}{\dfrac{28xy^2}{7xy}} = \frac{3x}{4y}$$

From these examples it is seen that the reduction of an algebraic fraction to its lowest terms really means finding the factors common to both the numerator and the denominator, and dividing them out. This is done by looking first at the coefficients, and then at each literal.

Reduction to lowest terms can be accomplished by an operation known as *canceling*. The last example can be rewritten as

$$\frac{21x^2y}{28xy^2} = \frac{3 \cdot 7 \cdot x \cdot x \cdot y}{4 \cdot 7 \cdot x \cdot y \cdot y}$$

The common terms between numerator and denominator can now be *canceled* to yield the fraction reduced to lowest terms. Thus,

$$\frac{3 \cdot 7 \cdot x \cdot x \cdot y}{4 \cdot 7 \cdot x \cdot y \cdot y} = \frac{3x}{4y}$$

Thus the rule:

REDUCTION OF FRACTIONS

To reduce an algebraic fraction to its lowest terms, cancel those factors that are common to both the numerator and the denominator.

From prior knowledge of division it is known that the quotient of two positive numbers is positive, and likewise the quotient of two negative numbers is also positive. Thus,

$$\frac{+a}{+b} = \frac{-a}{-b} = \frac{a}{b} = \text{a positive number}$$

It is also known that the quotient of a positive number and a negative number is negative. Thus,

$$\frac{-a}{b} = \frac{a}{-b} = -\frac{a}{b} = \text{a negative number}$$

Therefore:

SIGNS OF FRACTIONS

(1) If the signs of both the numerator and the denominator of a fraction are changed, the sign of the whole fraction remains unchanged.
(2) If the sign of either the numerator or the denominator alone is changed, the sign of the whole fraction is changed.

EXAMPLE 3.1. Remove the minus signs from inside the parentheses.
(a) $(-7/16)$
(b) $(-3/-8)$
(c) $(3a/-4b)$

SOLUTION.

(a) $\left(\dfrac{-7}{16}\right) = -\left(\dfrac{7}{16}\right)$ (using rule 1)

(b) $\left(\dfrac{-3}{-8}\right) = \left(\dfrac{3}{8}\right)$ (using rule 2)

(c) $\left(\dfrac{3a}{-4b}\right) = -\left(\dfrac{3a}{4b}\right)$ (using rule 1)

Exercises 3.1

1. Reduce the following to lowest terms:

(a) $\dfrac{2}{4}$ (b) $\dfrac{18}{24}$ (c) $\dfrac{64}{256}$ (d) $\dfrac{121}{165}$

(e) $\dfrac{3x^2}{x^4}$ (f) $\dfrac{I^2R}{2R}$ (g) $\dfrac{32x^3y}{56xy^2z}$ (h) $\dfrac{-35a^2bc}{-63ab^2c^3}$

(i) $\dfrac{30P^4R^3T}{42P^2RT^3}$ (j) $\dfrac{-10I^2R}{25IV}$ (k) $\dfrac{11xy}{-13z}$ (l) $\dfrac{1}{21x}$

2. Supply the missing terms in each of the following:

(a) $\dfrac{5}{8} = \dfrac{?}{48}$ (b) $\dfrac{3a}{4b} = \dfrac{18a^2b}{?}$ (c) $\dfrac{1}{y} = \dfrac{33xy^2}{?}$ (d) $\dfrac{?}{9a} = \dfrac{15ab}{45a^2b}$

3.2   MULTIPLICATION AND DIVISION

The multiplication of two algebraic fractions is accomplished by exactly the same method as that used for arithmetic fractions. Thus,

---

MULTIPLICATION OF FRACTIONS

To multiply two algebraic fractions, multiply the numerators to obtain the numerator of the product, and multiply the denominators to obtain the denominator of the product.

For example, the product of $2/3$ and $5/7$ is

$$\frac{2}{3} \cdot \frac{5}{7} = \frac{2 \cdot 5}{3 \cdot 7} = \frac{10}{21}$$

Similarly, the product of $2x/y$ and $a/3z$ is

$$\frac{2x}{y} \cdot \frac{a}{3z} = \frac{2x \cdot a}{y \cdot 3z} = \frac{2ax}{3yz}$$

After multiplication, the product could have common factors in the numerator and denominator. Thus, it is usually reduced to lowest terms by cancellation during multiplication. For instance,

$$\frac{2x}{3y} \cdot \frac{3xy}{5} = \frac{2 \cdot x \cdot \cancel{3} \cdot \cancel{x} \cdot \cancel{y}}{\cancel{3} \cdot \cancel{y} \cdot \cancel{5}} = \frac{2x^2}{5}$$

Fortunately the division of two algebraic fractions is also accomplished by exactly the same method as that used for arithmetic fractions. Thus,

---

DIVISION OF FRACTIONS

To divide two algebraic fractions, invert the divisor fraction and multiply.

For example, the quotient of $2/3$ and $5/7$ is

$$\frac{2}{3} \div \frac{5}{7} = \frac{2}{3} \cdot \frac{7}{5} = \frac{14}{15}$$

Similarly, the quotient of $2x/y$ and $a/3z$ is

$$\frac{2x}{y} \div \frac{a}{3z} = \frac{2x}{y} \cdot \frac{3z}{a} = \frac{6xz}{ay}$$

During the process of division the common factors between the numerator and the denominator are usually canceled out so that the quotient appears in its lowest terms. For example,

$$\frac{3ax}{y} \div \frac{4ax^2}{y^2} = \frac{3ax}{y} \cdot \frac{y^2}{4ax^2} = \frac{3y}{4x}$$

Exercises 3.2

1. Find the products of the following:

(a) $\dfrac{2}{3} \cdot \dfrac{1}{2}$

(b) $\dfrac{3a}{2} \cdot \dfrac{x}{4}$

(c) $\dfrac{2}{x} \cdot \dfrac{-3y}{x}$

(d) $\dfrac{3TR}{7} \cdot \dfrac{IR}{9a}$

(e) $\dfrac{iv}{3a} \cdot \dfrac{21ab}{i}$

(f) $\dfrac{-7x^2}{a^2y} \cdot \dfrac{3ab}{56xy}$

(g) $\dfrac{8mn^2}{p} \cdot \dfrac{7p^3}{12m^2n}$

(h) $\dfrac{3y}{-5q} \cdot \dfrac{-6pq}{9xy}$

(i) $\dfrac{1}{2} \cdot \dfrac{3}{8} \cdot \dfrac{4}{5}$

(j) $\dfrac{1}{3} \cdot \dfrac{3a}{y} \cdot \dfrac{7ay}{3x}$

(k) $\dfrac{-1}{x} \cdot \dfrac{y}{b} \cdot \dfrac{ab^2}{-4y}$

(l) $\dfrac{24x^3y^4}{9a^2b} \cdot \dfrac{81a^2b}{12x^2y^4}$

2. Find the quotients for each of the following:

(a) $\dfrac{1}{2} \div \dfrac{2}{3}$

(b) $\dfrac{3x}{4} \div \dfrac{3y}{8}$

(c) $\dfrac{-a}{b} \div \dfrac{a}{-2b}$

(d) $\dfrac{33x^2y}{56a} \div \dfrac{11x}{8ay}$

(e) $\dfrac{18a^2x^3}{125by} \div \dfrac{5by}{6a^2x^3}$

(f) $\dfrac{-6a^2b^2c^2}{11xyz} \div \dfrac{3abc}{22x^2y^2z^2}$

(g) $\dfrac{1}{3a} \div \dfrac{-1}{4x}$

(h) $\dfrac{\pi d^2}{4} \div \pi d$

(i) $\dfrac{1}{2}mv^2 \div 2gh$

(j) $\dfrac{1}{2\pi f_c} \div \dfrac{X_c}{f}$

(k) $\rho L \div \dfrac{\pi d^2}{4}$

(l) $\dfrac{T_0}{(T_0 + 20)} \div \dfrac{T_0}{(T + T_0)}$

UNIT 4
ALGEBRAIC FRACTIONS—II

4.1 LOWEST COMMON DENOMINATOR

In order to add or subtract fractions, it is necessary to change the form of each fraction so that each has the same denominator. For two common fractions, a *common denominator* is a number into which both the denominators will divide evenly. For example, the fractions $1/2$ and $3/5$ have a common denominator of 10, since both 2 and 5 will divide into 10 evenly.

Notice that 20 is also a common denominator. In fact, there are an infinite number of common denominators. Generally, the *lowest common denominator* (LCD) is of interest. The LCD is simply the *smallest* common denominator. For the fractions $1/2$ and $3/5$, 10 is clearly the LCD.

In order to find the LCD for two algebraic fractions, it is necessary to find a term into which both the denominator terms will divide evenly. In general, the LCD for an algebraic fraction will contain a numerical coefficient and one or more literals. For example, the LCD for $3/4x$ and b/y is $4xy$.

Notice that a common denominator can always be found by simply taking the product of the denominators. This technique may or may not provide an LCD! For example, $6xy$ is a common denominator for $p/2x$ and $q/3y$. The term $6xy$ is found by taking the product of $2x$ and $3y$ (the denominators of the fractions), and it is an LCD.

The term $6x^2y$ is a common denominator found by taking the product of the denominators of $a/2x$ and $b/3xy$; but $6x^2y$ is *not* an LCD. The LCD is $6xy$.

EXAMPLE 4.1. Find the LCD for the following fractions:
(a) $1/3$, $1/4$
(b) $2/5$, $5/8$
(c) $2/3$, $4/9$
(d) $a/2x$, $b/3y$
(e) $4/3a^2$, $x/2ab$

SOLUTION.
(a) 12 is the smallest number into which both 3 and 4 will divide evenly.
(b) 40 is the smallest number into which both 5 and 8 will divide evenly.
(c) 9 is the smallest number evenly divisible by both 3 and 9.
(d) The smallest term evenly divisible by both $2x$ and $3y$ is $6xy$.
(e) The smallest term evenly divisible by both $3a^2$ and $2ab$ is $6a^2b$.

The form of two or more fractions can be changed such that they will all have the same denominator by multiplying both numerator and denominator by the same term. Thus,

---

LCD

In order to express fractions in terms of their LCD:
(1) Find the LCD.
(2) Multiply both the numerator and the denominator of each fraction
by a term that will yield the LCD in the denominator.

For instance, the LCD of $a/3yz$ and $b/2xy$ is $6xyz$. Therefore multiply both the numerator and the denominator of the first fraction by the term $2x$ since $3xz \cdot 2x$ yields $6xyz$ (the LCD). The numerator and denominator of the second fraction are multiplied by the term $3z$ since $2xy \cdot 3z$ yields $6xyz$ (the LCD). Thus,

$$\frac{a}{3yz} \cdot \frac{2x}{2x} = \frac{2ax}{6xyz} \qquad \text{and} \qquad \frac{b}{2xy} \cdot \frac{3z}{3z} = \frac{3bz}{6xyz}$$

The two fractions are now expressed in a form such that they have the same LCD.

Notice that the first fraction was multiplied by the term $(2x/2x) = 1$, and thus the value of the fraction is unchanged even though it appears in a different form. Similarly, multiplication of the second fraction by the term $(2z/2z) = 1$ changes its form but leaves its value unchanged.

EXAMPLE 4.2. Express the following fractions in terms of their LCD's:
(a) $1/2$, $2/3$
(b) $2a/x$, $b/3y$

SOLUTION.
(a) LCD = 6. Multiply $1/2$ by $3/3$ to obtain 6 in the denominator. Thus,

$$\frac{1}{2} \cdot \frac{3}{3} = \frac{3}{6} \qquad \text{and} \qquad \frac{2}{3} \cdot \frac{2}{2} = \frac{4}{6}$$

(b) LCD = $3xy$. Multiply $2a/x$ by $3y/3y$ and $b/3y$ by x/x to obtain $3xy$ in the denominator. Thus,

$$\frac{2a}{x} \cdot \frac{3y}{3y} = \frac{6ay}{3xy} \qquad \text{and} \qquad \frac{b}{3y} \cdot \frac{x}{x} = \frac{bx}{3xy}$$

Exercises 4.1

Find the LCD for the following fractions and rewrite each fraction in terms of the LCD found:

1. $\dfrac{2}{3}$, $\dfrac{5}{7}$

2. $\dfrac{1}{2}$, $\dfrac{1}{3}$, $\dfrac{1}{4}$

3. $\dfrac{7}{8}$, $\dfrac{3}{16}$, $\dfrac{5}{24}$

4. $\dfrac{a}{2}, \dfrac{b}{3}, \dfrac{c}{5}$ 5. $\dfrac{3}{p}, \dfrac{4a}{3v}$ 6. $\dfrac{2a}{3x^2}, \dfrac{b}{5xy^2}$

7. $\dfrac{m}{2pv}, \dfrac{n}{iv}$ 8. $\dfrac{1}{R_1}, \dfrac{1}{R_2}$ 9. $\dfrac{1}{R_1}, \dfrac{1}{R_2}, \dfrac{1}{R_3}$

10. $\dfrac{V_1}{R_1}, \dfrac{V_2}{R_2}$ 11. $\dfrac{i_1}{G_1}, \dfrac{i_2}{G_2}$ 12. $\dfrac{V_1}{R_1}, \dfrac{V_2}{R_2}, \dfrac{V_3}{R_3}$

13. $\dfrac{i}{G_1}, \dfrac{1}{G_2}, \dfrac{1}{G_3}$ 14. $\dfrac{V_T}{V_1}, 7$ 15. $\dfrac{V_1}{R_1}, -3$

4.2   ADDITION AND SUBTRACTION

In order to add or subtract algebraic fractions, it is necessary to express them in terms of their LCD. Then, the sum (or difference) is found by writing the sum (or difference) of the numerators over the LCD.

For example, the LCD of $2/x$ and a/y is xy. Expressed in LCD form they become

$$\frac{2}{x} = \frac{2y}{xy} \qquad \text{and} \qquad \frac{a}{y} = \frac{ax}{xy}$$

The sum is then found as

$$\frac{2y}{xy} + \frac{ax}{xy} = \frac{2y + ax}{xy}$$

The difference is

$$\frac{2y}{xy} - \frac{ax}{xy} = \frac{2y - ax}{xy}$$

To summarize:

ADDITION OR SUBTRACTION

To add or subtract algebraic fractions:
(1) Express them in terms of their LCD.
(2) Add or subtract the numerators over the LCD.

EXAMPLE 4.3.   Add:

$$\frac{1}{b} \quad \text{and} \quad \frac{2}{d}$$

SOLUTION.   The LCD is bd. Thus,

$$\frac{1}{b} = \frac{d}{bd} \quad \text{and} \quad \frac{2}{d} = \frac{2b}{bd}$$

The sum is then

$$\frac{d}{bd} + \frac{2b}{bd} = \frac{d + 2b}{bd}$$

EXAMPLE 4.4.   Subtract:
$$\frac{1}{R_T} - \frac{1}{R_1}$$

SOLUTION.   The LCD is $R_1 R_T$. Thus,
$$\frac{1}{R_1} = \frac{R_T}{R_1 R_T}$$
and
$$\frac{1}{R_T} = \frac{R_1}{R_1 R_T}$$

The difference is then
$$\frac{R_1}{R_1 R_T} - \frac{R_T}{R_1 R_T} = \frac{R_1 - R_T}{R_1 R_T}$$

In the special case where the LCD of two fractions is found to be the product of the two denominators, it is possible to use a technique known as *cross multiplication* to aid in finding the sum or difference of the two fractions. Cross multiplication is accomplished as follows:

CROSS MULTIPLICATION

The sum or difference of two fractions can be found as:
(1) The LCD is the product of the two denominators.
(2) The numerator is the *first* numerator times the *second* denominator, plus or minus the *second* numerator times the *first* denominator.

Cross multiplication can be illustrated pictorially as follows:
$$\frac{a}{b} + \frac{x}{y} = \frac{ay + bx}{by}$$

EXAMPLE 4.5.   Use cross multiplication to add:
$$\frac{1}{R_a} + \frac{1}{R_b}$$

SOLUTION.   The denominator is $R_a R_b$. Cross multiplication yields $1 R_b + 1 \cdot R_a$ for the numerator. Thus,
$$\frac{1}{R_a} + \frac{1}{R_b} = \frac{R_b + R_a}{R_a R_b}$$

EXAMPLE 4.6.   Use cross multiplication for the difference:
$$\frac{2}{p} - \frac{v}{i}$$

SOLUTION.   The denominator is pi. Cross multiplication yields $2 \cdot i - pv$ for the numerator, since
$$\frac{2}{p} - \frac{v}{i} = \frac{2}{p} + \frac{(-v)}{i}$$
Thus,
$$\frac{2}{p} - \frac{v}{i} = \frac{2i - pv}{pi}$$

Algebraic expressions that contain both fractional terms and nonfractional terms are called *mixed expressions*. For instance, $(3+1/2)$, or $(2x - a/p)$, or $(2ax/3y + vi)$ are all mixed expressions.

Such expressions offer no difficulty since any term can be expressed as a fraction by simply dividing it by one (1); that is, $3 = 3/1$, $2x = 2x/1$, $vi = vi/1$, etc. Thus mixed expressions are simply treated as fractions, and cross multiplication works nicely.

For example,

$$3 + \frac{1}{2} = \frac{3}{1} + \frac{1}{2} = \frac{6+1}{2} = \frac{7}{2}$$

Also,

$$2x - \frac{a}{p} = \frac{2x}{1} - \frac{a}{p} = \frac{2xp - a}{p}$$

and

$$\frac{2ax}{3y} + vi = \frac{2ax}{3y} + \frac{vi}{1} = \frac{2ax + 3yvi}{3y}$$

Exercises 4.2

Find the indicated sums and differences of each of the following:

1. $\dfrac{2}{3} + \dfrac{5}{7}$

2. $\dfrac{5}{7} - \dfrac{2}{3}$

3. $\dfrac{1}{2} + \dfrac{1}{3} - \dfrac{1}{4}$

4. $\dfrac{a}{2} + \dfrac{b}{3} - \dfrac{c}{5}$

5. $\dfrac{3}{p} + \dfrac{4a}{3v}$

6. $\dfrac{2a}{3x^2} - \dfrac{b}{5xy^2}$

7. $\dfrac{m}{2pv} + \dfrac{n}{iv}$

8. $\dfrac{1}{R_1} + \dfrac{1}{R_2}$

9. $\dfrac{1}{R_1} + \dfrac{1}{R_2} + \dfrac{1}{R_3}$

10. $\dfrac{V_1}{R_1} - \dfrac{V_2}{R_2}$

11. $\dfrac{i_1}{G_1} + \dfrac{i_2}{G_2}$

12. $\dfrac{V_1}{R_1} + \dfrac{V_2}{R_2} - \dfrac{V_3}{R_3}$

13. $\dfrac{V_1}{I_1} + 6$

14. $\dfrac{V_T}{R_T} - 6$

15. $1 + \dfrac{3ab}{5p}$

16. $\dfrac{3a}{x} - \dfrac{5}{y} + \dfrac{1}{2}$

17. $\dfrac{v_2}{R} - \dfrac{7v_T}{R_T}$

18. $\dfrac{1}{R_1} + \dfrac{1}{R_2} - \dfrac{1}{R_T}$

19. $\dfrac{V_1^2}{R_1} + \dfrac{V_2^2}{R_2} + \dfrac{V_3^2}{R_3}$

20. $\dfrac{Q_1}{i_1} + \dfrac{Q_2}{-i_2}$

21. $\dfrac{v_1}{R_1} - \dfrac{-v_2}{R_2}$

The following problems deal with resistances connected in parallel.

22. Two resistors R_1 and R_2 are connected in *parallel* in Figure 4.1(a). They can be combined into a single equivalent resistance R_T as shown in Figure 4.1(b) by using the formula

$$\frac{1}{R_T} = \frac{1}{R_1} + \frac{1}{R_2}$$

Show that

$$R_T = \frac{R_1 R_2}{R_1 + R_2}$$

by adding the two fractions $1/R_1$ and $1/R_2$, and then taking the reciprocal of both sides of the equation.

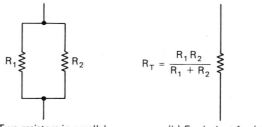

Figure 4.1   (a) Two resistors in parallel   (b) Equivalent for (a)

23. Using Figure 4.1, find R_T if:
    (a) $R_1 = 1 \text{ k}\Omega$, $R_2 = 2 \text{ k}\Omega$
    (b) $R_1 = 500 \ \Omega$, $R_2 = 500 \ \Omega$
    (c) $R_1 = 25 \ \Omega$, $R_2 = 15 \ \Omega$

24. Beginning with the formula

$$\frac{1}{R_T} = \frac{1}{R_1} + \frac{1}{R_2}$$

Problem 22 and Figure 4.1 show that

$$R_1 = \frac{R_2 R_T}{R_2 - R_T}$$

25. Use the results of Problem 24 to find R_1 if:
    (a) $R_2 = 10 \text{ k}\Omega$, $R_T = 2 \text{ k}\Omega$
    (b) $R_2 = 50 \ \Omega$, $R_T = 25 \ \Omega$
    (c) $R_2 = 100 \text{ k}\Omega$, $R_T = 80 \text{ k}\Omega$

26. What resistor must be connected in parallel with a 47 kΩ resistor to form an equivalent resistance of 33 kΩ? (*Hint:* Use the formula developed in Problem 24.)

27. The total equivalent resistance R_T of three resistors connected in parallel as shown in Figure 4.2 is given as

$$\frac{1}{R_T} = \frac{1}{R_1} + \frac{1}{R_2} + \frac{1}{R_3}$$

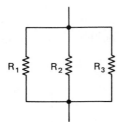

(a) Three parallel resistors

(b) Equivalent for (a)   **Figure 4.2**

Show that

$$R_T = \frac{R_1 R_2 R_3}{R_1 R_2 + R_1 R_3 + R_2 R_3}$$

28. Use the formula from Problem 27 to find R_T if:
    (a) $R_1 = 1 \text{ k}\Omega$, $R_2 = 2 \text{ k}\Omega$, $R_3 = 3 \text{ k}\Omega$
    (b) $R_1 = 50 \ \Omega$, $R_2 = 50 \ \Omega$, $R_3 = 50 \ \Omega$

29. An expression for the equivalent resistance R_T of N resistors connected in parallel is

$$\frac{1}{R_T} = \frac{1}{R_1} + \frac{1}{R_2} + \frac{1}{R_3} + \cdots + \frac{1}{R_N}$$

If the resistors all have equal values of R, show that

$$R_T = \frac{R}{N}$$

30. Use the formula developed in Problem 29 to find the equivalent resistance of:
    (a) Two 6.8 kΩ resistors in parallel
    (b) Three 39 kΩ resistors in parallel
    (c) Ten 500 Ω resistors in parallel

31. A 240 V source is connected to a resistance heating load that draws 100 A. What is the equivalent resistance of the load? If this load is actually composed of ten electric heaters connected in parallel, what is the resistance of each heater? (*Hint:* Use the formula in Problem 29.)

UNIT 5
POLYNOMIALS

5.1 ADDITION AND SUBTRACTION

A *polynomial* can be defined as an algebraic expression composed of two or more terms. Polynomials are frequently encountered in electric circuit problems, and it is essential to know how to apply the four basic operations $(+, -, \times, \div)$ to polynomials.

In finding the numerical values of expressions and in performing algebraic operations, there are two fundamental operations that have been used almost instinctively. These two operations are stated in the *commutative axiom of addition*, and the *associative axiom of addition*. An *axiom* is simply a statement assumed to be true and thus requires no proof.

The *commutative axiom of addition* states that "the sum of two numbers is independent of the order in which the numbers are added." In the form of an equation,

$$x + y = y + x \qquad \text{(commutative axiom)}$$

The validity of this axiom is made quite evident by substituting any two real numbers to test it. For example,

$$2 + 3 = 3 + 2$$
or
$$15 + 7 = 7 + 15$$
or
$$5 + (-3) = (-3) + 5$$

The last example illustrates how subtraction is really included in this axiom for addition.

The *associative axiom of addition* states that "the sum of three or more numbers is independent of grouping by parentheses." In equation form,

$$x + (y + z) = (x + y) + z \qquad \text{(associative axiom)}$$

The validity of this axiom is also easily illustrated by simply substituting real numbers. For example,

$$2 + (3 + 4) = (2 + 3) + 4$$
or
$$(-1) + (2 + 5) = [(-1) + 2] + 5$$

These two basic axioms for addition are summarized as:

ADDITION AXIOMS	
Commutative	$x + y = y + x$
Associative	$x + (y + z) = (x + y) + z$

The two axioms for addition can now be used along with prior knowledge of adding like algebraic terms to accomplish addition (or subtraction) of polynomials. Simply stated, the addition (or subtraction) of polynomials is accomplished by combining like terms. A number of examples will be of value.

EXAMPLE 5.1. Find the sums: (a) $3x, 2y, -x, 6, y$ (b) $2ab, b^2, 3ab, -b^2$ (c) $(2x-y), (2y+z)$	SOLUTION. Adding like terms: (a) $(3x-x)+(2y+y)+6=2x+3y+6$ (b) $(2ab+3ab)+(b^2-b^2)=5ab$ (c) $(2x-y)+(2y+z)=2x+(2y-y)+z=$ $2x+y+z$

Care must be taken when doing subtraction as the following example shows.

EXAMPLE 5.2. Subtract: (a) $(3x+y)-2x$ (b) $(2I_1+I_2)$ from (I_1-3I_2) (c) $\left(\frac{1}{2}V_1-\frac{1}{3}V_2\right)$ from $\left(-\frac{3}{4}V_1+\frac{2}{3}V_2\right)$	SOLUTION. (a) $(3x+y)-2x=3x+y-2x=x+y$ (b) $(I_1-3I_2)-(2I_1+I_2)=I_1-3I_2-2I_1-I_2$ $=-I_1-4I_2$ (c) $(-\frac{3}{4}V_1+\frac{2}{3}V_2)-(\frac{1}{2}V_1-\frac{1}{3}V_2)=$ $-\frac{3}{4}V_1+\frac{2}{3}V_2-\frac{1}{2}V_1+\frac{1}{3}V_2=-\frac{5}{4}V_1+V_2$

If more than two expressions are involved, it is sometimes easier to arrange all like terms in the same columns as in the solution to the next example.

EXAMPLE 5.3. Add: $(3I_1-I_2-4I_3)$ and $(-I_1+7I_2)+(-4I_1+6I_3)$	SOLUTION. $\begin{array}{rrr} 3I_1 & -I_2 & -4I_3 \\ -I_1 & 7I_2 & 0 \\ -4I_1 & 0 & 6I_3 \\ \hline -2I_1 & +6I_2 & +2I_3 \end{array}$

Exercises 5.1

1. Add the following:

(a) $R_1, 2R_2, 3R_1, -4R_2$
(b) $V_1, -V_2, 3V_2, 2V_1$
(c) $-I_a, 3I_b, -2I_b, 2I_a$
(d) $3i^2R, 5i^2R, -7i^2R$
(e) $V_a/6, -V_b/2, -V_a/9, 2V_b/3$

(f) $V^2/7, V^2/3, V^2/2, -V^2/6$
(g) $(i_1-i_2+i_3), (i_1+2i_2-3i_3)$
(h) $(x^2y+3x+2), (x^2y+x-4)$
(i) $(3xy^2-2y+5), (2y+7-x^2y)$
(j) $(at^2-3t+7), (5-4t-at^2)$

2. Subtract each of the following:

(a) $(3ax+7)-(2ax-3)$ (d) $(6I_1-3I_2)-(-I_1+4I_2)$

(b) $(5)-(2V_1+V_2)$ (e) $(36V_1-IR)$ from $(IR+7V_2)$

(c) $\dfrac{1}{2}-(V_1/4-3)$ (f) $(at^2-3t+7)-(5-4t-at^2)$

3. Simplify the following:

(a) $i_1^2R+i_2^2R-3+4i_2^2R-(i_1^2R-5)$

(b) $0.72ax-0.19by^2-0.11ax-by^2$

(c) $\dfrac{1}{2}t+3u-\dfrac{2}{3}s+\dfrac{u}{3}-\dfrac{s}{5}+\dfrac{3t}{4}-7t+s$

(d) $(2V_a-V_b)+(-3V_a+V_b-2V_c)-(3V_b+V_c)$

5.2 MULTIPLICATION

In the process of multiplication we instinctively make use of two fundamental concepts that are formally stated by the *commutative* and *associative* axioms of multiplication.

The *commutative axiom of multiplication* states that "the product of two numbers is independent of the order in which the numbers are multiplied." In equation form,

$$xy=yx \qquad \text{(commutative axiom)}$$

The validity is easily illustrated by using real numbers, That is,

$$2\cdot3=3\cdot2$$
or $$7(-2)=(-2)7$$

The *associative axiom of multiplication* states that "the product of three or more numbers is independent of grouping by parentheses." In equation form,

$$x(yz)=(xy)z \qquad \text{(associative axiom)}$$

Again, real numbers can be used to illustrate the validity.

$$2(3\cdot4)=(2\cdot3)4$$
or $$(-5)(2\cdot3)=[(-5)2]3$$

These two basic axioms for multiplication are summarized as:

MULTIPLICATION AXIOMS

Commutative $xy=yx$

Associative $x(yz)=(xy)z$

At this point, the commutative and associative axioms for both addition and multiplication have been stated (a total of four fundamental axioms). A fifth operation defined by the *distributive axiom* will provide the means to manipulate polynomial expressions.

The *distributive axiom* shows how to multiply a monomial and a polynomial (or two polynomials) and it is given in equation form as:

AXIOM

Distributive $x(y+z)=xy+xz$

To illustrate the validity of the distributive axiom, consider the following examples:

| EXAMPLE 5.4. Verify the distributive axiom for:

(a) $2(3+4)$
(b) $3(6-2)$ | SOLUTION.
(a) $2(3+4)=2\cdot3+2\cdot4$
or

$$2(7)=6+8$$

Thus,

$$14=14$$

(b) $3(6-2)=3\cdot6+3(-2)$
or

$$3(4)=18-6$$

Thus,

$$12=12$$ |

The distributive axiom can be used to find the product of a monomial and a polynomial. For example, the product of $(2a)$ and $(3t-s/4)$ is found as

$$(2a)\left(3t-\frac{s}{4}\right)=6at-\frac{as}{2}$$

A simple extension of the distributive axiom shows how to find the product of two polynomials. The rule is:

POLYNOMIAL MULTIPLICATION

To find the product of two polynomials: Multiply the second polynomial by *each term* in the first polynomial; then use the distributive axiom to complete the multiplication.

To illustrate, find the product of $(5+2)$ and $(6+3)$:

$$(5+2)(6+3)=5(6+3)+2(6+3)$$
$$=5\cdot6+5\cdot3+2\cdot6+2\cdot3$$
$$=30+15+12+6$$
$$=63$$

As a second illustration, multiply $(a+b)$ and $(x-y)$:

$$(a+b)(x-y)=a(x-y)+b(x-y)$$
$$=ax-ay+bx-by$$

The rule can of course be applied to polynomials having any number of terms. Take special care with minus signs.

EXAMPLE 5.5. Multiply:	SOLUTION.
(a) $(a+1)$ and $(x-9)$	(a) $(a+1)(x-9)=a(x-9)+1(x-9)$
(b) (i_1-i_2) and (v_1-v_2)	$\qquad = ax-9a+x-9$
(c) $(a+b+c)$ and $(x+y+z)$	(b) $(i_1-i_2)(v_1-v_2)$
	$\qquad = i_1(v_1-v_2)+(-i_2)$
	$\qquad \times(v_1-v_2)$
	$\qquad = i_1v_1-i_1v_2-i_2v_1+i_2v_2$
	(c) $(a+b+c)(x+y+z)$
	$\qquad = a(x+y+z)+b(x+y+z)$
	$\qquad +c(x+y+z)$
	$\qquad = ax+ay+az+bx+by$
	$\qquad + bz+cx+cy+cz$

Exercises 5.2

Multiply each of the following:

1. $I(R_1+R_2)$

2. $i(R_1+R_2-R_T)$

3. $(G_A+G_B+G_C)V_A$

4. $(R_v+R_m)I_m$

5. $V_A\left(\dfrac{1}{R_1}+\dfrac{1}{R_2}+\dfrac{1}{R_3}\right)$

6. $V^2\left(\dfrac{1}{R_A}+\dfrac{1}{R_B}+\dfrac{1}{R_C}\right)$

7. $(1+j)3$

8. $j(1-j)$

9. $j(2-3j)5$

10. $V_T\left(\dfrac{1}{4}+\dfrac{3}{8}+\dfrac{5}{9}\right)10^2$

11. $\dfrac{3j}{2}\left(\dfrac{a}{3}-\dfrac{jb}{2}\right)$

12. $(10^{-2}+3\times10^{-3})V_B$

13. $(1+j)(1-j)$

14. $(a+b)(x+y)$

15. $(v_1-v_2)(i_1+i_2)$

16. $(I_1-I_2)(R_1+R_2)$

17. $(V_A+V_B)\left(\dfrac{1}{R_1}+\dfrac{1}{R_2}\right)$

18. $(V_A + V_B)\left(\dfrac{1}{R_1} + \dfrac{1}{R_2} + \dfrac{1}{R_3}\right)$

19. $(R_1 + R_2 + R_3)(3I_1 - I_2 + 2I_3)$

20. $(V_A + V_B)(V_C - V_D)\left(\dfrac{1}{R_1}\right)$

21. $(I_1 + I_2)(I_3 - I_4)(R_1 + R_2)$

22. $(1 + j)(2 + j3)(j - 1)$

23. $(j + 2)\left(\dfrac{1}{2} - \dfrac{j}{3}\right)(2 - j)$

24. $(x - y)(x^2 + 2ax + y)$

25. $(3a^2 + 2ab + 4c)(a^2 - b^2 + c^2)$

5.3 DIVISION

The division of two monomials is accomplished by finding the quotient of their coefficients and then applying the Law of Exponents to their literal numbers, as has been seen previously. An extension of this fundamental rule will provide the means to divide a polynomial by a monomial. The rule is:

MONOMIAL DIVISION

To divide a polynomial by a monomial: First divide each term of the polynomial by the monomial; then take the algebraic sum of these quotients.

A number of examples will illustrate the application of this rule.

EXAMPLE 5.6. Divide:

$$(3x + 6y) \text{ by } (3)$$

SOLUTION. Dividing each term of $(3x + 6y)$ by (3), yields

$$\frac{3x + 6y}{3} = \frac{3x}{3} + \frac{6y}{3} = x + 2y$$

EXAMPLE 5.7. Divide:
(a) $(IR_1 + IR_2)$ by I
(b) $(9i^2 R_1 - 15i^2 R_2)$ by $(3i)$

(c) $\left(\dfrac{12v^2}{R_1} - \dfrac{4v^2}{R_2}\right)$ by $\left(\dfrac{2v}{R_T}\right)$

SOLUTION.

(a) $\dfrac{IR_1 + IR_2}{I} = \dfrac{IR_1}{I} - \dfrac{IR_2}{I} = R_1 - R_2$

(b) $\dfrac{9i^2 R_1 - 15i^2 R_2}{3i} = \dfrac{9i^2 R_1}{3i} - \dfrac{15i^2 R_2}{3i}$
$= 3iR_1 - 5iR_2$

(c) $\dfrac{\dfrac{12v^2}{R_1} - \dfrac{4v^2}{R_2}}{\dfrac{2v}{R_T}} = \left(\dfrac{12v^2}{R_1} - \dfrac{4v^2}{R_2}\right)\dfrac{R_T}{2v}$

$= \dfrac{12v^2 R_T}{2vR_1} - \dfrac{4v^2 R_T}{2vR_2}$

$= \dfrac{6vR_T}{R_1} - \dfrac{2vR_T}{R_2}$

There may be times when it will be necessary to determine a quotient by dividing one polynomial by another polynomial. The process is very similar to performing "long division" in an arithmetic problem, but there are some rules that will be of value. Thus,

---

POLYNOMIAL DIVISION

To divide one polynomial (dividend) by another polynomial (divisor):
(1) Arrange both polynomials according to descending powers of some common literal number.
(2) Divide the two first terms of each polynomial to obtain the first term of the quotient.
(3) Multiply the divisor by the first term of the quotient and subtract from the dividend to obtain a new second dividend.
(4) Repeat steps (2) and (3) until the remainder in step (3) is either zero or of lower degree than the divisor.

These four rules seem formidable, but a few examples will clarify the process.

EXAMPLE 5.8. Divide:

$$(x^2 + 7x + 12) \text{ by } (x + 3)$$

SOLUTION. The two terms are already arranged in descending powers of x. Thus we write

$$
\begin{array}{r}
x + 4 \\
x+3 \overline{\smash{)}\ x^2 + 7x + 12} \\
\underline{x^2 + 3x}(-) \\
4x + 12 \\
\underline{4x + 12}(-) \\
0
\end{array}
$$

EXAMPLE 5.9. Divide:

$$(-2 + 2t^2) \text{ by } (t - 1)$$

SOLUTION. First, rewrite in descending powers of t:

$$(2t^2 + 0t + 2) \div (t - 1)$$

Then divide

$$
\begin{array}{r}
2t + 2 \\
t-1 \overline{\smash{)}\ 2t^2 + 0t + 2} \\
\underline{2t^2 - 2t}(-) \\
2t - 2 \\
\underline{2t - 2}(-) \\
-0
\end{array}
$$

EXAMPLE 5.10.   Divide:

$$(2v^3 + 9v^2 + 7v + 15) \text{ by } (v + 3)$$

SOLUTION.

$$
\begin{array}{r}
(2v^2 + 3v - 2) \quad \text{(quotient)} \\
v + 3 \overline{) 2v^3 + 9v^2 + 7v + 15} \\
\underline{2v^3 + 6v^2(-)} \\
3v^2 + 7v \\
\underline{3v^2 + 9v(-)} \\
-2v + 15 \\
\underline{-2v - 6(-)} \\
21 \quad \text{(remainder)}
\end{array}
$$

This problem does not terminate in zero, but has a *remainder* of 21. Thus the solution can be written $(2v^2 + 3v - 2)$ with 21 remainder.

Exercises 5.3

Divide each of the following:

1. $(21x - 35xy)/(7x)$

2. $(9s^2t + 27st^2)$ by $(-3st)$

3. $(IR_1 + IR_2)$ by (IR_T)

4. $(15i^2R_1 - 35i^2R_2 + 20i^2R_T)$ by $(5i)$

5. $(6i^2R_1 - 7i^2R_2 + iR)$ by (iR)

6. $\left(\dfrac{5v^2}{R_1} + \dfrac{3v^2}{R_2} - \dfrac{6v^2}{R_3} \right)$ by $\left(\dfrac{4v}{R_A} \right)$

7. $\left(\dfrac{1}{R_1} + \dfrac{1}{R_2} + \dfrac{1}{R_3} \right)$ by $\left(\dfrac{v}{3} \right)$

8. $(4\pi r^2 - 7\pi r h) \div (\pi/4)$

9. $(3x^2y + 7xy - 11xy^2)$ by $\left(\dfrac{xy}{2} \right)$

10. $(ap^4 + bp^3 - cp^2 + 1) \div (p^2/2)$

11. $(x^2 + 3x + 2)$ by $(x + 1)$

12. $(-1 - x + 2x^2)$ by $(x - 1)$

13. $(9i^2 - 36)$ by $(3i + 6)$

14. $(9i^2 - 36)$ by $(3i - 6)$

15. $(3 + 10j + 3j^2)/(j + 3)$

16. $(3a^2 + 5ab - 2b^2) \div (3a - b)$

17. $(x^3 - 27)$ by $(x - 3)$

18. $(x^3 + 4x^2 + 3x - 2)$ by $(x^2 + 2x - 1)$

19. $(x^3 + y^3 + xy^2 + yx^2)$ by $(x^2 + y^2)$

20. $(x^2y^2 + by^2 - cx^2 - bc + 1)$ by $(y^2 - c)$

UNIT 6
KIRCHHOFF'S LAWS

6.1 KIRCHHOFF'S VOLTAGE LAW (KVL)

Up to this point the study of electric circuits has been limited to problems that can be solved by means of Ohm's Law. However, one can not advance beyond very simple electric circuits without utilizing Kirchhoff's Laws. *Kirchhoff's Voltage Law* (KVL) and *Kirchhoff's Current Law* (KCL) will provide the opportunity to apply some of the fundamentals of mathematics to the study of electric circuits. This will develop and strengthen mathematical skills, and at the same time it will greatly increase the understanding of electric circuits.

A law is a statement concerning observed experimental behavior. Kirchhoff's Voltage Law (KVL) deals with the voltages around a closed circuit and is stated as:

KVL

In a closed circuit, the algebraic sum of the source voltages must be equal to the algebraic sum of the voltage drops.

In Figure 6.1(a) the existence of a current I (shown by the dashed line) causes a *voltage drop* V_R as indicated. The current I in Figure 6.1(b) causes a voltage drop V_{R_1} across resistor R_1, and a voltage drop V_{R_2} across resistor R_2 as shown. Thus a *voltage drop* (or potential difference) exists across the terminals of a resistance any time there is an electric current in the resistance.

A source of electrical energy, such as chemical cell or an electronic power supply, is needed to produce an electric current in a resistance. In Figure 6.2, the battery has a *source voltage* of V_S, and it establishes an electric current I through the resistance R. Thus there is a *voltage drop* V_R across the terminals of the resistor. The application of KVL to this circuit simply requires the source voltage V_S be equal to the voltage drop V_R; or

$$V_S = V_R$$

The current I established by the source V_S in Figure 6.3 causes voltage drops V_{R_1} and V_{R_2}. Thus KVL requires

$$V_S = V_{R_1} + V_{R_2}$$

For example, if $V_S = 6$ V, $V_{R_1} = 4$ V, and $V_{R_2} = 2$ V, we obtain

$$6 = 4 + 2$$

It is easy to see how KVL can be applied to any number of resistors (N) connected in series with a source to yield

$$V_S = V_{R_1} + V_{R_2} + V_{R_3} + \cdots + V_{R_N} \tag{6.1}$$

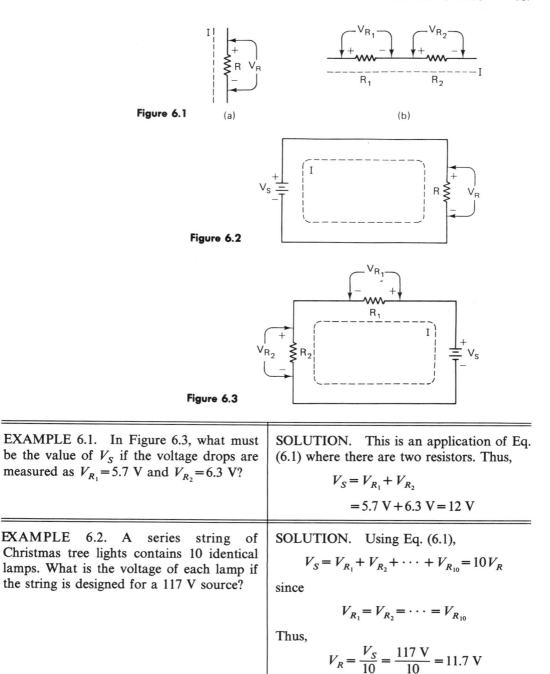

Figure 6.1   (a)   (b)

Figure 6.2

Figure 6.3

EXAMPLE 6.1. In Figure 6.3, what must be the value of V_S if the voltage drops are measured as $V_{R_1}=5.7$ V and $V_{R_2}=6.3$ V?	SOLUTION. This is an application of Eq. (6.1) where there are two resistors. Thus, $$V_S = V_{R_1} + V_{R_2}$$ $$= 5.7\text{ V} + 6.3\text{ V} = 12\text{ V}$$
EXAMPLE 6.2. A series string of Christmas tree lights contains 10 identical lamps. What is the voltage of each lamp if the string is designed for a 117 V source?	SOLUTION. Using Eq. (6.1), $$V_S = V_{R_1} + V_{R_2} + \cdots + V_{R_{10}} = 10 V_R$$ since $$V_{R_1} = V_{R_2} = \cdots = V_{R_{10}}$$ Thus, $$V_R = \frac{V_S}{10} = \frac{117\text{ V}}{10} = 11.7\text{ V}$$

EXAMPLE 6.3. The source voltage in Figure 6.3 is 100 V, and V_{R_1} is measured as $V_{R_1} = 37.5$ V. What must be the value of V_{R_2}?

SOLUTION. KVL requires

$$V_S = V_{R_1} + V_{R_2}$$

or $100 \text{ V} = 37.5 \text{ V} + V_{R_2}$

Thus,

$$V_{R_2} = 100 \text{ V} - 37.5 \text{ V} = 62.5 \text{ V}$$

If there is more than one voltage source in a closed circuit, KVL can be most easily applied by first combining the voltage sources to obtain an equivalent source voltage. For example, taking the algebraic sum of the two sources in Figure 6.4(a) yields the equivalent shown in Figure 6.4(b). In this case the voltage sources *aid* one another, and the equivalent is taken as the sum ($V_{eq} = V_{S_1} + V_{S_2}$).

In Figure 6.5(a), the sources *oppose* one another, and the equivalent is taken as the difference ($V_{eq} = |V_{S_1} - V_{S_2}|$). Choose V_{eq} to be a positive number, and then place the positive terminal of V_{eq} in agreement with the source having the larger magnitude. For example, if $V_{S_1} = 15$ V and $V_{S_2} = 10$ V, the equivalent is $V_{eq} = |V_{S_1} - V_{S_2}| = |15 \text{ V} - 10 \text{ V}| = 5$ V, and the positive terminal of V_{eq} is on top. On the other hand, if $V_{S_1} = 20$ V and $V_{S_2} = 30$ V, the equivalent is $V_{eq} = |20 \text{ V} - 30 \text{ V}| = 10$ V, and the positive terminal of V_{eq} is drawn on the bottom in Figure 6.5(b).

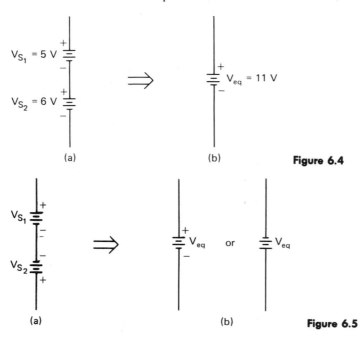

(a)   (b)   **Figure 6.4**

(a)   (b)   **Figure 6.5**

Notice carefully that in determining an equivalent source voltage we are simply *drawing* an equivalent representation of a circuit in order to simplify the application of KVL. We *cannot*, however, rearrange the original circuit.

EXAMPLE 6.4.   Apply KVL to the circuit in Figure 6.6(a) to verify the measured voltages indicated.

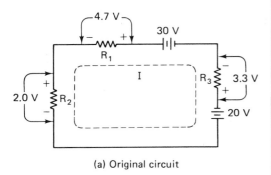

(a) Original circuit

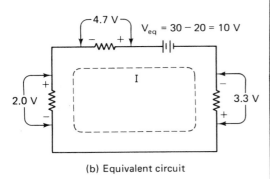

(b) Equivalent circuit

Figure 6.6

SOLUTION. The two voltage sources oppose one another, and they can be replaced with an equivalent equal to the difference between them, as shown in Figure 6.6(b). Applying KVL to Figure 6.6(b) then yields

$$V_{eq} = V_{R_1} + V_{R_2} + V_{R_3}$$

$$10\ V = 4.7\ V + 2.0\ V + 3.3\ V$$

The application of KVL to the series circuit in Figure 6.7(a) can be used to develop the law for combining series resistances. For this circuit, KVL states

$$V_S = V_{R_1} + V_{R_2} + V_{R_3}$$

But since this is a series circuit, the current I must be the same in all parts of the circuit. So, multiply both sides of the above equation by $1/I$, and use the distributive axiom to expand the right side of the equation as

$$V_S\left(\frac{1}{I}\right) = \left(V_{R_1} + V_{R_2} + V_{R_3}\right)\left(\frac{1}{I}\right) \qquad \frac{V_S}{I} = \frac{V_{R_1}}{I} + \frac{V_{R_2}}{I} + \frac{V_{R_3}}{I}$$

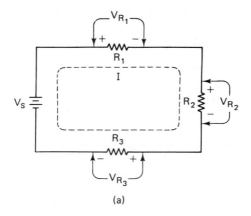

(a)

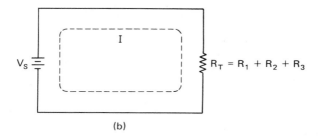

(b)

Figure 6.7

From Ohm's Law it is known that $R = V/I$, and thus each term in this equation is a resistance. In fact,

$$\frac{V_{R_1}}{I} = R_1 \qquad \frac{V_{R_2}}{I} = R_2 \qquad \frac{V_{R_3}}{I} = R_3$$

The term V_S/I must be equal to the total circuit resistance R_T, as shown in Figure 6.7(b). Making these substitutions,

$$R_T = R_1 + R_2 + R_3$$

These results extend immediately to a series circuit having any number of resistors (N), and the conclusion is

---

SERIES RESISTANCE

The total resistance of a series circuit is equal to the sum of the individual resistances.

$$R_T = R_1 + R_2 + R_3 + \cdots + R_N$$

EXAMPLE 6.5.   For the circuit in Figure 6.7(a), $V_S = 35$ V, $V_{R_1} = 3.3$ V, $V_{R_2} = 27$ V, and $I = 1$ mA. Determine V_{R_3}, R_1, R_2, R_3, and R_T.

SOLUTION.   Using KVL,
$$V_{R3} = V_S - V_{R_1} - V_{R_2}$$
$$= 35\ V - 3.3\ V - 27\ V = 4.7\ V$$

Ohm's Law yields
$$R_1 = \frac{V_{R_1}}{I} = \frac{3.3\ V}{1\ mA} = 3.3\ k\Omega$$

$$R_2 = \frac{V_{R_2}}{I_2} = \frac{27\ V}{1\ mA} = 27\ k\Omega$$

$$R_3 = \frac{V_{R_3}}{I_3} = \frac{4.7\ V}{1\ mA} = 4.7\ k\Omega$$

Thus $R_T = R_1 + R_2 + R_3 = (3.3 + 27 + 4.7)\ k\Omega$
$$= 35\ k\Omega$$

Exercises 6.1

1. Use Figure 6.3 to determine the following:
    (a) V_S if $V_{R_1} = 16$ V and $V_{R_2} = 7$ V
    (b) V_{R_1} if $V_S = 28$ V and $V_{R_2} = 11.5$ V
    (c) V_{R_2} if $V_S = 150$ V and $V_{R_1} = 68.2$ V
    (d) I if $V_S = 36$ V, $V_{R_1} = 11$ V and $R_1 = 3.3$ kΩ
    (e) I if $V_S = 12$ V, $V_{R_2} = 3.6$ V and $R_2 = 5.6$ kΩ
    (f) R_1 if $V_S = 15$ V, $V_{R_2} = 9$ V and $I = 6$ mA
    (g) R_2 if $V_S = 100$ V, $I = 5$ mA and $V_{R_1} = 2V_{R_2}$
    (h) V_{R_1} if $V_S = 230$ V, $I = 12$ A and $R_1 = 2R_2$
    (i) V_{R_2} if $V_S = 117$ V, $I = 125$ mA and $R_2 = 3/4\ R_1$
    (j) V_S if $I = 250$ mA, $R_1 = 100\ \Omega$ and $R_2 = 75\ \Omega$

2. Find the current I in a circuit composed of three resistors, 470 Ω, 330 Ω, and 150 Ω, connected in series with a 9 V battery.

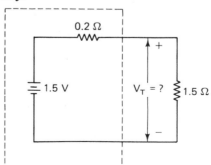

Figure 6.8

3. A 1.5 V cell having an internal resistance of 0.2 Ω is connected in series with a 1.5 Ω resistance, as shown in Figure 6.8. What is the terminal voltage V_T of the cell?

4. Draw the proper Veq in Figure 6.4 if:
   (a) $V_{S_1} = 6$ V and $V_{S_2} = 1.5$ V
   (b) $V_{S_1} = 25$ V and $V_{S_2} = 25$ V

5. Draw the proper Veq in Figure 6.5 if:
   (a) $V_{S_1} = 12$ V and $V_{S_2} = 3$ V
   (b) $V_{S_1} = 6$ V and $V_{S_2} = 18$ V
   (c) $V_{S_1} = 6$ V and $V_{S_2} = 6$ V

6. Find I, V_{R_1}, V_{R_2}, and V_{R_3} in Figure 6.9.

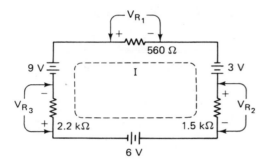

Figure 6.9

7. A 117 V source is connected to a motor drawing 9.8 A by means of two feeder wires. If the resistance of each wire is 0.15 Ω, what is the motor terminal voltage?

6.2 KIRCHHOFF'S CURRENT LAW (KCL)

A *node* can be defined as a connection between two or more circuit elements. The nodes in Figure 6.10 are labeled *A*, *B*, *C*, and *0*. Notice that node 0 is connected to *ground* and is the *reference node*. Kirchhoff's Current Law (KCL) deals with the electric current at a node, and can be stated as

---

KCL

The algebraic sum of the currents at any node is zero.

Another way of stating KCL is, "The sum of the currents *entering* a node must be equal to the sum of the currents *leaving* that node." This must be true since we obviously cannot gain or lose current at a node! For example, KVL

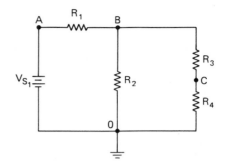

Figure 6.10

applied to the node in Figure 6.11 yields

$$I_1 + I_2 = I_3 + I_4$$
or
$$I_1 + I_2 - I_3 - I_4 = 0$$

EXAMPLE 6.6. Use KCL in Figure 6.11 to determine:

(a) I_1 if $I_2 = 3A$, $I_3 = 1$ A, and $I_4 = 4$ A.

(b) I_3 if $I_1 = 10$ mA, $I_2 = 15$ mA, and $I_4 = 25$ mA.

SOLUTION.

(a) $I_1 + I_2 = I_3 + I_4$

By substitution

$$I_1 + 3 = 1 + 4$$

Thus, $I_1 = 2$ A

(b) $I_1 + I_2 = I_3 + I_4$

Solving for I_3,

$$I_3 = I_1 + I_2 - I_4$$
$$= 10 \text{ mA} + 15 \text{ mA} - 25 \text{ mA} = 0$$

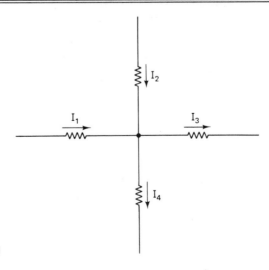

Figure 6.11

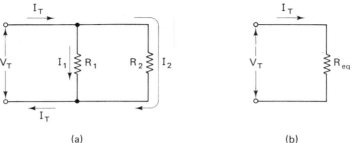

(a)   (b)   **Figure 6.12**

KCL can be used to develop the rule for combining parallel resistances. The resistors R_1 and R_2 in Figure 6.12(a) are connected in parallel, and KCL applied at either the top or bottom node yields

$$I_T = I_1 + I_2$$

However, since the two resistors are in parallel, the terminal voltage for each resistor is clearly V_T. We can then use Ohm's Law to determine

$$I_1 = \frac{V_T}{R_1}, \qquad I_2 = \frac{V_T}{R_2}$$

Substitution of these two relations into the KCL equation and using the distributive axiom yields

$$I_T = \frac{V_T}{R_1} + \frac{V_T}{R_2} = \left(\frac{1}{R_1} + \frac{1}{R_2} \right) V_T \tag{6.2}$$

Combining $1/R_1$ and $1/R_2$,

$$I_T = \left(\frac{R_1 + R_2}{R_1 R_2} \right) V_T$$

The equivalent resistance R_{eq} of R_1 and R_2 connected in parallel is shown in Figure 6.12(b); Ohm's Law requires

$$I_T = \frac{1}{R_{eq}} V_T$$

Comparing this with the above expression,

$$\frac{R_1 + R_2}{R_1 R_2} = \frac{1}{R_{eq}}$$

or
$$R_{eq} = \frac{R_1 R_2}{R_1 + R_2}$$

This is the well-known statement that "the equivalent resistance of two resistors in parallel is given as the product of the resistances divided by their sum."

Equation (6.2) can be written as

$$\frac{I_T}{V_T} = \left(\frac{1}{R_1} + \frac{1}{R_2} \right)$$

and it is easy to see that for N resistors in parallel, this becomes

$$\frac{1}{R_{eq}} = \frac{I_T}{V_T} = \frac{1}{R_1} + \frac{1}{R_2} + \frac{1}{R_3} + \cdots + \frac{1}{R_N} \tag{6.3}$$

Since $R_{eq} = V_T/I_T$, the following rule may be noted.

PARALLEL RESISTANCE

The equivalent resistance of N resistors connected in parallel can be determined from the equation

$$R_{eq} = \frac{1}{\dfrac{1}{R_1} + \dfrac{1}{R_2} + \dfrac{1}{R_3} + \cdots + \dfrac{1}{R_N}}$$

An interesting special case is the parallel connection of equal value resistances. In this case, $R_1 = R_2 = R_3 = \cdots = R_N$, and Eq. (6.3) becomes

$$\frac{I_T}{V_T} = \frac{1}{R} + \frac{1}{R} + \cdots + \frac{1}{R_N} = \frac{N}{R}$$

Thus,
$$R_{eq} = \frac{V_T}{I_T} = \frac{R}{N}$$

for N equal resistances in parallel.

EXAMPLE 6.7. Determine R_{eq} in Figure 6.12(b) if: (a) $R_1 = 3$ kΩ, $R_2 = 6$ kΩ (b) $R_1 = R_2 = 500$ Ω	**SOLUTION.** (a) $R_{eq} = \dfrac{R_1 R_2}{R_1 + R_2} = \dfrac{3 \text{ k}\Omega \times 6 \text{ k}\Omega}{3 \text{ k}\Omega + 6 \text{ k}\Omega} = 2 \text{ k}\Omega$ (b) $R_{eq} = \dfrac{R}{N} = \dfrac{500 \ \Omega}{2} = 250 \ \Omega$
EXAMPLE 6.8. Ten identical lamps are connected in parallel, and they draw 1.5 A from the source. What is the current in each lamp?	**SOLUTION.** Since the lamps are identical, they all have equal resistances; and since they are in parallel, they have the same terminal voltage. Thus they all have equal currents. So, $I = \dfrac{1.5 \text{ A}}{10} = 150 \text{ mA}$

Exercises 6.2

1. In Figure 6.11, determine:
   (a) I_1 if $I_2 = 3$ A, $I_3 = 4$ A, and $I_4 = 1$ A
   (b) I_2 if $I_1 = -2$ A, $I_3 = 1$ A, and $I_4 = 5$ A

    (c) I_3 if $I_1 = 1.5$ mA, $I_2 = 980$ μA, and $I_4 = -3$ mA

    (d) I_4 if $I_1 = 0.5$ mA, $I_2 = 450$ μA, and $I_3 = 4I_2$

2. Use Figure 6.12 to determine the following:

    (a) I_T if $I_1 = 33$ μA and $I_2 = 56$ μA

    (b) I_1 if $I_T = 6$ mA and $I_2 = 1$ mA

    (c) V_T if $I_T = 1.5$ mA, $I_1 = 500$ μA, and $R_2 = 56$ kΩ

    (d) I_T if $V_T = 12$ V, $R_1 = 10$ kΩ, and $R_2 = 5$ kΩ

    (e) I_2 if $V_T = 150$ V, $R_1 = 91$ kΩ and $I_T = 4$ mA

    (f) R_{eq} if $R_1 = 250$ Ω and $R_2 = 150$ Ω

    (g) R_{eq} if $R_1 = 80$ Ω and $R_2 = 60$ Ω

    (h) R_1 if $R_{eq} = 1.5$ kΩ and $R_2 = 5$ kΩ

    (i) R_2 if $R_{eq} = 50$ kΩ and $R_1 = 98$ kΩ

    (j) R_{eq} if $R_1 = R_2 = 33$ kΩ

3. What is the equivalent resistance of six 560 Ω resistors connected in parallel?

4. Begin with Eq. (6.3) and show that R_{eq} for three resistors ($N = 3$) is

$$R_{eq} = \frac{R_1 R_2 R_3}{R_1 R_2 + R_1 R_3 + R_2 R_3}$$

UNIT 7
FACTORING

7.1 COMMON FACTORS

The distribution axiom for multiplication provides the rule for multiplying two algebraic quantities. For example, the product of (a) and $(x+y)$ is given as

$$a(x+y)=ax+ay$$

The expression $(ax+ay)$ is clearly the product of the two quantities (a) and $(x+y)$, and these two quantities are defined as the *factors* of this expression. In fact, any time an algebraic expression can be written as the product of two or more quantities, these quantities are the *factors* of the expression, and the determination of these factors is called *factoring*. The distributive axiom will also provide the basis for factoring algebraic expressions.

Consider the algebraic expression $IR_1+IR_2+IR_3$. Each of the three terms in this expression contains the common factor I. The distributive axiom can then be used to factor the expression as

$$IR_1+IR_2+IR_3=I(R_1+R_2+R_3)$$

To factor the expression $2I^2R+6IV$, look for factors common to both terms. Notice that 2 is a common factor, and it is possible to rewrite the expression as

$$2(I^2R+3IV)$$

But, notice that I is also a common factor, and thus,

$$2I(IR+3V)$$

At this point there are no other factors common to both the terms in the parentheses, and the expression is therefore factored as

$$2I^2R+6IV=2I(IR+3V)$$

In general, the first step in factoring an algebraic expression involves removing the *greatest factor* common to all terms.

| EXAMPLE 7.1. Factor the following expressions:
 (a) $15x-5$
 (b) $3x^3-6x^2+9x$
 (c) $12V^2+6IV$
 (d) $2a^2bc^2-10ab+6ab^2c$ | SOLUTION.
 (a) $15x-5=5(3x-1)$
 The greatest common factor is 5.
 (b) $3x^2-6x^2+9x=3x(x^2-2x+3)$
 The greatest common factor is $3x$.
 (c) $12V^2+6IV=6V(2V+I)$
 The greatest common factor is $6V$.
 (d) $2a^2bc^2-10ab+6ab^2c=2ab(ac^2-5+3bc)$
 Thus $2ab$ is the greatest common factor. |

147

There is not always a factor common to all terms in an algebraic expression, but it might be possible to *group* the terms that do have common factors. For example, the expression $V_1 + 3IR_1 - 5V_1/V_T + 2IR_2$ can be factored by *grouping* terms with common factors as follows:

$$\left(V_1 - \frac{5V_1}{V_T}\right) + (3IR_1 + 2IR_2) = V_1\left(1 - \frac{5}{V_T}\right) + I(3R_1 + 2R_2)$$

EXAMPLE 7.2. Factor the following expression:

$$I_1R_1 + I_2R_2 + I_1R_2 + I_2R_1$$

SOLUTION. Grouping terms with a common I_1 and a common I_2 yields

$$(I_1R_1 + I_1R_2) + (I_2R_1 + I_2R_2)$$

Now, factor out I_1 from the first set of terms and I_2 from the second set.

$$I_1(R_1 + R_2) + I_2(R_1 + R_2)$$

Do not stop here, since $(R_1 + R_2)$ is a factor common to both terms. Thus,

$$(I_1 + I_2)(R_1 + R_2)$$

The distributive axiom can always be used to multiply out the factored form of an expression to verify that the factoring was carried out correctly. This should always be done. The factored expression in Example 7.2 can be multiplied out as

$$(I_1 + I_2)(R_1 + R_2) = I_1(R_1 + R_2) + I_2(R_1 + R_2)$$
$$= I_1R_1 + I_1R_2 + I_2R_1 + I_2R_2$$

and it is equivalent to the given expression.

Exercises 7.1

Factor each of the following:

1. $5R_1 + 10R_2$

2. $\dfrac{I_1}{10} + \dfrac{I_2}{20}$

3. $\dfrac{6}{R_1} + \dfrac{24}{R_1}$

4. $15R_1R_2 - 25R_1R_T$

5. $3IR_1 + 2IR_2 + 5IR_3$

6. $\dfrac{V}{R_1} + \dfrac{V}{R_2} - \dfrac{V}{R_3}$

7. $i^2R_1 + i^2R_2 + 3iv$

8. $6iv + v^2/R_1 - v^2/R_2$

9. $\pi r_1^2 - 3\pi d^2/4$

10. $as^2t - bst + cst^2$

11. $x^2y^3 - x^3y^2$

12. $4x^3 - 12x^2 + 28x$

13. $I_1(R_1 + R_2) - I_2(R_1 + R_2)$

14. $v_1\left(\dfrac{1}{R_1} + \dfrac{1}{R_2}\right) + v_2\left(\dfrac{1}{R_1} + \dfrac{1}{R_2}\right)$

15. $R_1(i_1 - i_2) + R_2(i_1 - i_2)$

16. $G_1(v_1-v_2)+G_2(v_1-v_2)+G_3(v_1-v_2)$

17. $ItV+V^2t/R+QV$

18. $3x^2(x+1)-4x(x+1)+6(x+1)$

19. $i^2R+Vi+P$

20. $I_1^2R_1+V_1I_1+V_2^2/R_2+V_2I_2$

21. $I_1R_1+I_1R_2+I_2R_1+I_2R_2$

22. $V_1/R_1+V_2/R_1+V_1/R_2+V_2/R_2$

23. $iv-ai+bv-ab$

24. $xy+x+ay+a$

25. $3a-at-st+3s$

26. $2aiR_1+3vR_2-3vR_1-2aiR_2$

27. v^2+1+v^3+v

28. $R_1i_1+i_1R_2+R_1i_2-i_3R_1+i_2R_2-i_3R_2$

29. $-R_Tv_1+v_1R_2+v_1R_1+R_1v_2+R_2v_2-R_Tv_2$

7.2 TRINOMIALS

A *trinomial* is an algebraic expression of the form

$$ax^2+bx+c$$

Frequently a trinomial can be factored into a product of two *binomials* in the form $(x+s)(x+t)$. To illustrate how this occurs, examine the following products:

Make the following observations:

1. The first term in the trinomial (x^2) is the product of the first term in each binomial (x). Thus in factoring, the two first terms in the binomials must be factors of the first term of the trinomial.

2. The third term in the trinomial (6) is the product of the second terms of the binomials (2 and 3). Thus when factoring, the second terms of the binomials must be factors of the third term in the trinomial.

3. The middle term in the trinomial is the algebraic sum of the products of the *outer terms* and the *inner terms* of the binomials. Thus the factors chosen as the second terms in the binomials must satisfy this requirement.

EXAMPLE 7.3. Factor:

$$x^2+5x+6$$

SOLUTION. The first terms in the factors are clearly x, and thus begin with

$$(x\quad)(x\quad)$$

The second terms must be chosen such that their product is 6 and their sum is 5. Possible choices are

$$6=6\cdot1\quad\text{or}\quad6=2\cdot3$$

Clearly the factors 2 and 3 have a sum of 5 (the coefficient of the middle term), and thus,

$$x^2 + 5x + 6 = (x + 2)(x + 3)$$

EXAMPLE 7.4. Factor: $$x^2 + 7x + 12$$	SOLUTION. The first two terms are x, and thus begin with $$(x \quad)(x \quad)$$ Possible factors for the second terms are $$12 = 12 \cdot 1 = 6 \cdot 2 = 4 \cdot 3$$ The choice of 4 and 3 yields the required sum of 7, and thus, $$x^2 + 7x + 12 = (x + 3)(x + 4)$$

Note that the middle term of the trinomial is the *algebraic sum* of the factors chosen for the third term. Therefore there is the possibility of minus signs appearing in the factors. For example,

$$x^2 + x - 6 = (x + 3)(x - 2)$$

In this case, the second terms of the factors must be chosen such that their *product* is (-6) and their *algebraic sum* is $(+1)$. The factors must therefore have *opposite* signs, and the larger factor must be positive. Clearly $(+3)$ and (-2) are the required factors, since $(3)(-2) = -6$ (the third term), and $(+3 - 2) = +1$ (the middle term).

In general:

1. If the third term of the trinomial is *positive*, the second terms of the factors must have *like* signs. The sign is the same as the sign of the middle term of the trinomial.

2. If the third term of the trinomial is *negative*, the second terms of the factors must have *opposite* signs.

EXAMPLE 7.5. Factor: $$x^2 - 5x + 6$$	SOLUTION. Begin with $$(x \quad)(x \quad)$$ The factors are $$6 = 1 \cdot 6 = (-1)(-6) = 2 \cdot 3 = (-2)(-3)$$ Clearly (-2) and (-3) are required since $$(-2) + (-3) = -5 \ldots \text{ (the middle term)}$$ Thus, $$x^2 - 5x + 6 = (x - 3)(x - 2)$$

If the coefficient of x^2 in the trinomial is not unity, the number of possible factors is increased, but the procedure remains the same.

EXAMPLE 7.6. Factor: $$2x^2 + 7x + 3$$	SOLUTION. Begin with factors for the term $2x^2$: $$(2x \quad)(x \quad)$$ The third-term factors are $3 = 3 \cdot 1$. The possibilities are: $$(2x+3)(x+1) \quad \text{and} \quad (2x+1)(x+3)$$ (1st choice) $\qquad$ (2nd choice) The first choice does not provide a $7x$ term, while the second choice does. Thus, $$(2x^2+7x+3)=(2x+1)(x+3)$$

When factoring trinomials, one should *always* verify the results by multiplying out the factors.

Exercises 7.2

1. Multiply each of the following:

   (a) $(x+1)(x+4)$ (b) $(x-1)(x+4)$ (c) $(x-2)(x-5)$

   (d) $(2x+1)(x-3)$ (e) $(i+6)(3i-9)$ (f) $(2v+2)(3v+3)$

   (g) $(2I+5)^2$ (h) $\left(\dfrac{V}{3}+2\right)\left(\dfrac{V}{2}-3\right)$ (i) $(0.7a-1.5)(0.1a+2)$

2. Factor and verify by multiplication:

   (a) x^2+5x+4 (b) $y^2+7y+10$ (c) $a^2+7a+12$

   (d) $x^2-7x+10$ (e) $p^2+30p+200$ (f) $Q^2-12Q+35$

   (g) $v^2-12v+36$ (h) $R^2-9R+14$ (i) i^2+4i-5

   (j) p^2-4 (k) $b^2-4b-21$ (l) $e^2+2e-24$

   (m) $z^2+5z-150$ (n) $\theta^2+2\theta-48$ (o) $i^2+1.5i+0.5$

   (p) $v^2+v+1/4$ (q) $2-3a+a^2$ (r) $j^2+4j-32$

   (s) $2x^2+5x-3$ (t) $3x^2+5x-2$ (u) $2i^2-3i+1$

   (v) $5j^2-13j-6$ (w) $3E^2-10E+3$ (x) $7v^2+37v+10$

   (y) $4p^2+13p+3$ (z) $6v^2+7v+2$ (aa) $15R^2-R-6$

   (bb) $8R^2-22R+15$ (cc) $21y^2+17y-30$ (dd) $4j^2-1$

7.3   DIFFERENCE OF TWO SQUARES—SUMMARY OF FACTORING

An algebraic expression of the form $(a^2 - b^2)$ is called the "difference of two squares." It is an important form since it is encountered frequently, and it is quite easy to factor. Multiplying $(a + b)$ and $(a - b)$ will yield

$$(a + b)(a - b) = a^2 + ab - ab - b^2 = a^2 - b^2$$

Thus,

DIFFERENCE OF SQUARES

The difference of squares of any two quantities is equal to the product of their sum and their difference.

EXAMPLE 7.7.   Factor the following:
(a) $x^2 - 4$
(b) $v^2 - 36$
(c) $9i^2 - 49$
(d) $1 - 81p^2$

SOLUTION.   Each is the difference of two squares.
(a) The square roots of the two terms are x and 2. Thus the product of the *sum* $(x + 2)$ and *difference* $(x - 2)$ of x and 2 is $(x + 2)(x - 2)$.
(b) The roots are v and 6. Thus
$$(v + 6)(v - 6).$$
(c) The roots are $3i$ and 7. Thus
$$(3i + 7)(3i - 7).$$
(d) The roots are 1 and $9p$. Thus
$$(1 + 9p)(1 - 9p).$$

Here is a summary of the procedures to be used in factoring algebraic expressions:

FRACTIONS

To factor an algebraic expression:
(1) Remove all monomial factors common to all terms.
(2) If there are two terms, try to factor as the difference of two squares.
(3) If there are three terms, try to factor as a trinomial.
(4) If there are four or more terms, try to factor by grouping.
(5) Apply the above steps to all new factors obtained until factoring is complete.

EXAMPLE 7.8. Factor:

$$3x^2 - 27$$

SOLUTION. First factor out a 3.

$$3x^2 - 27 = 3[x^2 - 9]$$

Recognize the factor $[x^2 - 9]$ as a difference of two squares. Thus,

$$3x^2 - 27 = 3[(x+3)(x-3)]$$

and the factoring is complete.

EXAMPLE 7.9. Factor:

$$4i^2 - 4i - 24$$

SOLUTION. First factor out a 4.

$$4i^2 - 4i - 24 = 4(i^2 - i - 6)$$

Factor the trinomial $(i^2 - i - 6)$ to obtain

$$4i^2 - 4i - 24 = 4(i+2)(i-3)$$

EXAMPLE 7.10. Factor:
(a) $i^4 - v^4$
(b) $3bx^2 - 3b + 3ab$
(c) $aiv + bciv - ap - bcp$

SOLUTION.
(a) $i^4 - v^4 = (i^2 + v^2)(i^2 - v^2)$

$$= (i^2 + v^2)(i + v)(i - v)$$

(b) $3bx^2 - 3b + 3ab = 3b[(x^2 - 1) + a]$

$$= 3b[(x+1)(x-1) + a]$$

(c) $(aiv + bciv) - (ap + bcp)$

$$= iv(a + bc) - p(a + bc)$$
$$= (iv - p)(a + bc)$$

For more complicated algebraic expressions, there are other special forms to look for (such as the sum and the difference of two cubes), or other special techniques (such as completing the square, and the factor theorem). For the interested reader, these topics are generally covered in a standard algebra textbook.

Exercises 7.3

1. Factor as a difference of two squares:

   (a) $x^2 - 25$

   (b) $16 - y^2$

   (c) $16i^2 - 9$

   (d) $\dfrac{v^2}{4} - \dfrac{1}{9}$

   (e) $0.01p^2 - 0.04$

   (f) $9Q^2 - 4$

   (g) $121x^4 - 49y^4$

   (h) $16a^2b^4p^6 - \dfrac{i^2R}{9}$

   (i) $\dfrac{p^4}{36} - (a-b)^4$

(j) $(i_1 - i_2)^2 R - 16$ (k) $(i_1 + i_2)^2 R - (i_1 - i_2)^2 R$ (l) $\dfrac{(v_1 - v_2)^2}{R} - \dfrac{(v_1 + v_2)^2}{9}$

2.   Factor and verify by multiplication:

(a) $5x^2 + 25x + 30$

(b) $2x^3 - 2x^2 - 4x$

(c) $90a + 33ai + 3ai^2$

(d) $6iv^2 - 6i$

(e) $4vi^2 - 36v$

(f) $175R - 7i^2 R$

(g) $ai^2 - 4a + bi^2 - 4b$

(h) $10a^2 bcxy - 6ab^2 c^2 z$

(i) $3x^3 y - 6x^2 y - 9xy$

(j) $2x^2(x+1) - (x^2 - 1)$

(k) $ax^2 - 9a + by^2 + by - 2b$

(l) $p^4 - 1$

(m) $bd + bci + adi + aci^2$

(n) $(i+v)^2 + i + v$

(o) $2as^2 + 2bs + ast + bt$

UNIT 8
ALGEBRAIC FRACTIONS—III

8.1   MULTIPLICATION AND DIVISION

It is sometimes possible to factor the numerator or the denominator (or both) of an algebraic fraction. When multiplying or dividing fractions written in factored form, it is easier to see how to simplify the expressions by cancellation of terms. Thus, always try to use factoring as an aid in simplifying algebraic expressions.

As an illustration, it is easy to see how to simplify the expression below, after factoring.

$$\frac{x^2+3x+2}{x^2-1}$$

Factor both the numerator and the denominator, and then cancel the *common term* to obtain

$$\frac{x^2+3x+2}{x^2-1}=\frac{\cancel{(x+1)}(x+2)}{\cancel{(x+1)}(x-1)}=\frac{x+2}{x-1}$$

EXAMPLE 8.1.   Simplify:

(a) $\dfrac{2IR+2R}{I^2-1}$

(b) $\dfrac{iR+3i+R+3}{i^2-2i-3}$

(c) $\dfrac{5v+35}{v^2+9v+14}$

SOLUTION.

(a) By factoring,

$$\frac{2IR+2R}{I^2-1}=\frac{2R\cancel{(I+1)}}{\cancel{(I+1)}(I-1)}=\frac{2R}{I-1}$$

(b) Factor and cancel terms,

$$\frac{iR+3i+R+3}{i^2-2i-3}=\frac{\cancel{(i+1)}(R+3)}{\cancel{(i+1)}(i-3)}=\frac{R+3}{i-1}$$

(c) $\dfrac{5\cancel{(v+7)}}{\cancel{(v+7)}(v+2)}=\dfrac{5}{v+2}$

When multiplying two algebraic fractions, it is much easier to factor the expressions *first*, cancel common terms, and *then* multiply. For instance, consider finding the product of the two following algebraic fractions:

$$\frac{7x^2-21x}{x^2-x-2}\quad\text{and}\quad\frac{x^2-4}{x^2-2x-3}$$

It is possible to first multiply according to the rules for operating on fractions and obtain

$$\frac{7x^2-21x}{x^2-x-2}\times\frac{x^2-4}{x^2-2x-3}=\frac{7x^4-21x^3-28x^2+84x}{x^4-3x^3-3x^2+7x+6}$$

We are now faced with the rather formidable task of trying to simplify this expression! A much easier way to find this product is to factor both expressions first, cancel terms, and then multiply. Thus,

$$\frac{7x^2-21x}{x^2-x-2} \times \frac{x^2-4}{x^2-2x-3} = \frac{7x\cancel{(x-3)}}{\cancel{(x-2)}(x+1)} \times \frac{(x+2)\cancel{(x-2)}}{\cancel{(x-3)}(x+1)} = \frac{7x(x+2)}{(x+1)^2}$$

Since the rule for dividing algebraic fractions is to "invert and multiply," it makes sense to use factoring in an effort to cancel terms *first*, and then carry out the division process. For instance, the quotient of

$$\frac{iv+3v}{v^2-49} \quad \text{and} \quad \frac{i^2v+3iv}{2v^2-14v}$$

is found by factoring first.* Thus,

$$\frac{iv+3v}{v^2-49} \div \frac{i^2v+3iv}{2v^2-14v} = \frac{v\cancel{(i+3)}}{(v+7)\cancel{(v-7)}} \times \frac{2v\cancel{(v-7)}}{iv\cancel{(i+3)}} = \frac{2v}{i(v+7)}$$

Exercises 8.1

1. Factor and simplify:

   (a) $\dfrac{R_1 R_2 + R_1}{R_1 R_2 - R_1}$

   (b) $\dfrac{i + vi}{i^2 - 2i}$

   (c) $\dfrac{3v^2 - 12}{3v^2 + 6v}$

   (d) $\dfrac{9i + 6i^2}{4i^2 - 9}$

   (e) $\dfrac{R^2 - 4}{2R^2 + R - 6}$

   (f) $\dfrac{-2 + v + v^2}{6 - 9v + 3v^2}$

   (g) $\dfrac{T_0 T + T_0^2}{20T_0 + T_0^2}$

   (h) $\dfrac{2s^2 + 3s}{6s^2 + 13s + 6}$

   (i) $\dfrac{t^2 + at + bt + ab}{t^2 - at + bt - ab}$

2. Find the product and simplify:

   (a) $\dfrac{R_{20} T_0}{T_0 + 20} \times \dfrac{T + T_0}{T_0}$

   (b) $I_T \times \dfrac{R_1 R_2}{R_1 + R_2} \times \dfrac{1}{R_1}$

   (c) $\dfrac{R_1 R_2 + R_1 R_3}{R_T} \times \dfrac{R_T R_2}{R_2 + R_3}$

   (d) $\dfrac{1}{q\mu_n N + q\mu_p N} \times \dfrac{qN}{3}$

   (e) $\dfrac{R_1 + 4}{R_2 + 2} \times \dfrac{3iR_2 + 6i}{R_1^2 - 16}$

   (f) $\dfrac{2v}{3v + 3iR} \times \dfrac{vi + i^2 R}{2}$

   (g) $\dfrac{3R - 6}{R_1^2 - 4} \times \dfrac{R_1 + 2}{2R_2 + 4} \times \dfrac{R_2 + 2}{R - 2}$

   (h) $\dfrac{\theta^2 - \theta - 6}{\alpha^2 + 5\alpha + 6} \times \dfrac{\alpha + 2}{\theta - 3}$

   (i) $\dfrac{s^2 + as + bs + ab}{t^2 + at + bt + ab} \times \dfrac{at + t^2}{bs + s^2}$

---

*You might invert and multiply first, and then try to simplify.

3. Find the quotient and simplify:

(a) $\dfrac{R_1 R_2}{R_1 + R_2} \div \dfrac{R_2}{I_T}$

(d) $\dfrac{3x^2 - 9x + 6}{5x^2 + 35x + 60} \div \dfrac{x^2 - 3x + 2}{5x^2 + 20x}$

(b) $(i^2 R - 3iv) \div \dfrac{1}{v}$

(e) $\dfrac{1 - x^2}{3x^3} \div \dfrac{9x + a}{9x}$

(c) $\dfrac{(i^2 - 4)}{(v^2 - 9)} \div \dfrac{(i + 2)}{(v - 3)}$

(f) $\dfrac{10i - 15}{6} \div \dfrac{28i - 42}{12}$

4. Simplify:

(a) $\dfrac{i^2 R_T}{R_1 + R_2} \times \dfrac{iR_1 + iR_2}{R_T} \div \dfrac{i^2 R_1 - i^2 R_T}{R_1 R_T}$

(b) $\dfrac{R_1 R_2 R_3}{R_1 R_2 + R_1 R_3 + R_2 R_3} \div \dfrac{R_1 R_2}{R_1 (R_2 + R_3) + R_2 R_3}$

(c) $\dfrac{x^2 - 4}{x^2 + 4x + 1} \times \dfrac{-3 - 2x + x^2}{2x + x^2} \div \dfrac{x^2 - 5x + 6}{x^2 + 3x}$

8.2   ADDITION AND SUBTRACTION

As seen previously, it is necessary to find the LCD (lowest common denominator) in order to add or subtract algebraic fractions. For algebraic fractions having polynomial expressions in their denominators, the task of finding the LCD can often be simplified by first factoring the denominators. For instance, to find the LCD for the two fractions

$$\frac{a}{3x + 6} \quad \text{and} \quad \frac{b}{5x + 10}$$

factor the denominators first to obtain

$$\frac{a}{3(x + 2)} \quad \text{and} \quad \frac{b}{5(x + 2)}$$

The LCD is then the product of 3, 5, and $(x + 2)$, or $15(x + 2)$.

Notice that the LCD is the simplest expression into which both denominators will divide evenly. In general, the LCD is found as the product of all different factors appearing in the denominators.

EXAMPLE 8.2. Find the LCD:

(a) $\dfrac{a}{3x^2+9x}$, $\dfrac{b}{x^2+4x+3}$

(b) $\dfrac{a}{i^2+i-2}$, $\dfrac{b}{i^2-4}$

SOLUTION. Factor each denominator first:

(a) $\dfrac{a}{3x(x+3)}$, $\dfrac{b}{(x+3)(x+1)}$

The LCD is then the product:

$$3x(x+3)(x+1)$$

(b) $\dfrac{a}{(i+2)(i-1)}$, $\dfrac{b}{(i+2)(i-2)}$

The LCD is then,

$$(i+2)(i-1)(i-2)$$

The addition and subtraction of fractions having polynomial denominators are accomplished in exactly the same fashion as with simple fractions. But, make use of factoring to determine the LCD.

EXAMPLE 8.3. Find the sum:

$$\frac{5}{3s+6}+\frac{3a}{10s+20}$$

SOLUTION. The fractions are first written in factored form as

$$\frac{5}{3(s+2)}+\frac{3a}{10(s+2)}$$

Clearly the LCD is $3\times10(s+2)$. Rewrite the fractions over the LCD and then add.

$$\frac{5}{3(s+2)}\times\frac{10}{10}+\frac{3a}{10(s+2)}\times\frac{3}{3}$$

$$=\frac{50}{30(s+2)}+\frac{9a}{30(s+2)}=\frac{50+9a}{30(s+2)}$$

EXAMPLE 8.4. Find the difference:

$$\frac{v-2}{R_1+1}-\frac{v+7}{R_1-1}$$

SOLUTION. The denominators cannot be factored, and cross multiplication is used.

$$\frac{(v-2)(R_1-1)-(v+7)(R_1+1)}{(R_1+1)(R_1-1)}$$

$$=\frac{-9R_1-2v-5}{(R_1+1)(R_1-1)}$$

EXAMPLE 8.5. Find the sum:

$$\frac{x+5}{x^2+3x+2}+\frac{x-7}{x^2+x-2}$$

SOLUTION. Factoring,

$$\frac{x+5}{(x+1)(x+2)}+\frac{x-7}{(x+2)(x-1)}$$

Rewriting over the LCD $\{(x+1)(x+2)(x-1)\}$ and adding,

$$\frac{(x+5)(x-1)+(x-7)(x+1)}{(x+1)(x+2)(x-1)}$$

$$=\frac{2x^2-2x-12}{(x+1)(x+2)(x-1)}$$

Notice that the numerator can be factored, and then simplified by cancellation.

$$\frac{2x^2-2x-12}{(x+1)(x+2)(x-1)}=\frac{2(x-3)\cancel{(x+2)}}{(x+1)\cancel{(x+2)}(x-1)}$$

$$=\frac{2(x-3)}{(x+1)(x-1)}$$

Exercises 8.2

1. Find the LCD:

(a) $\dfrac{1}{R_1}, \dfrac{1}{R_1+R_2}$

(b) $\dfrac{a}{3i+6}, \dfrac{b}{i+2}$

(c) $\dfrac{V_1}{R_1R_2+R_1R_3}, \dfrac{V_2}{R_2+R_3}$

(d) $\dfrac{1}{R_1}, \dfrac{1}{R_2}, \dfrac{1}{R_3}$

(e) $\dfrac{V_1}{R_1R_2}, \dfrac{V_2}{R_1R_2+2R_2}$

(f) $\dfrac{3}{x^2-1}, \dfrac{4a}{x^2+3x+2}$

2. Simplify:

(a) $\dfrac{1}{R_A}+\dfrac{1}{R_B}-\dfrac{1}{R_C}$

(b) $\dfrac{R_1R_2}{R_1+R_2}+\dfrac{1}{R_1}$

(c) $\dfrac{V_1}{3}+\dfrac{V_2}{8}-\dfrac{V_3}{6}$

$\dfrac{V_A}{?}+\dfrac{V_B}{3}-\dfrac{V_T}{6a}$

$\dfrac{?_A}{R_1R_3}-\dfrac{V_B}{R_TR_3+R_TR_2}$

(f) $\dfrac{1}{x-3}-\dfrac{x-3}{x^2-4}$

(g) $\dfrac{A}{s+7}-\dfrac{B}{s}+\dfrac{C}{s^2+7s}$

(h) $\dfrac{3}{x^2-1}-\dfrac{4a}{x^2+3x+2}$

(i) $\dfrac{a}{2i_1-4i_2}-\dfrac{b}{3i_1-6i_2}$

(j) $\dfrac{V_1}{V_1-V_2}-\dfrac{V_2}{V_1+V_2}$

(k) $\dfrac{1}{i^2-iR}+\dfrac{1}{iR-i^2}+\dfrac{1}{i-R}$

(q) $\dfrac{1}{i^2+3i+2}+\dfrac{1}{i^2+5i+6}$

(l) $\dfrac{I-2}{I-1}+\dfrac{I+2}{I+1}$

(r) $\dfrac{1}{v^2-v-6}-\dfrac{1}{v^2-4v+3}$

(m) $\dfrac{I-2}{I-1}-\dfrac{I+2}{I+1}$

(s) $\dfrac{(x+3)}{x^2+2x-3}-\dfrac{(x+7)}{x^2+4x-21}$

(n) $\dfrac{1}{(R_1+R_2)^2}+\dfrac{1}{R_1^2-R_2^2}$

(t) $\dfrac{1}{(R_1+2)(R_2+3)}+\dfrac{1}{(R_2+3)(R_3+4)}$

(o) $\dfrac{x}{x-y}-\dfrac{2xy}{x^2-y^2}$

(u) $\dfrac{1}{v^2+3v+2}+\dfrac{1}{v^2+v-2}-\dfrac{2}{v^2-1}$

(p) $\dfrac{1}{R_1+R_2}+\dfrac{2}{R_1-R_2}-\dfrac{3R_1}{R_1^2-R_2^2}$

8.3   MIXED EXPRESSIONS

Algebraic expressions containing both integers and fractions are called *mixed expressions*. For instance, $1+R/2$, and $3I-4V/R$ are mixed expressions. Before applying the operations of algebra to a mixed expression, it is usually desirable to rewrite the entire expression as a fraction.

An integer can be rewritten as a fraction by multiplying both the numerator and the denominator by the same quantity. This is essentially the same as multiplying by a quantity $\dfrac{Q}{Q}=1$, and clearly the value of the integer is unchanged. For instance, the integer 2 can be rewritten as a fraction in many different ways:

$$2\cdot\frac{x}{x}=\frac{2x}{x}\qquad\left(\text{multiplying by }\frac{x}{x}\right)$$

$$2\cdot\frac{3a}{3a}=\frac{6a}{3a}\qquad\left(\text{multiplying by }\frac{3a}{3a}\right)$$

$$2\cdot\frac{(x^2-1)}{(x^2-1)}=\frac{2(x^2-1)}{x^2-1}\qquad\left(\text{multiplying by }\frac{x^2-1}{x^2-1}\right)$$

Since any value for Q $\left(\text{in }\dfrac{Q}{Q}=1\right)$ can be selected, choose a value of Q consistent with a desired LCD. In order to simplify an expression such as $x+(a)/(y+1)$, for example, multiply the x term by $(y+1)/(y+1)$. Thus

$$x+\frac{a}{y+1}=x\frac{(y+1)}{(y+1)}+\frac{a}{y+1}=\frac{x(y+1)+a}{y+1}$$

EXAMPLE 8.6.   Express as a fraction:

$$R + \frac{1}{R_1 + j\omega L}$$

SOLUTION.   Multiply the term R by the quantity

$$\frac{R_1 + j\omega L}{R_1 + j\omega L}$$

Thus,

$$\frac{R(R_1 + j\omega L)}{(R_1 + j\omega L)} + \frac{1}{R_1 + j\omega L} = \frac{R(R_1 + j\omega L) + 1}{R_1 + j\omega L}$$

EXAMPLE 8.7.   Express as a fraction:

$$I_T - \frac{(V_1 - V_2)}{R} - 2I_2$$

SOLUTION.   The LCD is R, and thus multiply both the first and last terms by R/R. Thus,

$$I_T \frac{R}{R} - \frac{(V_1 - V_2)}{R} - 2I_2 \frac{R}{R}$$
$$= \frac{I_T R - V_1 + V_2 - 2I_2 R}{R}$$

EXAMPLE 8.8.   Express as a fraction:

$$I_T - \frac{3V_2}{R_1 + R_2} - \frac{V_3}{R_3}$$

SOLUTION.   The desired LCD is $R_3(R_1 + R_2)$. Thus, multiply the first term by

$$\frac{R_3(R_1 + R_2)}{R_3(R_1 + R_2)}$$

Multiply the second term by R_3/R_3, and multiply the third term by

$$\frac{(R_1 + R_2)}{(R_1 + R_2)}$$

Thus,

$$I_T \frac{R_3(R_1 + R_2)}{R_3(R_1 + R_2)}$$

$$- \frac{3V_2 R_3}{R_3(R_1 + R_2)} - \frac{V_3(R_1 + R_2)}{R_3(R_1 + R_2)}$$

$$= \frac{I_T R_3(R_1 + R_2) - 3V_2 R_3 - V_3(R_1 + R_2)}{R_3(R_1 + R_2)}$$

$$= \frac{(I_T R_3 - V_3)(R_1 + R_2) - 3V_2 R_3}{R_3(R_1 + R_2)}$$

Exercises 8.3

1. Express as fractions:

(a) $1 + \dfrac{1}{R}$

(b) $1 + \dfrac{1}{R_1} + \dfrac{1}{R_2}$

(c) $\dfrac{T}{T_0} + 1$

(d) $\dfrac{V}{V_T} - 1$

(e) $\dfrac{9}{5}\,°C + 32$

(f) $I_1 + \dfrac{I_T R_0}{R_0 + R_x}$

(g) $12 + \dfrac{V}{R_1} + \dfrac{V}{R_2}$

(h) $I_1 + I_2 + \dfrac{V}{R_1 + R_2}$

(i) $R + j\omega L - \dfrac{j}{\omega C}$

2. Perform the indicated operations and simplify:

(a) $\left(1 + \dfrac{1}{R_1}\right)\left(1 + \dfrac{1}{R_2}\right)$

(f) $\left(\dfrac{V^2}{4R} + 2\right) \div \left(VI + \dfrac{1}{8}\right)$

(b) $\left(1 + \dfrac{1}{R_1}\right) \div \left(1 + \dfrac{1}{R_2}\right)$

(g) $\left(10^3 + \dfrac{R_1 R_2}{R_1 + R_2}\right)\left(10^{-3} - \dfrac{V}{10^3}\right)$

(c) $\dfrac{R_0 T}{T_0} + R_0$

(h) $\left(\dfrac{V^2}{R} - 2\right) \div \left(10^{-3} R + \dfrac{200}{R}\right)$

(d) $\left(12 - \dfrac{V}{R}\right)\left(IR + \dfrac{V}{6}\right)$

(i) $\left(\dfrac{i_1 + i_2}{1 - i_1 i_2} - i_2\right) \div \left(1 + \dfrac{i_2(i_1 + i_2)}{1 - i_1 i_2}\right)$

(e) $\left(x + 1 - \dfrac{2}{x+2}\right)\left(2x - \dfrac{4}{x+1}\right)$

8.4   FRACTIONAL EQUATIONS

In order to solve an algebraic equation containing fractional expressions, it is generally desirable to eliminate all fractions first. The KCL equation

$$30 \text{ mA} = \frac{V}{2 \text{ k}\Omega} + \frac{V}{3 \text{ k}\Omega}$$

can be solved for V by first of all eliminating the fractions on the right side. This is done by multiplying *both sides* of the equation by the LCD of the fractions. This is referred to as "clearing the equation of fractions." Thus, multiply both sides by the LCD, which is 2 k$\Omega \times$ 3 k$\Omega =$ 6 kΩ.

$$30 \text{ mA} \times 6 \text{ k}\Omega = \frac{V}{2 \text{ k}\Omega} \times 6 \text{ k}\Omega + \frac{V}{3 \text{ k}\Omega} \times 6 \text{ k}\Omega$$

$$180 = 3V + 2V$$

$$180 = 5V$$

$$V = \frac{180}{5} = 36\,V$$

(d) $\dfrac{}{2}$

(e) $\dfrac{}{R_1 R_2 +}$

In general,

CLEARING FRACTIONS

To clear an equation of fractions, multiply both sides of the equation by the LCD.

EXAMPLE 8.9.   Solve the equation for V:

$$I = \left(\frac{1}{R_1} + \frac{1}{R_2} \right) V$$

SOLUTION. Multiplying both sides by the LCD $R_1 R_2$,

$$IR_1 R_2 = R_1 R_2 \left(\frac{1}{R_1} + \frac{1}{R_2} \right) V$$

$$= \left(\frac{R_1 R_2}{R_1} + \frac{R_1 R_2}{R_2} \right) V$$

$$= (R_2 + R_1) V$$

Then,

$$V = \frac{IR_1 R_2}{R_1 + R_2}$$

EXAMPLE 8.10.   Solve for x:

$$\frac{x}{x+1} = \frac{2}{3}$$

SOLUTION. Clearing of fractions is accomplished by multiplying both sides by the LCD $3(x+1)$. Thus,

$$\frac{x}{(x+1)} \times 3(x+1) = \frac{2}{3} \times 3(x+1)$$

$$3x = 2(x+1)$$

Then,

$$3x = 2x + 2$$

$$x = 2$$

The result is checked by substituting the solution $(x=2)$ back into the original equation:

$$\frac{2}{2+1} = \frac{2}{3}$$

and the result checks.

An algebraic fraction that has a fraction for its numerator, or for its denominator, or for both, is called a *complex fraction*. For example,

$$\frac{\dfrac{x}{3}}{2}, \quad \frac{5}{\dfrac{v}{2}}, \quad \frac{\dfrac{v}{3}}{\dfrac{R}{5}}$$

are all complex fractions.

The simplification of complex fractions is accomplished by first reducing the numerator and denominator fractions separately. Then, invert the denominator and multiply it by the numerator.

COMPLEX FRACTIONS

To simplify complex fractions:
(1) First reduce the numerator and denominator fractions.
(2) Then invert the denominator and multiply with the numerator.

EXAMPLE 8.11.   Simplify:

(a) $\dfrac{\dfrac{a}{b}}{\dfrac{x}{y}}$

(b) $\dfrac{4 + V/3}{R/2 + 3}$

SOLUTION.

(a) The numerator and denominator are in lowest terms; thus invert and multiply:

$$\frac{\dfrac{a}{b}}{\dfrac{x}{y}} = \frac{a}{b} \times \frac{y}{x} = \frac{ay}{bx}$$

(b) Operate on the numerator and denominator separately to obtain

$$\frac{4 + V/3}{R/2 + 3} = \frac{\dfrac{3 \times 4 + V}{3}}{\dfrac{R + 2 \times 3}{2}} = \frac{\dfrac{12 + V}{3}}{\dfrac{R + 6}{2}}$$

Then invert and multiply:

$$\frac{\dfrac{12 + V}{3}}{\dfrac{R + 6}{2}} = \frac{12 + V}{3} \times \frac{2}{R + 6} = \frac{2(12 + V)}{3(R + 6)}$$

The rule for finding the equivalent resistance of two resistors connected in parallel involves a complex fraction as the following example shows.

EXAMPLE 8.12.   Begin with

$$R_T = \cfrac{1}{\cfrac{1}{R_1} + \cfrac{1}{R_2}}$$

and show that

$$R_T = \frac{R_1 R_2}{R_1 + R_2}$$

SOLUTION.

$$R_T = \cfrac{1}{\cfrac{1}{R_1} + \cfrac{1}{R_2}} = \cfrac{\cfrac{1}{1}}{\cfrac{R_2 + R_1}{R_1 R_2}} = \frac{1}{1} \times \frac{R_1 R_2}{R_1 + R_2}$$

Thus,

$$R_T = \frac{R_1 R_2}{R_1 + R_2}$$

An important type of problem that leads to a fractional equation is called a *rate problem*. The word *rate* suggests the involvement of time. In fact, rate can be thought of as the accomplishment of something in a certain period of time, and it can be stated that

(Rate of doing something) × (time) = (fractional part of something done)

For example, if a certain task can be done in 5 hours, 1/5 the task can be done in one hour. This fraction is the "rate of doing work." In general, the rate of working is $1/x$ if the total task can be accomplished in x hours.

EXAMPLE 8.13.   An electronics technician wires a system in 14 hours, and his partner can do it in 12 hours. How long would it take to do the job if the two men work together?

SOLUTION.   The rate at which each works is established as

$$\text{Technician rate} = \frac{1}{14}$$

$$\text{Partner rate} \quad = \frac{1}{12}$$

Now, if the whole job takes x hours, use (rate) × (time) = (fractional part) to obtain

$$\text{Work by Technician} = \frac{1}{14} x$$

$$\text{Work by Partner} \quad = \frac{1}{12} x$$

The whole job (1) is equal to the sum of the fractional parts. Thus

$$\frac{1}{14} x + \frac{1}{12} x = 1$$

The LCD is 84. Thus,

$$\frac{6}{6} \cdot \frac{1}{14} x + \frac{7}{7} \cdot \frac{1}{12} x = 1$$

$$\frac{6x + 7x}{84} = 1$$

$$13x = 84$$

$$x = \frac{84}{13} = 6\frac{6}{13} \text{ hr}$$

Exercises 8.4

1. Solve and check:

(a) $\dfrac{v}{2} + \dfrac{v}{3} = 5$

(b) $i - \dfrac{1}{8} = \dfrac{7}{6}$

(c) $\dfrac{z}{4} - 1 = z$

(d) $\dfrac{V}{5} - \dfrac{V}{6} = \dfrac{V}{2} - \dfrac{1}{15}$

(e) $\dfrac{V-1}{6} = \dfrac{3+V}{4}$

(f) $\dfrac{7}{R} + \dfrac{3}{R} = 5$

(g) $\dfrac{2}{R} + \dfrac{1}{8} = \dfrac{1}{2}$

(h) $\dfrac{i+2}{3i} = \dfrac{5-i}{4} + \dfrac{i}{4}$

(i) $\dfrac{1}{2}(\theta - 4) - \dfrac{2}{3}(1 + \theta) = 0$

(j) $\dfrac{5}{R-1} - \dfrac{7}{R-2} + \dfrac{2}{R-3} = 0$

(k) $\dfrac{3}{2x-1} + \dfrac{1}{2} = \dfrac{1}{4x-2}$

(l) $\dfrac{v+1}{v-1} - \dfrac{v-1}{v+1} = 0$

(m) $36 = 60\dfrac{1 \text{ k}\Omega}{1 \text{ k}\Omega + R_S}$

(n) $3 \text{ mA} = \dfrac{38}{R_{TH} + 6 \text{ k}\Omega}$

(o) $10\dfrac{500}{700 + 500} = 10\dfrac{R_3}{900 + R_3}$

2. Simplify the following:

(a) $\dfrac{\dfrac{3}{a}}{\dfrac{5}{5}}$

(b) $\dfrac{\dfrac{R}{2}}{6}$

(c) $\dfrac{\dfrac{v}{5}}{\dfrac{R}{2}}$

(d) $\dfrac{v+1}{\dfrac{3}{8}}$

(e) $\dfrac{\dfrac{a}{3}}{R+5}$

(f) $\dfrac{\dfrac{(i+1)}{(i-2)}}{\dfrac{(i+1)}{3}}$

3. Solve and check:

(a) $500 = \dfrac{1}{\dfrac{1}{1000} + \dfrac{1}{R_1}}$

(b) $1 \, \mu F = \dfrac{1}{\dfrac{1}{5 \, \mu F} + \dfrac{1}{C}}$

(c) $10 = \dfrac{\dfrac{6}{R_1} + 20 \times 10^{-3}}{\dfrac{1}{R_1} + \dfrac{1}{1 \, k\Omega}}$

(d) $1 \, k\Omega = \dfrac{1}{\dfrac{1}{3 \, k\Omega} + \dfrac{1}{6 \, k\Omega} + \dfrac{1}{R}}$

(e) $5 \, mA = \dfrac{10 \, mA \dfrac{2 \, k\Omega \times R_1}{2 \, k\Omega + R_1}}{R_1}$

(f) $10 = \dfrac{12 - V_T}{\dfrac{V_T}{500}}$

4. Solve the following problems and check your solution in each case:

   (a) Given

   $$\frac{1}{R_T} = \frac{1}{R_1} + \frac{1}{R_2} + \frac{1}{R_3}$$

   solve for (i) R_T, (ii) R_1.

   (b) Two thirds of a certain number added to 3 is equal to 9. Find the number.

   (c) Find a number such that the difference between 9/5 the number and 1/4 the number is 7.

   (d) The internal resistance of a circuit is increased by 25% in order to obtain a total resistance of 1250 Ω. Find the original resistance.

   (e) Find the consecutive integers such that one-third the smaller exceeds one-fifth the larger by 3.

   (f) The sum of two numbers is 100. When the larger is divided by the smaller, the quotient is 4 with a remainder of 5. Find the two numbers.

   (g) The cross-sectional area of a piece of electrical conduit is given by $A = \dfrac{\pi d^2}{4}$, where d is the pipe diameter. If a second pipe has 3 times the cross-sectional area, what is the ratio of the two diameters?

   (h) An equation for the resistance R of a material at any temperature T is

   $$R = R_{20}[1 + \alpha(T - 20)]$$

   where R_{20} is the resistance at 20°C, and $\alpha = \dfrac{1}{T_0 + 20}$, which is the "temperature coefficient of resistance" (given in tables). Begin with the equation

   $$R = R_{20} + R_{20}\left(\frac{T - 20}{T_0 + 20}\right)$$

   and develop the given equation.

5. Solve the following rate problems:

(a) A worker can complete a certain job in 7 days, and the same job requires 10 days for his apprentice. How many days would be required if the two men work together?

(b) A certain task requires 3 min 15 sec for worker A, and 2 min 50 sec for worker B. How long will it take to complete the task if A and B work together?

(c) Three workers with equal abilities can each accomplish a certain task in 25 hours. If they work together, how long will it take?

(d) A tank fitted with two fill pipes can be filled in 4 hr 30 min by one pipe alone, and in 3 hr 40 min by the other pipe alone. How long will it take to fill the tank if both fill pipes are used? (*Hint:* The rate for the first pipe is either $\dfrac{1}{4\frac{1}{2}\ \text{hr}} = \dfrac{2}{9\ \text{hr}}$, or $\dfrac{1}{270\ \text{min}}$.)

(e) A tank can be filled by an intake pipe in 23 min, and it can be emptied by a drain pipe in 67 min. Beginning with an empty tank, how long will it take to fill if both pipes are open?

(f) A storage battery can supply an emergency lighting load for 3 hr 10 min, or an emergency laboratory load for 1 hr 15 min. How long could the battery supply both loads? (*Hint:* The rate of discharge for the lighting load is either $\dfrac{1}{3\frac{1}{6}\ \text{hr}}$ or $\dfrac{1}{190\ \text{min}}$.)

(g) A chemical cell will supply a certain load for 28 hr, and it takes 12 hr to recharge the cell. If the discharged cell is left connected to its load while being recharged, how long will it take to recharge completely?

(h) A satellite power source must discharge into its load no more than 50% of its capacity in 1 1/2 hr (the time of one revolution around the earth). While in the sun, the source is recharged by means of solar cells. If the time in the sun is taken to be 2/3 hr, what must be the recharge rate in order to completely recharge the source at the end of each revolution? The time chart will be an aid in solving this problem (see Figure 8.1).

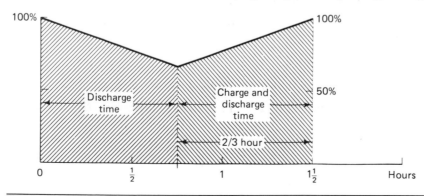

Figure 8.1

UNIT 9
TWO SIMULTANEOUS EQUATIONS

9.1 INTRODUCTION—GRAPHING

In many electric circuit problems there is more than one unknown quantity. The application of KCL to an electric circuit could provide a *node* equation such as

$$40 = 10V_A + 5V_B$$

where the two unknown quantities to be determined are V_A and V_B. Similarly, KVL applied to an electric circuit might lead to a *mesh (loop)* equation such as

$$10 = 3I_1 - 2I_2 - 4I_3$$

where the unknown quantities are I_1, I_2, and I_3.

An attempt to solve the above node equation ($40 = 10V_A + 5V_B$) leads to an infinite number of possible sets of V_A and V_B that will satisfy the equation. For example, $V_A = 0$ and $V_B = 8$ will satisfy the equation. So will all of the following sets of (V_A, V_B): (1,6), (2,4), and (−1,10). The graph of this node equation is the straight line shown in Figure 9.1, and clearly any point on the graph will provide a value of V_A and a value of V_B satisfying the equation.

Normally, when solving an electric circuit problem, we will be interested in finding a single (*unique*) value of V_A and its corresponding value of V_B for one particular circuit. In order to do this, we must have a second node equation, such as $0 = 2V_A - V_B$. Since the graph of this equation is also a straight line, there are an infinite number of sets of (V_A, V_B) that will satisfy this equation. For example: (0,0), (1,2), etc.

These two node equations are graphed together in Figure 9.2. Notice that the two lines intersect at a single point where $V_A = 2$ and $V_B = 4$. This set of values of

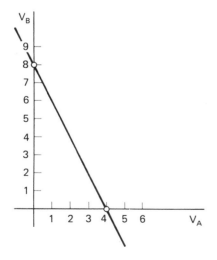

V_A	−2	−1	0	1	2	3	4	
V_B	12	10	8	6	4	2	0	

$$40 = 10 V_A + 5 V_B$$

Figure 9.1

V_A and V_B (2,4) satisfies both equations *simultaneously*, and provides a *unique* solution for V_A and V_B. Since the two node equations

$$40 = 10V_A + 5V_B$$

$$0 = 2V_A - V_B$$

are satisfied *simultaneously* by a unique set of values for V_A and V_B, they are called *simultaneous equations*. In general,

---

SIMULTANEOUS EQUATIONS

When two or more algebraic equations are satisfied simultaneously by a unique set of values of the unknowns, they are called *simultaneous equations*.

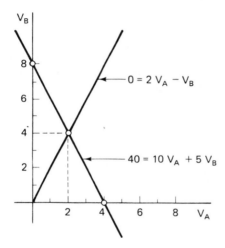

Figure 9.2

Notice that it required *two* equations to find a unique solution for a problem with *two* unknowns. In a similar way, it can be shown that it requires *three* equations to find a unique solution for *three* unknowns; and in general,

---

UNIQUE SOLUTION

It will be necessary to solve N simultaneous equations to determine a unique solution for a problem having N unknowns.

EXAMPLE 9.1.   Find a solution for the simultaneous equations

$$y = 2x + 4$$
$$y = x + 3$$

by graphing.

SOLUTION.   Both equations are straight lines, and they are easily graphed by locating the x and y intercepts. Thus,

for $y = 2x + 4$
$$\begin{array}{c||c|c} x & 0 & -2 \\ \hline y & 4 & 0 \end{array}$$

for $y = x + 3$
$$\begin{array}{c||c|c} x & 0 & -3 \\ \hline y & 3 & 0 \end{array}$$

The lines are graphed in Figure 9.3, and the single point of the intersection occurs at $x = -1$, $y = 2$, which is the unique solution. The solution can be checked by substituting $(-1, 2)$ into the two equations:

$$y = 2x + 4 \qquad y = x + 3$$
$$2 = 2(-1) + 4 \qquad 2 = -1 + 3$$

and the solution is seen to be valid.

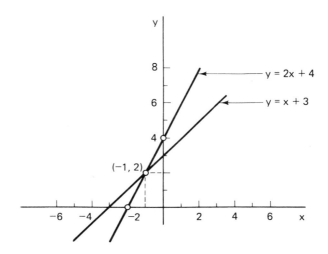

Figure 9.3

It is possible to have two equations in two unknowns for which no solution exists. Consider the two equations

$$y = x + 3$$
$$y = x$$

which are graphed together in Figure 9.4. Notice that the two lines are *parallel*. Thus the lines have no intersection, and there is no unique solution. That is, there is

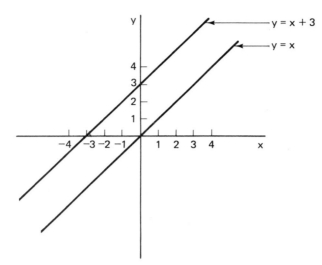

Figure 9.4

no set of values for x and y that will satisfy both equations simultaneously (you might try to find a set). Two such equations having no intersection (and therefore no unique solution) are said to be *inconsistent*. Conversely, intersecting lines that have a unique solution are said to be *consistent*.

Exercises 9.1

Graph the following sets of equations and determine a unique solution if one exists. Check the solutions by substituting values into the original equations.

1. $y = x$
   $y = 2x$

2. $y = x + 1$
   $y = 2x + 1$

3. $y = -x$
   $y = x - 1$

4. $y = 2x - 1$
   $y = 3x + 2$

5. $y = -x + 3$
   $y = -3x + 1$

6. $2x - y = 0$
   $x + y = -1$

7. $y = 3x + 2$
   $y = 3x - 1$

8. $3y + 2x = 6$
   $2y + 3x = 2$

9. $\dfrac{x}{2} + \dfrac{y}{3} = 4$
   $x + y = 1$

10. $V_A = V_B - 1$
    $V_A = 2V_B + 3$

11. $V_A - 2V_B = 0$
    $V_A = V_B + 1$

12. $2V_A - V_B = 6$
    $V_A - \dfrac{1}{2}V_B = -2$

13. $I_1 + I_2 = 0$
    $I_1 - I_2 = 0$

14. $I_1 + I_2 = 0$
    $I_1 - I_2 = 2$

15. $3 = 4I_1 - 3I_2$
    $0 = -3I_1 + 9I_2$

9.2   SOLUTION BY SUBSTITUTION

In order to obtain a graphical solution for two simultaneous equations, it is necessary to plot the two lines with great care. Even so, the results obtained may not be sufficiently accurate when the solutions are not integers. That is to say, numerical values such as 17/32 or 7.96 are not as easy to determine from a graph as are values such as 3 or 5.

It is possible to find a solution for two simultaneous equations by combining the two equations in such a way as to eliminate one of the unknowns. The resulting equation is solved for its unknown, and this value is then substituted into either of the original equations to find the second unknown.

The method to be used here is called *elimination by substitution*, and the procedure is:

ELIMINATION BY SUBSTITUTION

To solve two simultaneous equations using elimination by substitution:
(1) Solve either equation for one of the unknowns.
(2) Substitute this value into the other equation and solve for the remaining unknown.
(3) Use the solution in either of the given equations to solve for the other unknown.
(4) Check the solution by substituting values into *both* original equations.

Let's look at an example.

EXAMPLE 9.2.   Solve by substitution:

$$y = x$$
$$y = 2x + 1$$

SOLUTION. The first equation is solved for y in terms of x. Substituting $y = x$ into the second equation yields

$$(x) = 2x + 1$$

Solving for x,

$$x = -1$$

Substituting this value back into the first equation gives a solution for y as

$$y = x = -1$$

As a check:

$$y = x \qquad y = 2x + 1$$
$$-1 = -1 \qquad -1 = 2(-1) + 1$$

and the solution $(-1, -1)$ is seen to be valid.

Notice that both equations in Example 9.2 are solved for y. In this case we could simply equate the expression for y from each equation to obtain $x = 2x + 1$. The solution then follows immediately.

EXAMPLE 9.3.   Solve by substitution:

$$40 = 10V_A + 5V_B$$

$$0 = 2V_A - V_B$$

SOLUTION. Solve the second equation for V_B to obtain

$$V_B = 2V_A$$

Substitute this value of V_B into the first equation:

$$40 = 10V_A + 5(2V_A)$$

Solve this equation for V_A:

$$40 = 10V_A + 10V_A = 20V_A$$

Thus $V_A = 2$.

Substitute this solution for V_A into the given equation:

$$0 = 2V_A - V_B$$

to obtain

$$0 = 2(2) - V_B$$

Then, $V_B = 4$

As a check:

$$40 = 10V_A + 5V_B \qquad 0 = 2V_A - V_B$$

$$40 = 10(2) + 5(4) \qquad 0 = 2(2) - 4$$

and the solution $(2, 4)$ is valid.

If we attempt to use elimination by substitution to solve two equations that are inconsistent (they have no intersection point and therefore no solution, such as two parallel lines), we will obtain an *inequality*. That is, we will end up with an equation which is *not true*. As an example, consider the two equations graphed in Figure 9.4. The equations are

$$y = x$$
$$y = x + 3$$

If we substitute the first equation into the second equation to eliminate the variable y, we obtain

$$(x) = x + 3$$

But this leads to $0 = 3$!!!—which is nonsense. Thus we immediately know that there is no solution for these two equations, and the equations are said to be *inconsistent*.

EXAMPLE 9.4.   Solve by substitution:

$$x+y=1$$
$$2y=-2x-6$$

SOLUTION.   Solving the first equation for x,

$$x=1-y$$

Substituting this into the second equation to eliminate x,

$$2y=-2(1-y)-6$$

Thus, $2y=-2+2y-6$

and, $0=-8???$

Therefore, these two equations are inconsistent.

Rewriting the two equations in Example 9.4 in the standard slope-intercept form ($y=mx+b$), we obtain

$$y=-x+1$$
$$y=-x-3$$

Notice that both equations have the *same slope* ($m=-1$), and by inspection we can say the equations are inconsistent. There is no solution.

When solving simultaneous equations, care must be taken to ensure that there are indeed N equations for N unknowns. Graphs of the two equations

$$y=2x+1$$
$$2y-4x=2$$

will reveal that they are exactly the same straight line. Even though it appears that there are two equations, there is in fact only one equation written in two different forms. This can be shown by dividing the second equation by 2 to obtain

$$y-2x=1$$

and then transposing terms to obtain

$$y=2x+1$$

An attempt to solve simultaneously a single equation written in two different forms (thinking that these are in fact two independent equations), will lead to an *identity*. Let's use the above example to illustrate. Substituting $(2x+1)$ for y in the equation $2y-4x=2$ to eliminate y yields

$$2(2x+1)-4x=2$$

Simplifying,

$$4x+2-4x=2$$
$$2=2$$

The result is a valid statement that simply verifies that the first equation is *identical* with the second equation. Clearly there can be no unique solution since we in fact are dealing with only *one* equation, and we must have *two* equations when dealing with *two* unknowns.

EXAMPLE 9.5.   Solve by substitution:

$$y = \frac{x}{3} + \frac{1}{2}$$

$$y = 2x - 2$$

SOLUTION.   It will generally be easier to clear all fractions first. Thus, rewrite the first equation as (multiply both sides by 6)

$$6y = 2x + 3$$

Substituting $(2x - 2)$ for y to eliminate x,

$$6(2x - 2) = 2x + 3$$
$$12x - 12 = 2x + 3$$
$$10x = 15$$
$$x = \frac{3}{2}$$

Then,

$$y = 2x - 2 = 2(3/2) - 2 = 1$$

As a check:

$$y = \frac{x}{3} + \frac{1}{2} \qquad y = 2x - 2$$

$$1 = \frac{\frac{3}{2}}{3} + \frac{1}{2} \qquad 1 = 2\left(\frac{3}{2}\right) - 2$$

$$1 = 1 \qquad\qquad 1 = 1$$

and the solution is verified.

Exercises 9.2

Use elimination by substitution to determine a unique solution if one exists. Check all solutions by substituting values into the original equations. If a solution does not exist, state the reason (inconsistent or identity).

1. $y = x$
   $y = \frac{1}{2}x$

2. $y = x$
   $y = x + 3$

3. $y = -x$
   $y = 2x + 1$

4. $y = x + 1$
   $y = -x - 1$

5. $y = x + 1$
   $y = 2x + 1$

6. $y = 2x - 1$
   $y = 3x + 2$

7. $y = 2x - 1$
   $y = 2x + 4$

8. $2x - y = 0$
   $x + y = -1$

9. $\dfrac{x}{2} + \dfrac{y}{3} = 4$
   $2x + 2y = 2$

10. $I_1 + I_2 = 0$
    $I_1 - I_2 = 0$

11. $4I_1 - 3I_2 = 3$
    $-I_1 + 3I_2 = 0$

12. $12 = 10I_1 - 7I_2$
    $0 = -7I_1 + 12I_2$

13. $\dfrac{I_1}{2} - \dfrac{I_2}{3} = 7$
    $3I_1 - 2I_2 = 42$

14. $6 = \dfrac{V_A}{5} - \dfrac{V_B}{8}$
    $0 = -\dfrac{V_A}{8} + \dfrac{V_B}{2}$

15. $0 = \dfrac{V_1}{5} - \dfrac{V_2}{7}$
    $0 = -\dfrac{V_1}{7} + \dfrac{V_2}{2}$

16. $2 = \left(\dfrac{1}{2} + \dfrac{1}{3}\right)V_A - \dfrac{1}{3}V_B$
    $0 = -\dfrac{1}{3}V_A + \left(\dfrac{1}{3} + \dfrac{1}{4}\right)V_B$

17. $V_1 = \dfrac{V_2}{2} + 1$
    $2V_1 - V_2 = 2$

18. $6 = \dfrac{V_1}{R_1} - \dfrac{V_2}{R_2}$
    $V_1 = 2V_2$

19. $5 = R_1 I_1 - R_2 I_2$
    $0 = -R_2 I_1 + R_3 I_2$

20. $10 = R_1 I_1 - R_2 I_2$
    $I_1 = 2I_2$

9.3   SOLUTION BY ADDITION

It is sometimes easier to find a solution for two simultaneous equations by either adding or subtracting the two equations in order to eliminate one of the variables. This method is called *elimination by addition (or subtraction)* and in general:

ELIMINATION BY ADDITION

To solve two simultaneous equations using elimination by addition (or subtraction):
(1) Multiply each equation by a suitable number such that one of the variables will be eliminated when two equations are added (subtracted).
(2) Solve this equation for the remaining unknown.
(3) Use this solution in either of the given equations to solve for the other unknown.
(4) Check the solution by substituting values into *both* original equations.

EXAMPLE 9.6. Solve using elimination by addition:

$$y + x = 3$$
$$y - x = 1$$

SOLUTION. Step (1) is not necessary, since the variable x can be eliminated by simply adding the two equations. Thus,

$$y + x = 3$$
$$\underline{y - x = 1(+)}$$
$$y = 2$$

Then substituting this value for y into the first equation yields

$$(2) + x = 3$$
$$x = 1$$

Thus the solution is

$$x = 1, y = 2$$

As a check, substitute these values into the given equations

$$y + x = 3 \quad y - x = 1$$
$$2 + 1 = 3 \quad 2 - 1 = 1$$

and the solution is verified.

Let's look at an example where it will be necessary to multiply both equations by a suitable constant in order to eliminate a variable.

EXAMPLE 9.7. Solve, using elimination by addition:

$$2x + 3y = 1$$
$$3x + 4y = 2$$

SOLUTION. There are two choices: (1) multiply the first equation by 3, multiply the second equation by 2, and subtract the eliminate x; or (2) multiply the first equation by 4, multiply the second equation by 3, and subtract to eliminate y. Choose the first alternative:

$$(\times 3)[2x + 3y = 1] \rightarrow 6x + 9y = 3$$
$$(\times 2)[3x + 4y = 2] \rightarrow \underline{6x + 8y = 4(-)}$$
$$y = -1$$

Substituting this value for y into the first given equation yields

$$2x + 3(-1) = 1$$
$$2x = 4$$
$$x = 2$$

Thus the solution is
$$x = 2, y = -1$$

As a check:

$$2x + 3y = 1 \qquad\qquad 3x + 4y = 2$$
$$2(2) + 3(-1) = 1 \qquad 3(2) + 4(-1) = 2$$

and the solution is valid.

Equations having literal coefficients instead of numerical coefficients can be handled in exactly the same manner.

EXAMPLE 9.8. Solve using elimination by addition:

$$ax + y = 0$$
$$x + by = 3$$

SOLUTION. Multiply the second equation by (a) and subtract the first equation to eliminate x.

$$(xa)[x + by] = 3 \rightarrow ax + aby = 3a$$
$$\underline{ax + y = 0(-)}$$
$$aby - y = 3a$$

Factoring, $\qquad (ab - 1)y = 3a$

Thus, $\qquad\qquad y = \dfrac{3a}{ab - 1}$

Now, substitute this value of y into the first given equation to determine a value for x.

$$ax + \left(\frac{3a}{ab - 1}\right) = 0$$
$$ax = -\frac{3a}{ab - 1}$$
$$x = \frac{-3}{ab - 1}$$

To check: First,
$$ax + y = 0$$

Substituting,

$$a\left(\frac{-3}{ab - 1}\right) + \left(\frac{3a}{ab - 1}\right) = 0$$

Clearing fractions,

$$-3a + 3a = 0 \qquad 0 = 0$$

Second, $x + by = 3$

Substituting, $\left(\dfrac{-3}{ab-1}\right)+b\left(\dfrac{3a}{ab-1}\right)=3$

Clearing fractions, $-3+3ab=3(ab-1)$
$$3(ab-1)=3(ab-1)$$

and the solution is valid.

At this point, let's summarize the key points involved in solving two simultaneous equations.

1. There must be *two* equations to solve for *two* unknowns.

2. If the graphs of the two equations intersect, there is a unique solution and the equations are *consistent*.

3. If the two equations are parallel lines (same slope), they do not intersect, and there is no solution. The equations are said to be *inconsistent*. An attempt to solve these two equations simultaneously will lead to an *inequality*.

4. If the graphs of the two equations are the same, essentially there is only one equation (two are required), and there is no unique solution. An attempt to solve simultaneously will lead to an *identity*.

Exercises 9.3

Use elimination by addition (or subtraction) to determine a unique solution if one exists. Check all solutions by substituting values into the original equations. If a solution does not exist, state the reason (inconsistent or identity).

1. $y=x$
   $y=2x$

2. $y=-x$
   $y=x-1$

3. $y+x=3$
   $y+3x=1$

4. $2x-y=0$
   $3x+3y=-3$

5. Do Example 9.7 using the second alternative.

6. $3y+2x=6$
   $2y+3x=2$

7. $\dfrac{x}{2}+\dfrac{y}{3}=4$
   $2x+2y=2$

8. $3x-y+2=0$
   $3x-y-1=0$

9. $V_A-V_B=-1$
   $V_A-2V_B=3$

10. $2V_A-V_B=6$
    $V_A-\dfrac{1}{2}V_B=-2$

11. $6=\dfrac{V_A}{5}-\dfrac{V_B}{8}$
    $0=\dfrac{-V_A}{8}+\dfrac{V_B}{2}$

12. $6 = \dfrac{V_1}{R_1} - \dfrac{V_2}{R_2}$

$\dfrac{V_1}{2} - V_2 = 0$

13. $\dfrac{I_1}{2} - \dfrac{I_2}{3} = 6$

$3I_1 - 2I_2 = 36$

14. $4I_1 - 3I_2 = 3$

$\dfrac{-I_1}{3} + I_2 = 0$

15. $12 = 10I_1 - 7I_2$

$0 = -7I_1 + 12I_2$

16. $10 = R_1 I_1 - R_2 I_2$

$0 = -I_1 + 2I_2$

17. $ax - y = 1$

$x + by = 0$

18. $1 \text{ mA} = \left(\dfrac{1}{2 \text{ k}\Omega} + \dfrac{1}{1 \text{ k}\Omega} \right) V_A - \left(\dfrac{1}{1 \text{ k}\Omega} \right) V_B$

$-2 \text{ mA} = -\left(\dfrac{1}{1 \text{ k}\Omega} \right) V_A + \left(\dfrac{1}{1 \text{ k}\Omega} + \dfrac{1}{2 \text{ k}\Omega} \right) V_B$

19. $6 = (2 \text{ k}\Omega + 1 \text{ k}\Omega) I_1 - (2 \text{ k}\Omega) I_2$

$0 = -(2 \text{ k}\Omega) I_1 + (2 \text{ k}\Omega + 4 \text{ k}\Omega) I_2$

20. $6 = 3 \text{ k}\Omega I_1 - 2 \text{ k}\Omega I_2$

$-6 = -2 \text{ k}\Omega I_1 + 4 \text{ k}\Omega I_2$

UNIT 10
THREE SIMULTANEOUS EQUATIONS

10.1 INTRODUCTION—GRAPHING

It has been stated previously that it requires N equations to solve for N unknowns. Thus, three *independent* equations can be solved simultaneously for three unknown quantities. The *independent* means that there must in fact be three distinct equations; that is, none of the equations can be derived from the other equations by simply changing forms (this is similar to the *identity* case when dealing with two equations). Furthermore, in order to have a *unique solution*, the three equations must have a single point of intersection on a 3-dimensional graph. Three independent equations having a unique solution are said to be *consistent* (otherwise they are said to be *inconsistent*).

A 3-dimensional graph is drawn on a coordinate system having three mutually perpendicular axes, as shown in Figure 10.1. The x-axis can be thought of as the "right-left" axis, the y-axis is the "in-out" axis, and the z-axis is the "up-down" axis. The origin O is the point of intersection of the three axes.

A point can be uniquely located in the 3-dimensional space of Figure 10.1 by specifying a value of x, a value of y, and a value of z as (x,y,z). For example, the unique points $(2,1,1)$ and $(-3,-2,2)$ are graphed in Figure 10.2.

Any equation of the form $Ax + By + Cz = D$ represents a plane that can be graphed in the 3-dimensional space of Figure 10.1. For example, the plane $x + y + z = 3$ is graphed in Figure 10.3.

The graphs of two equations in 3-dimensional space will form two planes. If the planes are not parallel, then the points at which they intersect will form a straight line. For example, $x + y = 1$ and $z = 2$ are graphed in Figure 10.4.

Now in order for three equations in three unknowns to have a unique solution, their 3-dimensional graphs must intersect at a common point, and the coordinates of this point is the solution. The intersection of two of the planes forms a line, and the point where this line pierces the third plane is the unique solution point.

EXAMPLE 10.1. Find a graphical solution by finding the intersection in 3-dimensional space of the three planes:

$$x = 2$$
$$y = 1$$
$$z = 3$$

SOLUTION.
(1) $x = 2$ is a plane parallel to the y-z plane.
(2) $y = 1$ is a plane parallel to the x-z plane.
(3) $z = 3$ is a plane parallel to the x-y plane.

The common point $(2,1,3)$ is shown in Figure 10.5. Note that the planes form three sides of a box with $(2,1,3)$ at one corner.

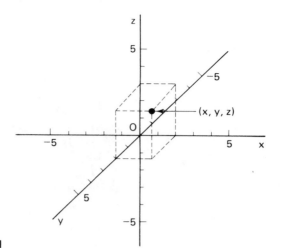

Figure 10.1

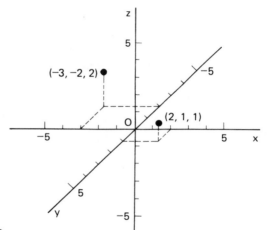

Figure 10.2

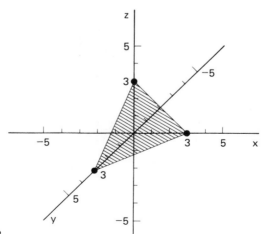

Figure 10.3

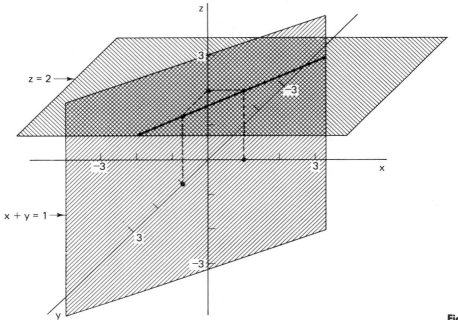

z = 2 →

x + y = 1 →

Figure 10.4

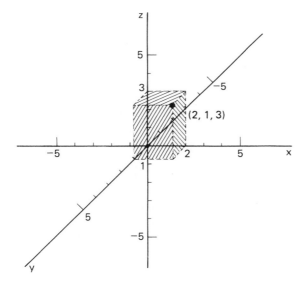

(2, 1, 3)

Figure 10.5

In general, finding a graphical solution for three simultaneous equations is not practical, and we will develop more useful techniques in the following sections.

Exercises 10.1

1. Plot the following points on a 3-dimensional coordinate system:

    (a) $(0,0,0)$
    (b) $(0,0,1)$
    (c) $(-1,0,0)$
    (d) $(0,3,0)$
    (e) $(-1,1,0)$
    (f) $(3,0,-2)$
    (g) $(0,-5,6)$
    (h) $(5,6,7)$
    (i) $(-5,-5,-5)$
    (j) $(10,-10,5)$

2. Plot the following planes on a 3-dimensional coordinate system:

    (a) $x=5$ (b) $y=3$ (c) $z=6$ (d) $x=-4$
    (e) $x+y=2$ (a plane parallel to the z-axis)
    (f) $x+z=4$ (a plane parallel to the y-axis)
    (g) $y-z=3$ (a plane parallel to the x-axis)
    (h) $x+y+z=2$
        (*Hint:*   With $x=0$, plot the line in the y-z plane.
                 With $y=0$, plot the line in the x-z plane.
                 With $z=0$, plot the line in the x-y plane.)
    (i) $x+y-z=5$
    (j) $2x+y+3z=6$

3. Find a graphical solution:

    (a) $x+z=2,\ x=z,\ y=0$
    (b) $x=-1,\ y=z,\ y+z=4$
    (c) $z=5,\ y+x=6,\ y=2x$

10.2 SOLUTION BY ELIMINATION

In general it will be much easier to find a solution for three simultaneous equations by algebraic techniques rather than by graphing. In fact, the same process of elimination by substitution or addition developed for two simultaneous equations can be used. The basic idea is to combine the three equations in a way that will provide two equations in two unknowns. The solution of two equations in two unknowns can then be used. The procedure is:

ELIMINATION

To solve three simultaneous equations by elimination:
(1) Choose a variable to be eliminated.
(2) Combine two of the equations to eliminate the chosen variable.
(3) Combine two *different* equations to eliminate the chosen variable.
(4) Solve these two equations in two unknowns using previously developed techniques.
(5) Substitute into an original equation to solve for the eliminated variable.
(6) Check the solution by substitution into all three original equations.

An example will best illustrate the procedure.

EXAMPLE 10.2. Solve by elimination:
(a) $x + y + z = 3$
(b) $y - z = 1$
(c) $2x + y + z = 2$

SOLUTION. Choose to eliminate the variable z. Adding (a) and (b),

$$x + 2y = 4$$

Then adding (b) and (c),

$$2x + 2y = 3$$

Now solve these equations simultaneously for x and y. By subtraction,

$$\begin{array}{r} 2x + 2y = 3 \\ x + 2y = 4 (-) \\ \hline x = -1 \end{array}$$

Then substituting (-1) for x,

$$(-1) + 2y = 4$$

and thus,

$$y = \frac{5}{2}$$

We now substitute $5/2$ for y into the original equation (b) to obtain

$$\frac{5}{2} - z = 1$$

Thus,

$$z = \frac{3}{2}$$

The solution is then

$$x = -1, y = \frac{5}{2}, z = \frac{3}{2}$$

As a check:

$x+y+z=3$	$y-z=1$	$2x+y+z=2$
$-1+\frac{5}{2}+\frac{3}{2}=3$	$\frac{5}{2}-\frac{3}{2}=1$	$2(-1)+\frac{5}{2}+\frac{3}{2}=2$

and the solution is valid.

Note that the solution could have been accomplished by eliminating either x or y, instead of z, from the original equations.

EXAMPLE 10.3.   Solve by elimination:

(a) $\dfrac{x}{2}+y=1$

(b) $y-3z=0$

(c) $x+\dfrac{z}{3}=0$

SOLUTION.   Choose to eliminate the variable y. Subtracting (b) from (a),

$$\frac{x}{2}+y=1$$
$$y-3z=0 \; (-)$$
$$\overline{\frac{x}{2}+3z=1}$$

Use equation (c) as it is since it does not contain the variable y. Thus,

$$x+\frac{z}{3}=0$$

Now, solve these two equations simultaneously for x and z.

$$\frac{x}{2}+3z=1$$
$$x+\frac{z}{3}=0$$

Multiply the first equation by 2 and subtract the second equation.

$$(x2)\left[\frac{x}{2}+3z=1\right] \rightarrow x+6z=2$$
$$x+\frac{z}{3}=0 \; (-)$$
$$\overline{\frac{17}{3}z=2}$$

and

$$z=\frac{6}{17}$$

Substituting $z = 6/17$ into $x + z/3 = 0$ yields

$$x + \frac{\frac{6}{17}}{3} = 0$$

$$x = -\frac{2}{17}$$

Now, substituting $z = 6/17$ into the original equation $y - 3z = 0$ yields

$$y - 3\left(\frac{6}{17}\right) = 0$$

$$y = \frac{18}{17}$$

The solution is thus,

$$x = -\frac{2}{17}, \qquad y = \frac{18}{17}, \qquad z = \frac{6}{17}$$

As a check:

$$\frac{x}{2} + y = 1 \qquad\qquad y - 3z = 0 \qquad\qquad x + \frac{z}{3} = 0$$

$$\frac{-\frac{2}{17}}{2} + \frac{18}{17} = 1 \qquad \frac{18}{17} - 3\left(\frac{6}{17}\right) = 0 \qquad \frac{-2}{17} + \frac{\frac{6}{17}}{3} = 0$$

$$\frac{17}{17} = 1 \qquad\qquad\qquad 0 = 0 \qquad\qquad\qquad 0 = 0$$

When solving by elimination, great care must be taken to always use *all three* of the given equations. Otherwise, the solution will not be unique.

Exercises 10.2

1. Check to see if the given solutions are valid for the given equations. If a solution is not valid, find the correct solution.

(a) $x + y + z = 5$
$y + z = 3$
$x + y - z = 0$

$\left(2, \frac{1}{2}, \frac{5}{2}\right)$

(b) $x + y + z = 5$
$y - z = 3$
$x + y - z = 0$

$\left(\frac{1}{3}, \frac{11}{3}, \frac{2}{3}\right)$

(c) $10 = 3I_1 - 2I_2 - I_3$
$0 = -2I_1 + 4I_2 - 2I_3$
$0 = -I_1 - 2I_2 + 5I_3$

$(I_1 = 10, I_2 = 7.5, I_3 = 5)$

(d) $-0.01 = 4I_1 - 3I_2 - I_3$
   $-0.02 = -3I_1 + 7I_2$
   $0.02 = -I_1 + 3I_3$
   $I_1 = -5 \times 10^{-3}$
   $I_2 = -5 \times 10^{-3}$
   $I_3 = 5 \times 10^{-3}$

(e) $6 = 3I_1 - 2I_2$
   $0 = -4I_1 + 12I_2 - 2I_3$
   $-3 = -I_2 + 3I_3$
   $(I_1 = 2,\ I_2 = 1,\ I_3 = -1)$

2. Find a unique solution and check by substitution:

(a) $x + y + z = 1$
   $2x + y + z = 0$
   $x - y + z = 0$

(f) $I_1 + I_2 = 1$
   $2I_1 + I_2 + I_3 = 1$
   $2I_2 - I_3 = 1$

(b) $x - y + z = 0$
   $-x + y + z = 2$
   $x + y - z = 0$

(g) $V_A + \dfrac{V_B}{2} + \dfrac{V_C}{3} = 1$
   $\dfrac{V_B}{2} + \dfrac{V_C}{2} = 0$
   $\dfrac{V_A}{2} + \dfrac{V_B}{2} = 0$

(c) $x - y + z = 0$
   $-x + y + z = 0$
   $x + y - z = 3$

(h) $V_A - \dfrac{V_B}{2} - \dfrac{V_C}{3} = \dfrac{1}{6}$
   $-\dfrac{V_A}{2} + V_B - \dfrac{V_C}{3} = 0$
   $-\dfrac{V_A}{3} - \dfrac{V_B}{2} + V_C = 0$

(d) $2I_1 - I_3 = 2$
   $I_2 + I_3 = 0$
   $I_1 + 2I_2 = 0$

(i) $x + y - z = A$
   $x - y + z = B$
   $-x + y + z = C$

(e) $3I_1 + I_2 - 2I_3 = 4$
   $I_1 - I_2 - 2I_3 = 0$
   $-I_2 + I_3 = 0$

UNIT 11
DETERMINANTS

11.1 2×2 DETERMINANTS

In the previous sections we have seen how to find solutions for sets of two or three simultaneous equations by substitution and by elimination. These techniques are satisfactory for simple equations, but they may become unwieldy if the coefficients are other than small integers. The solution of simultaneous equations will be necessary when finding a solution for an electric circuit using mesh equations or node equations, and the coefficients will not in general be simple integers. For example, it might be necessary to find the following set of mesh equations:

$$10 = 8.6 \text{ k}\Omega \, I_1 - 3.9 \text{ k}\Omega \, I_2$$
$$0 = -3.9 \text{ k}\Omega \, I_1 + 11.7 \text{ k}\Omega \, I_2 - 2.7 \text{ k}\Omega \, I_3$$
$$-12 = -2.7 \text{ k}\Omega \, I_2 + 9.5 \text{ k}\Omega \, I_3$$

The problem of finding a simultaneous solution for any number of equations is covered quite thoroughly in the study of *matrix algebra*.* One technique used is called the *method of determinants*. The use of determinants provides a simple systematic procedure for solving simultaneous equations, and we will begin by applying this method to find a solution for two simultaneous equations.

Given the two mesh equations

$$V_1 = AI_1 + BI_2 \tag{11.1}$$
$$V_2 = CI_1 + DI_2 \tag{11.2}$$

it is necessary to solve for the variables I_1 and I_2. Note that the voltages V_1 and V_2 are voltage *sources*, and they are considered as *constants*. The *coefficients* of the variables A, B, C, and D are circuit resistances. The coefficients and the constants are simply numbers (the values depend on the particular circuit), and these are the quantities that will be manipulated in order to determine solutions for the variables I_1 and I_2.

A very concise way to express a set of simultaneous equations is to write them in *matrix form*. For example, the mesh equations (11.1) and (11.2) can be written in matrix form as

$$\begin{bmatrix} V_1 \\ V_2 \end{bmatrix} = \begin{bmatrix} A & B \\ C & D \end{bmatrix} \begin{bmatrix} I_1 \\ I_2 \end{bmatrix} \tag{11.3}$$

(Constants matrix) (Coefficients matrix) (Variables matrix)

Each of the quantities contained within a pair of brackets [ ] is called a *matrix*. Any set of simultaneous equations written in matrix form will have three matrices

---

*For example, see F. E. Hohn, *Matrix Algebra* (New York: Macmillan), 1967.

—the *constants* matrix, the *coefficients* matrix, and the *variables* matrix. Clearly, a matrix is simply a rectangular array of quantities, and any matrix is composed of horizontal *rows* and vertical *columns*. For example, the coefficients matrix in Eq. (11.3) consists of the two rows $\begin{bmatrix} A & B \end{bmatrix}$ and $\begin{bmatrix} C & D \end{bmatrix}$ and the two columns

$$\begin{bmatrix} A \\ C \end{bmatrix} \quad \text{and} \quad \begin{bmatrix} B \\ D \end{bmatrix}$$

It is said to be a 2×2 matrix (2 rows and 2 columns), or a *second-order* matrix. In general, the *order* of any square matrix is equal to the number of rows (or columns) in the matrix.

EXAMPLE 11.1. Write the following simultaneous equations in matrix form:

(a) $x + y = 3$
       $-3x + 2y = 6$

(b) $6 = I_1 - I_2 + I_3$
     $0 = I_1 + 2I_2$
     $5 = -I_1 + 3I_2 + 4I_3$

SOLUTION.

(a) $\begin{bmatrix} 3 \\ 6 \end{bmatrix} = \begin{bmatrix} 1 & 1 \\ -3 & 2 \end{bmatrix} \begin{bmatrix} x \\ y \end{bmatrix}$

(b) $\begin{bmatrix} 6 \\ 0 \\ 5 \end{bmatrix} = \begin{bmatrix} 1 & -1 & 1 \\ 1 & 2 & 0 \\ -1 & 3 & 4 \end{bmatrix} \begin{bmatrix} I_1 \\ I_2 \\ I_3 \end{bmatrix}$

Note carefully that the coefficient of I_3 in the second equation is zero.

All of the information needed to find a solution for a set of simultaneous equations written in matrix form is contained within the constants matrix and the coefficients matrix. In order to find a solution, we must be able to determine the numerical value of a matrix.

The numerical value of a square matrix is called its *determinant*. Note carefully that a matrix has a determinant (a unique numerical value) only if it is square—it must have the same number of rows and columns. A determinant is distinguished by using brackets of the form | | instead of [ ], and we will use the symbol Δ. For example, the determinant of the coefficients matrix in Eq. (11.3) is written as

$$\Delta = \begin{vmatrix} A & B \\ C & D \end{vmatrix} \tag{11.4}$$

In order to calculate the determinant of a 2×2 (second-order) matrix, take the product of the elements in the *forward diagonal* and subtract from it the product of the elements in the *reverse diagonal*. For example, the determinant below is evaluated as

$$\Delta \overset{(+)}{=} \begin{vmatrix} A & B \\ C & D \end{vmatrix} \overset{(-)}{=} AD - BC$$

(Reverse   (Forward
diagonal)   diagonal)

2×2 DETERMINANT

To calculate the determinant of a 2×2 matrix:
Take the product of the elements in the *forward diagonal* and subtract from it the product of the elements in the *reverse diagonal*.

EXAMPLE 11.2. Find the determinants:

(a) $\begin{vmatrix} 1 & 2 \\ 3 & 4 \end{vmatrix}$

(b) $\begin{vmatrix} 2 & -1 \\ 3 & 2 \end{vmatrix}$

SOLUTION.   Find the product of the two numbers in the forward diagonal and subtract the product of the two numbers in the reverse diagonal.

(a) $\Delta = \begin{vmatrix} 1 & 2 \\ 3 & 4 \end{vmatrix} = (1)(4) - (2)(3) = -2$

(b) $\Delta = \begin{vmatrix} 2 & -1 \\ 3 & 2 \end{vmatrix} = (2)(2) - (-1)(3) = 7$

Needless to say, one must take great care with *signs* when working with determinants.

Now that we know how to find the determinant of a 2×2 matrix, how can determinants be used to solve two simultaneous equations? It will be necessary to find the determinants of three matrices, and these determinants can then be used to calculate the two unknowns.

Continuing with Eq. (11.3), the first determinant comes from the coefficient matrix and is defined as

$$\Delta = \begin{vmatrix} A & B \\ C & D \end{vmatrix}$$

The second determinant is found by replacing the *first column* of the coefficient matrix with the constants. This is defined as Δ_1, and

$$\Delta_1 = \begin{vmatrix} V_1 & B \\ V_2 & D \end{vmatrix}$$

The third determinant is found by replacing the *second column* of the coefficient matrix with the constants. This is defined as Δ_2, and

$$\Delta_2 = \begin{vmatrix} A & V_1 \\ C & V_2 \end{vmatrix}$$

The values of the unknowns I_1 and I_2 are then easily calculated from these three determinants as follows:

$$I_1 = \frac{\Delta_1}{\Delta} \tag{11.5a}$$

$$I_2 = \frac{\Delta_2}{\Delta} \tag{11.5b}$$

The procedure is quite easy and can be summarized as:

TWO EQUATIONS

To solve two simultaneous equations using determinants:

(1) Write the two equations in matrix form

$$\begin{bmatrix} V_1 \\ V_2 \end{bmatrix} = \begin{bmatrix} A & B \\ C & D \end{bmatrix} \begin{bmatrix} I_1 \\ I_2 \end{bmatrix}$$

(2) Evaluate Δ, Δ_1, and Δ_2.

(3) Calculate the unknowns as

$$I_1 = \frac{\Delta_1}{\Delta} \qquad I_2 = \frac{\Delta_2}{\Delta}$$

(4) Check the solution by substitution.

EXAMPLE 11.3. Solve, using determinants:

$$x + y = 2$$
$$x - y = 1$$

SOLUTION.

(1) In matrix form,

$$\begin{bmatrix} 2 \\ 1 \end{bmatrix} = \begin{bmatrix} 1 & 1 \\ 1 & -1 \end{bmatrix} \begin{bmatrix} x \\ y \end{bmatrix}$$

(2) $\Delta = \begin{vmatrix} 1 & 1 \\ 1 & -1 \end{vmatrix} = (1)(-1) - (1)(1) = -2$

$\Delta_1 = \begin{vmatrix} 2 & 1 \\ 1 & -1 \end{vmatrix} = (2)(-1) - (1)(1) = -3$

$\Delta_2 = \begin{vmatrix} 1 & 2 \\ 1 & 1 \end{vmatrix} = (1)(1) - (2)(1) = -1$

(3) $x = \dfrac{\Delta_1}{\Delta} = \dfrac{-3}{-2} = \dfrac{3}{2}$

$y = \dfrac{\Delta_2}{\Delta} = \dfrac{-1}{-2} = \dfrac{1}{2}$

(4) Check:

$$x + y = 2 \qquad x - y = 1$$

$$\frac{3}{2} + \frac{1}{2} = 2 \qquad \frac{3}{2} - \frac{1}{2} = 1$$

EXAMPLE 11.4. Solve, using determinants:

$$6.83I_1 - 5.16I_2 = 11.4$$
$$-11.32I_1 + 1.73I_2 = 6.51$$

SOLUTION. After a little practice, it is possible to write the determinants directly.

$$\Delta = \begin{vmatrix} 6.83 & -5.16 \\ -11.32 & 1.73 \end{vmatrix}$$
$$= (6.83)(1.73) - (-5.16)(-11.32)$$
$$= -46$$

$$\Delta_1 = \begin{vmatrix} 11.4 & -5.16 \\ 6.51 & 1.73 \end{vmatrix}$$
$$= (11.4)(1.73) - (-5.16)(6.51)$$
$$= 53.3$$

$$\Delta_2 = \begin{vmatrix} 6.83 & 11.4 \\ -11.32 & 6.51 \end{vmatrix}$$
$$= (6.83)(6.51) - (11.4)(-11.32)$$
$$= 173.5$$

Then,

$$I_1 = \frac{\Delta_1}{\Delta} = \frac{53.3}{-46.6} = -1.14$$

$$I_2 = \frac{\Delta_2}{\Delta} = \frac{173.5}{-46.6} = -3.72$$

As a check:

$$6.83I_1 - 5.16I_2 = 11.4$$
$$(6.83)(-1.14) - (5.16)(-3.72) = 11.4$$
$$-7.79 + 19.19 = 11.4$$
$$11.4 = 11.4$$
$$-11.32I_1 + 1.73I_2 = 6.51$$
$$(-11.32)(-1.14) + (1.73)(-3.72) = 6.51$$
$$12.9 - 6.43 = 6.51$$
$$6.47 \cong 6.51$$

Exercises 11.1

Use determinants to find a solution and check your results.

1. $x - y = 0$
   $2x - y = 0$

2. $y - x = 1$
   $y - 2x = 1$

3. $x + y = 0$
   $x - y = 1$

4. $2x - y = 1$
   $3x - y = -2$

5. $2x - y = 0$

    $x + y = -1$

6. $\dfrac{x}{2} + \dfrac{y}{3} = 5$

    $x + y = 1$

7. $4I_1 - 3I_2 = 3$

    $-I_1 + 3I_2 = 0$

8. $12 = 10I_1 - 7I_2$

    $0 = -7I_1 + 12I_2$

9. $6 = \dfrac{V_A}{5} - \dfrac{V_B}{8}$

    $0 = \dfrac{-V_A}{8} + \dfrac{V_B}{2}$

10. $2 = \left(\dfrac{1}{2} + \dfrac{1}{3}\right)V_A - \dfrac{1}{3}V_B$

    $0 = -\dfrac{1}{3}V_A + \left(\dfrac{1}{3} + \dfrac{1}{4}\right)V_B$

11. $6 = 3I_1 + 2I_2$

    $2 = 2I_1 + 3I_2$

12. $V_A = \dfrac{V_B}{3} + \dfrac{1}{2}$

    $V_A = 2V_B - 2$

13. $6 = \dfrac{V_1}{R_1} - \dfrac{V_2}{R_2}$

    $0 = -V_1 + 2V_2$

14. $10 = R_1I_1 - R_2I_2$

    $0 = -I_1 + 2I_2$

11.2   3×3 DETERMINANTS

Determinants can also be used to find a solution for a set of three simultaneous equations. Consider the following three mesh equations obtained by applying KVL to an electric circuit.

$$V_1 = AI_1 + BI_2 + CI_3 \tag{11.6a}$$

$$V_2 = DI_1 + EI_2 + FI_3 \tag{11.6b}$$

$$V_3 = GI_1 + HI_2 + JI_3 \tag{11.6c}$$

These three equations can be rewritten in *matrix form* as

$$
\underbrace{\begin{bmatrix} V_1 \\ V_2 \\ V_3 \end{bmatrix}}_{\text{(Constants)}} = \underbrace{\begin{bmatrix} A & B & C \\ D & E & F \\ G & H & J \end{bmatrix}}_{\text{(Coefficients)}} \underbrace{\begin{bmatrix} I_1 \\ I_2 \\ I_3 \end{bmatrix}}_{\text{(Variables)}} \tag{11.7}
$$

Notice that the coefficient matrix is a 3×3 or third-order matrix (3 rows and 3 columns). Since it is a square matrix, it must have a determinant, and it will be necessary to learn how to calculate the determinant of a third-order matrix.

Finding the determinant of the 3×3 matrix below is straightforward and can be summarized as follows:

$$
\Delta = \begin{vmatrix} A & B & C \\ D & E & F \\ G & H & J \end{vmatrix}
$$

3×3 DETERMINANTS

To calculate the determinant of a third-order matrix:

(1) To the right of the determinant, copy the first two columns.
(2) Take the product of the three elements in each of the three *forward diagonals*, and subtract the product of the three elements in each of the three *reverse diagonals*.

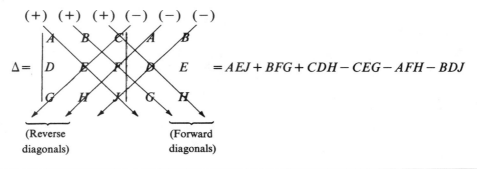

$$\Delta = \begin{vmatrix} A & B & C \\ D & E & F \\ G & H & J \end{vmatrix} \begin{matrix} A & B \\ D & E \\ G & H \end{matrix} = AEJ + BFG + CDH - CEG - AFH - BDJ$$

(Reverse diagonals)   (Forward diagonals)

EXAMPLE 11.5. Evaluate the determinant:

$$\Delta = \begin{vmatrix} 1 & 2 & 1 \\ 0 & -1 & 3 \\ -2 & 4 & 0 \end{vmatrix}$$

SOLUTION.

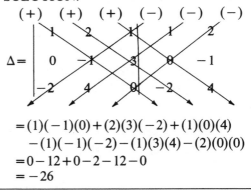

$$= (1)(-1)(0) + (2)(3)(-2) + (1)(0)(4) \\ - (1)(-1)(-2) - (1)(3)(4) - (2)(0)(0)$$
$$= 0 - 12 + 0 - 2 - 12 - 0$$
$$= -26$$

There are other methods for finding the determinant of a 3×3 matrix, and they will be considered in the next section.

In order to use determinants to solve three simultaneous equations, it is necessary to find four different determinants. Using Eq. (11.7), first write the determinant of the coefficient matrix as

$$\Delta = \begin{vmatrix} A & B & C \\ D & E & F \\ G & H & J \end{vmatrix}$$

The remaining three determinants are written by replacing one of the columns of the coefficients matrix with the constants (similar to the case with two equations). Thus,

$$\Delta_1 = \begin{vmatrix} V_1 & B & C \\ V_2 & E & F \\ V_3 & H & J \end{vmatrix} \qquad \Delta_2 = \begin{vmatrix} A & V_1 & C \\ D & V_2 & F \\ G & V_3 & J \end{vmatrix} \qquad \Delta_3 = \begin{vmatrix} A & B & V_1 \\ D & E & V_2 \\ G & H & V_3 \end{vmatrix}$$

(First column)   (Second column)   (Third column)

Then, the three unknowns are calculated as:

$$I_1 = \frac{\Delta_1}{\Delta} \qquad I_2 = \frac{\Delta_2}{\Delta} \qquad I_3 = \frac{\Delta_3}{\Delta} \qquad (11.8)$$

The procedure can be summarized as

THREE EQUATIONS

To solve three simultaneous equations using determinants:
(1) Write the equations in matrix form:

$$\begin{bmatrix} V_1 \\ V_2 \\ V_3 \end{bmatrix} = \begin{bmatrix} A & B & C \\ D & E & F \\ G & H & J \end{bmatrix} \begin{bmatrix} I_1 \\ I_2 \\ I_3 \end{bmatrix}$$

(2) Evaluate Δ, Δ_1, Δ_2, and Δ_3.
(3) Calculate the unknowns as

$$I_1 = \frac{\Delta_1}{\Delta}, I_2 = \frac{\Delta_2}{\Delta}, I_3 = \frac{\Delta_3}{\Delta}$$

(4) Check the solution by substitution.

EXAMPLE 11.6. Solve using determinants:

$$6 = 3I_1 - 2I_2$$
$$0 = -2I_1 + 6I_2 - I_3$$
$$-3 = -I_2 + 3I_3$$

SOLUTION. First, write the three equations in matrix form. Note carefully the two zero coefficients.

$$\begin{bmatrix} 6 \\ 0 \\ -3 \end{bmatrix} = \begin{bmatrix} 3 & -2 & 0 \\ -2 & 6 & -1 \\ 0 & -1 & 3 \end{bmatrix} \begin{bmatrix} I_1 \\ I_2 \\ I_3 \end{bmatrix}$$

Then evaluate the four determinants:

$$\Delta = \begin{vmatrix} 3 & -2 & 0 \\ -2 & 6 & -1 \\ 0 & -1 & 3 \end{vmatrix}$$

$$= (3)(6)(3) + (-2)(-1)(0) + (0)(-2)(-1)$$
$$- (0)(6)(0) - (-2)(-2)(3) - (3)(-1)(-1)$$
$$= 54 + 0 + 0 - 0 - 12 - 3 = 39$$

$$\Delta_1 = \begin{vmatrix} 6 & -2 & 0 \\ 0 & 6 & -1 \\ -3 & -1 & 3 \end{vmatrix}$$

$$= (6)(6)(3) + (-2)(-1)(-3) + (0)(0)(-1)$$
$$\quad - (0)(6)(-3) - (-2)(0)(3) - (6)(-1)(-1)$$
$$\quad + 108 - 6 + 0 - 0 - 0 - 6$$
$$= 96$$

$$\Delta_2 = \begin{vmatrix} 3 & 6 & 0 \\ -2 & 0 & -1 \\ 0 & -3 & 3 \end{vmatrix}$$

$$= (3)(0)(3) + (6)(-1)(0) + (0)(-2)(-3)$$
$$\quad - (0)(0)(0) - (6)(-2)(3) - (3)(-1)(-3)$$
$$= 0 + 0 + 0 - 0 + 36 - 9$$
$$= 27$$

$$\Delta_3 = \begin{vmatrix} 3 & -2 & 6 \\ -2 & 6 & 0 \\ 0 & -1 & -3 \end{vmatrix}$$

$$= (3)(6)(-3) + (-2)(0)(0) + (6)(-2)(-1)$$
$$\quad - (6)(6)(0) - (-2)(-2)(-3) - (3)(-1)(0)$$
$$= -54 + 0 + 12 - 0 + 12 + 0$$
$$= -30$$

(3) The unknowns are then:

$$I_1 = \frac{\Delta_1}{\Delta} = \frac{96}{39} = 2.46$$

$$I_2 = \frac{\Delta_2}{\Delta} = \frac{27}{39} = 0.693$$

$$I_3 = \frac{\Delta_3}{\Delta} = \frac{-30}{39} = -0.769$$

(4) As a check:

$$6 = 3I_1 - 2I_2$$
$$6 = 3(2.46) - 2(0.693)$$
$$6 = 7.38 - 1.38$$
$$6 = 6$$
$$0 = -2I_1 + 6I_2 - I_3$$
$$0 = -2(2.46) + 6(0.693) - (-0.769)$$
$$0 = -4.92 + 4.16 + 0.769$$
$$0 = 0$$

$$\begin{vmatrix} -3 = -I_2 + 3I_3 \\ -3 = -(0.693) + 3(-0.769) \\ -3 = -0.693 - 2.307 \\ -3 = -3 \end{vmatrix}$$

Exercises 11.2

1. Evaluate the following determinants:

(a) $\begin{vmatrix} 1 & 1 & 1 \\ 0 & 1 & 0 \\ -1 & 0 & -2 \end{vmatrix}$ 　(b) $\begin{vmatrix} 1 & 2 & 3 \\ 3 & 2 & 1 \\ 1 & 2 & -3 \end{vmatrix}$ 　(c) $\begin{vmatrix} 1 & 3 & 3 \\ 0 & 1 & 0 \\ 2 & 1 & 1 \end{vmatrix}$

(d) $\begin{vmatrix} 1 & 2 & 1 \\ 0 & a & 0 \\ 1 & 3 & -1 \end{vmatrix}$ 　(e) $\begin{vmatrix} A & 1 & 1 \\ B & 2 & 2 \\ C & 3 & 4 \end{vmatrix}$ 　(f) $\begin{vmatrix} A & 2 & -3 \\ B & 3 & 4 \\ -A & -2 & 3 \end{vmatrix}$

2. Solve, using determinants:

(a) $\begin{aligned} x + y + z &= 5 \\ y + z &= 3 \\ x + y - z &= 0 \end{aligned}$ 　(b) $\begin{aligned} I_1 + I_2 + I_3 &= 5 \\ I_2 - I_3 &= 3 \\ I_1 + I_2 - I_3 &= 0 \end{aligned}$

(c) $\begin{aligned} 10 &= 3I_1 - 2I_2 - I_3 \\ 0 &= -2I_1 + 4I_2 - 2I_3 \\ 0 &= -I_1 - 2I_2 + 5I_3 \end{aligned}$ 　(d) $\begin{aligned} -0.01 &= 4I_1 - 3I_2 - I_3 \\ -0.02 &= -3I_1 + 7I_2 \\ 0.02 &= -I_1 + 3I_3 \end{aligned}$

(e) $\begin{aligned} r - s + t &= 0 \\ -r + s + t &= 2 \\ r + s - t &= 0 \end{aligned}$ 　(f) $\begin{aligned} a - b + c &= 0 \\ -a + b + c &= 0 \\ a + b - c &= 0 \end{aligned}$

(g) $\begin{aligned} 1 &= V_A + \frac{V_B}{2} + \frac{V_C}{3} \\ 0 &= \frac{V_B}{3} + \frac{V_C}{2} \\ 0 &= \frac{V_A}{2} + \frac{V_B}{2} \end{aligned}$ 　(h) $\begin{aligned} V_A + V_B - V_C &= A \\ V_A - V_B + V_C &= B \\ -V_A + V_B + V_C &= C \end{aligned}$

11.3 EXPANSION BY MINORS

The evaluation of *second-order* (2 rows and 2 columns) and *third-order* (3 rows and 3 columns) determinants can be carried out using forward and reverse diagonals. However, this technique will not work for *higher order* determinants. For example, the solution of four simultaneous equations in four unknowns requires the evaluation of five *fourth-order* (4 rows and 4 columns) determinants. A general procedure that can always be used to calculate the determinant of any square matrix is called *expansion by minors*.

Each entry in a determinant can be uniquely specified according to its row and column numbers. The rows and columns can be numbered as shown in the following determinant.

$$\text{(Row)} \begin{cases} 1 \\ 2 \\ 3 \end{cases} \overset{\overset{\displaystyle\text{(Column)}}{\overset{\displaystyle 1 \quad 2 \quad 3}{\frown}}}{\begin{vmatrix} A & B & C \\ D & E & F \\ G & H & J \end{vmatrix}} \qquad (11.9)$$

Using the row number first and then the column number, each entry can be uniquely located. For example, the 1-1 element is A, the 1-3 element is C, the 2-3 element is F, and so on.

Every element in a determinant has a *minor*. The minor of any element is simply the determinant remaining when the row and column of that element are lined out. For example, the minor of A (the 1-1 element) is found by crossing out row 1 and column 1.

$$\begin{vmatrix} A & B & C \\ D & E & F \\ G & H & J \end{vmatrix} \qquad \text{minor of } A = \begin{vmatrix} E & F \\ H & J \end{vmatrix}$$

The minors of D, E, and J, found by crossing out the appropriate row and column, are:

$$\text{Minor of } D = \begin{vmatrix} B & C \\ H & J \end{vmatrix} \qquad \text{Minor of } E = \begin{vmatrix} A & C \\ G & J \end{vmatrix} \qquad \text{Minor of } J = \begin{vmatrix} A & B \\ D & E \end{vmatrix}$$

Try it. Notice that each minor is itself a determinant, but it is one order *smaller* than the original determinant! The rules for the procedure known as *expansion by minors* are given below.

EXPANSION BY MINORS

To calculate the determinant of a matrix by *expansion by minors*:
(1) Choose a row (or column) for expansion.
(2) Attach the appropriate *sign* (+ or −) to each element in the chosen row (column) as follows:

   (+) if row number + column number are *even*.

   (−) if row number + column number are *odd*. Or use this "checkerboard" arrangement:

$$\begin{vmatrix} + & - & + & \cdots \\ - & + & - & \cdots \\ + & - & + & \cdots \\ \vdots & \vdots & \vdots & \end{vmatrix}$$

(3) Multiply each element in the chosen row (or column) along with its appropriate sign by its minor and take the algebraic sum of these products to find the value of the determinant.

The determinant in Eq. (11.9) can be evaluated by expansion by minors. Since there are three elements in each row or column, it will be necessary to find the determinants of three second-order matrices. The evaluation can be done by expanding along any row or column, and Example 11.7 illustrates expansion along row 1.

EXAMPLE 11.7. Evaluate by expansion along row 1.

$$\Delta = \begin{vmatrix} A & B & C \\ D & E & F \\ G & H & J \end{vmatrix} \leftarrow \text{Row 1}$$

SOLUTION.

(1) The three elements in the first row are A, B, and C, and their appropriate signs are:

(2) $+A$, since $1+1=2$ (even)
$-B$, since $1+2=3$ (odd)
$+C$, since $1+3=4$ (even)

or use the checkerboard.

(3) Multiplying each *signed* element in the first row by its minor and adding algebraically yields:

$$\Delta = +A \begin{vmatrix} E & F \\ H & J \end{vmatrix} + (-B) \begin{vmatrix} D & F \\ G & J \end{vmatrix} + C \begin{vmatrix} D & E \\ G & H \end{vmatrix}$$
$$= A(EJ - HF) - B(DJ - FG) + C(DH - EG)$$
$$= AEJ - AFH - BDJ + BFG + CDH - CEG$$

Compare this result with that obtained by expansion along diagonals obtained in the previous section.

Example 11.7 illustrates the proper technique, but a numerical answer is usually more satisfying, as illustrated in the next example.

EXAMPLE 11.8. Evaluate by expansion along row 1.

$$\begin{vmatrix} 1 & 2 & 3 \\ 3 & 1 & 0 \\ 1 & 0 & 1 \end{vmatrix} \leftarrow \text{Row 1}$$

SOLUTION.

(1) The elements in row 1 are $1, 2, 3$.

(2) The elements with appropriate signs are: $+1, -2, +3$.

(3) Expanding:

$$\Delta = 1 \begin{vmatrix} 1 & 0 \\ 0 & 1 \end{vmatrix} - 2 \begin{vmatrix} 3 & 0 \\ 1 & 1 \end{vmatrix} + 3 \begin{vmatrix} 3 & 1 \\ 1 & 0 \end{vmatrix}$$
$$= 1(1-0) - 2(3-0) + 3(0-1)$$
$$= 1 - 6 - 3 = -8$$

In order to show that any row or column can be chosen for expansion, the determinant in Example 11.8 will be evaluated by expanding along column 2.

EXAMPLE 11.9. Evaluate by expansion along column 2.	SOLUTION.

EXAMPLE 11.9.   Evaluate by expansion along column 2.

$$\begin{vmatrix} 1 & 2 & 3 \\ 3 & 1 & 0 \\ 1 & 0 & 1 \end{vmatrix}$$

$\uparrow$Column 2

SOLUTION.

(1) The elements in column 2 are 2, 1, 0.

(2) The signed elements are $-2, +1, -0$.

(3) Expanding:

$$\Delta = (-2)\begin{vmatrix} 3 & 0 \\ 1 & 1 \end{vmatrix} + 1\begin{vmatrix} 1 & 3 \\ 1 & 1 \end{vmatrix} - 0\begin{vmatrix} 1 & 3 \\ 3 & 0 \end{vmatrix}$$

$$= (-2)(3-0) + 1(1-3) - 0$$

$$= -6 - 2 = -8$$

In comparing Examples 11.8 and 11.9, notice that the latter was simpler since it is only necessary to evaluate two second-order matrices. This is a result of the zero element in column 2. Clearly the presence of zeros in the expansion row or column will reduce the computational labor, and you should choose the expansion row or column accordingly.

Expansion by minors is a technique that can be used to evaluate determinants of any order. However, the physical labor required increases rapidly as the order of the determinant increases. For example, to find the value of a 6×6 determinant using this method requires the computation of 720 terms! For this reason, one might look for techniques to simplify the determinant (next section), or perhaps utilize a good digital computer program. In any case, the computational labor involved can be minimized by expansion along a row or column containing zeros. Therefore you should always evaluate by expansion along a row or column containing the maximum number of zeros.

As one final example of expansion by minors, consider a fourth-order determinant.

EXAMPLE 11.10.   Evaluate by expanding along row 4:

$$\Delta = \begin{vmatrix} 2 & 1 & 3 & -1 \\ -1 & -2 & 1 & 0 \\ 0 & 1 & -2 & 2 \\ 1 & 0 & 0 & 0 \end{vmatrix} \leftarrow \text{Row 4}$$

SOLUTION.

(1) The elements in row 4 are: 1, 0, 0, 0.

(2) The *signed* elements in row 4 are:
$-1, +0, -0, +0$.

(3) Expanding:

$$\Delta = (-1) \begin{vmatrix} 1 & 3 & -1 \\ -2 & 1 & 0 \\ 1 & -2 & 2 \end{vmatrix}$$

$$+0 \begin{vmatrix} 2 & 3 & -1 \\ -1 & 1 & 0 \\ 0 & -2 & 2 \end{vmatrix}$$

$$-0 \begin{vmatrix} 2 & 1 & -1 \\ -1 & -2 & 0 \\ 0 & 1 & 2 \end{vmatrix}$$

$$+0 \begin{vmatrix} 2 & 1 & 3 \\ -1 & -2 & 1 \\ 0 & 1 & -2 \end{vmatrix}$$

$$= - \begin{vmatrix} 1 & 3 & -1 \\ -2 & 1 & 0 \\ 1 & -2 & 2 \end{vmatrix}$$

$$= (1)(1)(2) + (3)(0)(1) + (-1)(-2)(-2)$$
$$\quad - (-1)(1)(1) - (3)(-2)(2) - (1)(-2)(0)$$
$$= 2 + 0 - 4 + 1 + 12 + 0$$
$$= 11$$

Exercises 11.3

1. Check the results of Example 11.7 by expanding along:
   (a) Row 2 (b) Row 3   (c) Column 1   (d) Column 2
   (e) Column 3

2. Check the results of Example 11.8 by expanding along:
   (a) Row 2   (b) Row 3   (c) Column 1   (d) Column 3

3. Evaluate by expansion by minors. Check by expansion along a different row or column.

   (a) $\begin{vmatrix} 1 & 2 & 1 \\ 0 & -1 & 3 \\ -2 & 4 & 0 \end{vmatrix}$ (b) $\begin{vmatrix} 3 & -2 & 0 \\ -2 & 6 & -1 \\ 0 & -1 & 3 \end{vmatrix}$

   (c) $\begin{vmatrix} 6 & -2 & 0 \\ 0 & 6 & -1 \\ -3 & -1 & 3 \end{vmatrix}$ (d) $\begin{vmatrix} -3 & 6 & 0 \\ 2 & 0 & -1 \\ 0 & -3 & 3 \end{vmatrix}$

   (e) $\begin{vmatrix} 3 & 6 & -2 \\ -2 & 0 & 6 \\ 0 & -3 & -1 \end{vmatrix}$ (f) $\begin{vmatrix} 1 & 1 & 1 \\ 0 & 1 & 0 \\ -1 & 0 & -2 \end{vmatrix}$

(g) $\begin{vmatrix} 1 & 2 & 3 \\ 1 & 2 & -3 \\ 3 & 2 & 1 \end{vmatrix}$ (h) $\begin{vmatrix} 1 & 3 & 2 \\ 0 & 1 & 0 \\ 2 & 1 & 1 \end{vmatrix}$

(i) $\begin{vmatrix} 1 & 2 & 1 \\ 0 & a & 0 \\ 1 & 3 & -1 \end{vmatrix}$ (j) $\begin{vmatrix} A & 1 & 1 \\ B & 2 & 2 \\ C & 3 & 4 \end{vmatrix}$

(k) $\begin{vmatrix} 1 & 2 & 0 & 1 \\ 3 & 1 & 0 & 2 \\ 0 & 1 & 1 & 3 \\ 1 & 0 & 0 & 1 \end{vmatrix}$ (l) $\begin{vmatrix} 1 & -1 & 0 & 3 \\ -1 & 2 & 1 & 2 \\ 1 & 0 & 1 & 0 \\ -1 & -2 & -1 & 0 \end{vmatrix}$

11.4 OPERATIONS ON DETERMINANTS

The mechanics of evaluating a determinant are not difficult, and the most bothersome aspect is simply doing the arithmetic correctly. Clearly, it is easier to evaluate a determinant having numerous zero elements since the arithmetic is simplified. There are a number of fundamental operations that can be applied to the rows or columns in a determinant, and they can be used to generate zero elements, or otherwise simplify the determinant.

First of all, notice that it is possible to factor out a term common to all elements in any row (column). For example:

(k is common to all elements in column 1.)

(x is common to all elements in row 1.)

$$\begin{vmatrix} kA & B & C \\ kD & E & F \\ kG & H & J \end{vmatrix} = k \begin{vmatrix} A & B & C \\ D & E & F \\ G & H & J \end{vmatrix} \quad \text{and} \quad \begin{vmatrix} xA & xB & xC \\ D & E & F \\ G & H & J \end{vmatrix} = x \begin{vmatrix} A & B & C \\ D & E & F \\ G & H & J \end{vmatrix}$$

EXAMPLE 11.11. Evaluate:

$$\Delta = \begin{vmatrix} 5 & 2 & 3 \\ 15 & 1 & 0 \\ 5 & 0 & 1 \end{vmatrix}$$

SOLUTION. First, factor a 5 out of column 1.

$$\Delta = 5 \begin{vmatrix} 1 & 2 & 3 \\ 3 & 1 & 0 \\ 1 & 0 & 1 \end{vmatrix} = 5[1+0+0-3-0-6]$$

$$= 5[-8]$$

$$= -40$$

Interchanging any two rows (columns) of a determinant will simply change the algebraic sign of its value. For example:

(Interchanging columns)

(Interchanging rows)

$$\begin{vmatrix} A & B \\ C & D \end{vmatrix} = - \begin{vmatrix} B & A \\ D & C \end{vmatrix} \quad \text{and} \quad \begin{vmatrix} A & B \\ C & D \end{vmatrix} = - \begin{vmatrix} C & D \\ A & B \end{vmatrix}$$

EXAMPLE 11.12.   Show that:

(a) $\begin{vmatrix} 1 & 2 \\ 3 & 4 \end{vmatrix} = -\begin{vmatrix} 2 & 1 \\ 4 & 3 \end{vmatrix}$

and

(b) $\begin{vmatrix} 1 & 2 \\ 3 & 4 \end{vmatrix} = -\begin{vmatrix} 3 & 4 \\ 1 & 2 \end{vmatrix}$

SOLUTION.   The original determinant is evaluated as

$$\Delta = \begin{vmatrix} 1 & 2 \\ 3 & 4 \end{vmatrix} = 4 - 6 = -2$$

(a) Interchanging columns:

$$\begin{vmatrix} 2 & 1 \\ 4 & 3 \end{vmatrix} = 6 - 4 = 2$$

(b) Interchanging rows:

$$\begin{vmatrix} 3 & 4 \\ 1 & 2 \end{vmatrix} = 6 - 4 = 2$$

Now a very useful property that is widely used to simplify determinants—that is, it is possible to multiply each element in a row (column) by some constant and add the result to the corresponding element in any other row (column) without changing the value of the determinant. For example, multiplying each element in row 1 by k and adding them to the elements in row 2, gives

$$\begin{vmatrix} A & B & C \\ D & E & F \\ G & H & J \end{vmatrix} = \begin{vmatrix} A & B & C \\ (D+kA) & (E+kB) & (F+kC) \\ G & H & J \end{vmatrix} \begin{matrix} \leftarrow \text{Row 1} \\ \leftarrow \text{Row 2} \\ \ \end{matrix}$$

Or, multiplying each element in column 1 by $-k$ and adding them to the elements in column 3, gives

$$\begin{vmatrix} A & B & C \\ D & E & F \\ G & H & J \end{vmatrix} = \begin{vmatrix} A & B & (C-kA) \\ D & E & (F-kD) \\ G & H & (J-kG) \end{vmatrix}$$

This property provides a very useful technique for generating zeros in the rows or columns of the matrix, and this in turn makes it much easier to calculate its determinant.

EXAMPLE 11.13.   Use the above procedure to generate zeros in the first column of

$$\Delta = \begin{vmatrix} 1 & 2 & 1 \\ 1 & 3 & 0 \\ -2 & 1 & 1 \end{vmatrix}$$

and then calculate the determinant.

SOLUTION.

(1) Copy the first *row* unchanged:

$$\Delta = \begin{vmatrix} 1 & 2 & 1 \end{vmatrix}$$

(2) Subtract the first row from the second row:

$$\Delta = \begin{vmatrix} 1 & 2 & 1 \\ (1-1) & (3-2) & (0-1) \end{vmatrix}$$

$$= \begin{vmatrix} 1 & 2 & 1 \\ 0 & 1 & -1 \end{vmatrix}$$

(3) Multiply the first row by 2 and add to the third row:

$$\Delta = \begin{vmatrix} 1 & 2 & 1 \\ 0 & 1 & -1 \\ (2-2) & (4+1) & (2+1) \end{vmatrix}$$

$$= \begin{vmatrix} 1 & 2 & 1 \\ 0 & 1 & -1 \\ 0 & 5 & 3 \end{vmatrix}$$

The resulting determinant is easier to evaluate by expansion along the first column:

$$\Delta = 1 \begin{vmatrix} 1 & -1 \\ 5 & 3 \end{vmatrix}$$
$$= 1(3+5)$$
$$= 8$$

Make the following observations regarding determinants:

1. If all the elements in any row (column) are zeros, the value of the determinant is zero. This is obvious since expansion by minors along the row (column) of zeros yields zero.

$$\begin{vmatrix} A & B & C \\ D & E & F \\ 0 & 0 & 0 \end{vmatrix} = 0 \quad \text{and} \quad \begin{vmatrix} 0 & B & C \\ 0 & E & F \\ 0 & H & J \end{vmatrix} = 0$$

2. If the elements in any row (column) are equal to or proportional to the corresponding elements in any other row (column), the value of the determinant is zero. This is true since the addition of one row (column) to another row (column) after multiplication by the appropriate constant will generate a row (column) of zeros.

$$\begin{vmatrix} A & B & C \\ kA & kB & kC \\ D & E & F \end{vmatrix} = 0 \qquad \begin{vmatrix} kA & A & C \\ kD & D & F \\ kG & G & J \end{vmatrix} = 0$$

EXAMPLE 11.14.   Evaluate:

(a) $\Delta_1 = \begin{vmatrix} 1 & 2 & 1 \\ 0 & 3 & -2 \\ 0 & 0 & 0 \end{vmatrix}$

(b) $\Delta_2 = \begin{vmatrix} 3 & 1 & 6 \\ 1 & 1 & 5 \\ 6 & 2 & 12 \end{vmatrix}$

SOLUTION.
(a) Expanding along row 3 yields:

$$\Delta_1 = 0 \begin{vmatrix} 2 & 1 \\ 3 & -2 \end{vmatrix} - 0 \begin{vmatrix} 1 & 1 \\ 0 & -2 \end{vmatrix} + 0 \begin{vmatrix} 1 & 2 \\ 0 & 3 \end{vmatrix}$$

$$= 0$$

(b) It is zero since row 3 is twice row 1. To verify, multiply row 1 by two and subtract from row 3 to obtain:

$$\Delta_2 = \begin{vmatrix} 3 & 1 & 6 \\ 1 & 1 & 5 \\ (6-6) & (2-2) & (12-12) \end{vmatrix}$$

$$= \begin{vmatrix} 3 & 1 & 6 \\ 1 & 1 & 5 \\ 0 & 0 & 0 \end{vmatrix} = 0$$

At this point, a summary of the properties of determinants will be helpful.

PROPERTIES OF DETERMINANTS

For any determinant:
(1) It is possible to factor out any term common to all elements in a row (column).
(2) Interchanging any two rows (columns) changes the algebraic sign of the determinant.
(3) It is possible to multiply each element in a row (column) by a constant and add to the corresponding element in another row (column) without changing the value of the determinant.
(4) The value of a determinant is zero if:
  (a) All elements in any row (column) are zero.
  (b) The elements in any row (column) are equal to or proportional to the corresponding elements in any other row (column).

These properties are quite valuable in simplifying the calculation of the numerical value of a determinant. They can also be used to test the coefficients matrix of a set of simultaneous equations to see whether or not a solution exists. Recall that the solutions for the simultaneous equation

$$\begin{bmatrix} V_1 \\ V_2 \\ V_3 \end{bmatrix} = \begin{bmatrix} A & B & C \\ D & E & F \\ G & H & J \end{bmatrix} \begin{bmatrix} I_1 \\ I_2 \\ I_3 \end{bmatrix}$$

are given according to Cramer's Rule as

$$I_1 = \frac{\Delta_1}{\Delta} \qquad I_2 = \frac{\Delta_2}{\Delta} \qquad I_3 = \frac{\Delta_3}{\Delta}$$

If the determinant of the coefficients matrix is zero (if $\Delta = 0$), then there are no solutions! In this case, either the equations are *inconsistent* (no solution exists), or they are *dependent* (there are not three equations). In any case, when solving electric circuit problems, evaluate the coefficients matrix first; and if it is zero, stop, since no solution exists for the given set of equations.

Exercises 11.4

1. Use row or column factoring as an aid in evaluating the following determinants. Check the result by evaluating the original determinant *without* factoring.

(a) $\begin{vmatrix} 5 & 10 \\ 1 & 3 \end{vmatrix}$
(b) $\begin{vmatrix} 2 & 5 \\ 4 & -1 \end{vmatrix}$
(c) $\begin{vmatrix} 3 & -9 \\ 6 & 2 \end{vmatrix}$

(d) $\begin{vmatrix} -3 & -21 \\ -7 & -49 \end{vmatrix}$
(e) $\begin{vmatrix} 1 & 5 & 3 \\ 0 & 10 & 2 \\ -1 & 15 & 0 \end{vmatrix}$
(f) $\begin{vmatrix} 2 & 4 & 6 \\ 0 & 2 & -1 \\ 1 & 6 & 0 \end{vmatrix}$

(g) $\begin{vmatrix} -3 & 1 & 6 \\ 9 & 0 & 2 \\ 12 & 6 & 3 \end{vmatrix}$
(h) $\begin{vmatrix} 6 & -18 & 24 \\ -12 & 9 & -48 \\ 4 & 27 & 3 \end{vmatrix}$

2. Make use of the properties of determinants to generate zero elements and then evaluate the following determinants. Notice that there are many different ways to evaluate a given determinant. Some are easier than others, and you should attempt to find a "best method."

(a) $\begin{vmatrix} 1 & 2 & 1 \\ 0 & -1 & 3 \\ -2 & 4 & 0 \end{vmatrix}$
(b) $\begin{vmatrix} 0 & -1 & 3 \\ 3 & -2 & 0 \\ -2 & 6 & -1 \end{vmatrix}$
(c) $\begin{vmatrix} 6 & 0 & -3 \\ -2 & 6 & -1 \\ 0 & -1 & 3 \end{vmatrix}$

(d) $\begin{vmatrix} -3 & 6 & 0 \\ 4 & 0 & -2 \\ 0 & -3 & 3 \end{vmatrix}$
(e) $\begin{vmatrix} 5 & 10 & 5 \\ 0 & V & 0 \\ 1 & 3 & -1 \end{vmatrix}$
(f) $\begin{vmatrix} A & 1 & 2 \\ B & 2 & 4 \\ C & 3 & 6 \end{vmatrix}$

(g) $\begin{vmatrix} 1 & -1 & 0 & 3 \\ -1 & 2 & 1 & 2 \\ 1 & 0 & 1 & 0 \\ -1 & -2 & -1 & 0 \end{vmatrix}$
(h) $\begin{vmatrix} 3 & 1 & 9 & -2 \\ 1 & 3 & 3 & 1 \\ 7 & 0 & 21 & 3 \\ -2 & 4 & -6 & 5 \end{vmatrix}$

3. Determine whether the following sets of equations have solutions by evaluating the coefficients matrix in each case. Do not solve.

(a) $2 = 2x + 1$
$3 = -6x - 3y$

(b) $7 = 6I_1 - 2I_2$
$5 = -2I_1 + 4I_2$

(c) $12 = 10I_1 - 4I_2$
$0 = -4I_1 + 11I_2 - 3I_3$
$6 = -3I_2 + 9I_3$

(d) $10^{-2} = \dfrac{4}{11} V_A - \dfrac{2}{11} V_B$

$10^{-3} = -\dfrac{2}{11} V_A + \dfrac{7}{11} V_B$

(e) $5 \times 10^{-3} = \left(\dfrac{1}{2k} + \dfrac{1}{3k} \right) V_A - \dfrac{1}{2k} V_B - \dfrac{1}{3k} V_C$

$0 = -\dfrac{1}{2k} V_A + \left(\dfrac{1}{2k} + \dfrac{1}{3k} + \dfrac{1}{4k} \right) V_B - \dfrac{1}{4k} V_C$

$6 \times 10^{-3} = -\left(\dfrac{10}{6k} \right) V_B + \dfrac{1}{1k} V_B + \dfrac{2}{3k} V_C$

UNIT 12
EXPONENTS

12.1 LAWS OF EXPONENTS

Two previously encountered rules that tell how to handle exponents when finding the product or quotient of two algebraic terms are called *Laws of Exponents*. These are:

$$a^m \cdot a^n = a^{m+n} \tag{12.1}$$
$$a^m \div a^n = a^{m-n} \tag{12.2}$$

There are three other Laws of Exponents that will be useful from time to time. These are discussed below—Eqs. (12.3), (12.4), and (12.5).

Consider first of all the problem of raising an *exponential term* (a term having an exponent) to some power. For example, $(x^2)^3$ or $(a^{-2})^2$, or in general, $(a^m)^n$. The term $(a^m)^n$ can be rewritten as

$$(a^m)^n = (a^m)(a^m)(a^m) \ldots \text{ taken } n \text{ times}$$
$$= a^{m+m+m+\cdots} \text{ taken } n \text{ times}$$
$$= a^{mn}$$

Therefore, a third Law of Exponents can be stated as:

$$(a^m)^n = a^{mn} \tag{12.3}$$

EXAMPLE 12.1 Evaluate:	SOLUTION. Using Eq. (12.3):
(a) $(x^2)^3$	(a) $(x^2)^3 = x^{2\cdot3} = x^6$
(b) $(2^3)^4$	(b) $(2^3)^4 = 2^{3\cdot4} = 2^{12} = 4096$
(c) $(R^2)^{-2}$	(c) $(R^2)^{-2} = R^{2(-2)} = R^{-4}$

A fourth Law of Exponents can be used to evaluate a product raised to some power; that is, $(ab)^n$. This term can be written as

$$(ab)^n = (ab)(ab)(ab) \cdots \text{ taken } n \text{ times}$$
$$= (a \cdot a \cdot a \cdots n \text{ times})(b \cdot b \cdot b \cdots n \text{ times})$$
$$= a^n b^n$$

Thus the rule is

$$(ab)^n = a^n b^n \tag{12.4}$$

A similar rule can be established for the quotient of two terms raised to a power; that is, $(a/b)^n$. Write

$$\left(\frac{a}{b}\right)^n = \left(\frac{a}{b}\right)\left(\frac{a}{b}\right)\left(\frac{a}{b}\right) \cdots \text{ taken } n \text{ times}$$
$$= \frac{a \cdot a \cdot a \cdots n \text{ times}}{b \cdot b \cdot b \cdots n \text{ times}}$$
$$= \frac{a^n}{b^n}$$

The rule is then

$$\left(\frac{a}{b}\right)^n = \frac{a^n}{b^n} \qquad\qquad (12.5)$$

EXAMPLE 12.2. Evaluate:
(a) $(2x^2)^2$
(b) $-(x^3)^2$
(c) $(-x^3)^2$
(d) $\left(\dfrac{R}{5}\right)^3$
(e) $\left(\dfrac{2i^2}{p}\right)^3$

SOLUTION.
Using Eq. (12.4):
(a) $(2x^2)^2 = 2^2x^4 = 4x^4$
(b) $-(x^3)^2 = -x^{3\cdot 2} = -x^6$
(c) $(-x^3)^2 = (-x^3)(-x^3) = x^6$
(Careful with sign!)
Using Eq. (12.5):
(d) $\left[\dfrac{R}{5}\right]^3 = \dfrac{R^3}{5^3} = \dfrac{R^3}{125}$
(e) $\left[\dfrac{2i^2}{p}\right]^3 = \dfrac{2^3i^{2\cdot 3}}{p^3} = \dfrac{8i^6}{p^3}$

Here is a summary of the Laws of Exponents as discussed so far.

EXPONENTS

Laws of Exponents:

$$a^m \cdot a^n = a^{m+n} \qquad a^m/a^n = a^{m-n}$$

$$(a^m)^n = a^{mn} \qquad (ab)^n = a^n b^n$$

$$\left(\frac{a}{b}\right)^n = \frac{a^n}{b^n}$$

Exercises 12.1

Use the Laws of Exponents to evaluate or simplify each of the following:

1. $x^2 \cdot x^3$
2. $m^4 \cdot m$
3. $i^3 \cdot i^0$
4. $R \cdot R^2 \cdot R^3$
5. $v^2 \cdot v^{-1}$
6. $(2^3)(2^4)$
7. $(-3)^2(3)^3$
8. $(-3)^2(-3)^3$
9. $-(5)(5)^2$
10. $(2x^2)^3$
11. $(-2x^2)^3$
12. $(ai^2)^2$
13. $(mn)(mn)^2(mn)^4$
14. $(iv)(i^2v)^2(iv^2)$
15. $\left(\dfrac{v}{R}\right)\left(\dfrac{v^2}{R}\right)$
16. $\left(\dfrac{P}{i^2}\right)\left(\dfrac{P}{i}\right)^2$
17. $\dfrac{ai^2R}{biv}$
18. $\dfrac{abR^2}{a^2bR}$

19. $\dfrac{x^2y^2z^3}{a^2xyz^2}$ 20. $\dfrac{(3a^2b)^2}{(2ab)^3}$ 21. $\dfrac{(2a^2xy^3)^3}{-8(ay)^5}$

22. $\left(\dfrac{3V^2}{R}\right)(2I^2R)^2$ 23. $\dfrac{(ax^2)^3(by)^3}{abxy}$ 24. $\left(\dfrac{a}{xy^2}\right)^2\left(\dfrac{-b}{x}\right)^3\left(\dfrac{x^2y^2}{a^2b}\right)$

12.2   FRACTIONAL EXPONENTS

In a previous section, the Law of Exponents $a^m \cdot a^n = a^{m+n}$ was used to define the square root and the cube root. Specifically,

$$a^1 = a^{1/2} \cdot a^{1/2}$$

where $a^{1/2}$ is the square root of a, and

$$a^1 = a^{1/3} \cdot a^{1/3} \cdot a^{1/3}$$

where $a^{1/3}$ is the cube root of a.

This idea can easily be extended to the general case of n factors. That is,

$$a^1 = a^{1/n} \cdot a^{1/n} \cdot a^{1/n} \cdots \text{ with } n \text{ terms}$$

In this case, since there are n terms, the nth root is defined as

$$a^{1/n} \text{ is the } n\text{th root of } a$$

The radical sign can also be used to express a root, and thus,

---

nth ROOT

The nth principal root of a is

$$a^{1/n} = \sqrt[n]{a}$$

Calculations of the square root, cube root, or nth root of a number can sometimes be done by inspection (e.g., $\sqrt[3]{8} = 2$), but the electronic hand calculator can always be used to determine a root by using the y^x function. You should determine how to evaluate the y^x function on your own calculator, and verify the following examples.

EXAMPLE 12.3
(a) Find the 4th principal root of 16.
(b) Express the 7th root of 128 in exponential form, and in radical form. Evaluate.
(c) Find the 5th principal root of 12.

SOLUTION.
(a) By inspection,
$$16 = \sqrt[4]{16} = 2$$

Check:
$$2 \cdot 2 \cdot 2 \cdot 2 = 16$$

(b) Exponential: $128^{1/7}$
Radical: $\sqrt[7]{128}$
By inspection,
$128 = \sqrt{128} = 2$
Check:

$$2 \cdot 2 \cdot 2 \cdot 2 \cdot 2 \cdot 2 \cdot 2 = 128$$

(c) Using an electronic calculator,
$\sqrt[5]{12} = 1.644$
Check:

$$1.644^5 \cong 12$$

At this point, the Laws of Exponents can be used to evaluate expressions having integers as exponents (3^2, $2x^4$, etc.) and expressions having fractional exponents of the form ($a^{1/2}$, $b^{1/3}$, or $x^{1/n}$). But, what must be done in order to evaluate a term such as $2^{3/2}$?

The Law of Exponents that tells how to raise an exponent to a power can be used to write

$$a^{m/n} = (a^{1/n})^m = (a^m)^{1/n}$$

Thus, to evaluate $a^{m/n}$, there are three choices:

1. Raise the nth root of a to the m power.
2. Find the nth root of a raised to the m power.
3. Calculate the ratio $m/n = x$, and use the electronic calculator to determine a^x.

In general:

FRACTIONAL EXPONENTS

For any quantity having a fractional exponent such as $a^{m/n}$:
(1) The numerator (m) is the power to which the quantity must be raised.
(2) The denominator (n) is the root that must be taken.
(3) The following three notations are equivalent:

$$a^{m/n} = \sqrt[n]{a^m} = (\sqrt[n]{a})^m$$

(4) The hand calculator can be used to find the numerical value of $a^{m/n} = a^x$.

Again, the evaluation of terms having fractional exponents can sometimes be done by inspection. In general, the electronic hand calculator can be used to determine the numerical values of coefficients, and the Laws of Exponents can be applied to the literal terms.

EXAMPLE 12.4. Evaluate:	SOLUTION.
(a) $9^{3/2}$	(a) $9^{3/2} = (^2\sqrt{9}\,)^3 = 3^3 = 27$
(b) $(8x^3)^{2/3}$	$\qquad = {}^2\sqrt{9^3} = \sqrt{729} = 27$
(c) $16^{0.25}$	$\qquad = 9^{1.5} = 27$
	(b) $(8x^3)^{2/3} = (^3\sqrt{8x^3}\,)^2 = (2x)^2 = 4x^2$
	$\qquad = 8^{2/3}x^{3\cdot2/3} = 4x^2$
	(c) $16^{0.25} = 16^{1/4} = {}^4\sqrt{16} = 2$
	$\qquad = 16^{0.25} = 2$

All the Laws of Exponents can be used to evaluate expressions containing fractional exponents. The next example illustrates the application of the Laws of Exponents to a number of different problems.

EXAMPLE 12.5. Evaluate:	SOLUTION.
(a) $i^{1/2}\cdot i^{1/4}$	(a) $i^{1/2+1/4} = i^{3/4}$
(b) $x^{3/2} \div x^{3/4}$	(b) $x^{3/2}\cdot x^{4/3} = x^{(3/2+4/3)} = x^2$
(c) $(v^{2/3})^4$	(c) $v^{(2/3)\cdot4} = v^{8/3}$
(d) $(27x^6)^{1/3}$	(d) $^3\sqrt{27x^6} = 3x^2$
(e) $\left[\dfrac{1}{36v^4}\right]^{0.5}$	(e) $\left[\dfrac{1}{36v^4}\right]^{1/2} = \sqrt{\dfrac{1}{36v^4}} = \dfrac{\sqrt{1}}{\sqrt{36v^4}} = \dfrac{1}{6v^2}$

Exercises 12.2

1. Express in radical form:

   (a) $a^{1/5}$ (b) $y^{1/6}$ (c) $z^{2/3}$ (d) $Q^{6/7}$

   (e) $v^{4/3}$ (f) $x^{1\frac{1}{2}}$ (g) $a^{1\frac{1}{3}}$ (h) $p^{1.5}$

   (i) $2x^{1/2}$ (j) $(2x)^{1/2}$ (k) $3y^{2/3}$ (l) $(3y)^{2/3}$

   (m) $4x^{1.25}$ (n) $(i+2)^{3/2}$ (o) $3(x-y)^{2/3}$ (p) $(27a^3b^6)^{1\frac{1}{3}}$

2. Express in exponential form:

   (a) $\sqrt[4]{b}$ (b) $\sqrt[7]{y}$ (c) $\sqrt[2]{p^4}$ (d) $\sqrt[3]{z^5}$

(e) $2\sqrt{y}$ (f) $\sqrt[3]{3x}$ (g) $11\sqrt[3]{y^2}$ (h) $\dfrac{\sqrt{x}}{\sqrt{y}}$

(i) $\dfrac{\sqrt[3]{a}}{2\sqrt[3]{b}}$ (j) $\dfrac{2\sqrt[4]{x^3}}{\sqrt[4]{y^3}}$ (k) $\dfrac{\sqrt[3]{P}}{\sqrt[4]{R}}$ (l) $\dfrac{3a\sqrt{x}}{5\sqrt[3]{b}}$

3. Evaluate:

(a) $49^{1/2}$ (b) $27^{1/3}$ (c) $64^{1/6}$ (d) $64^{5/6}$

(e) $4^{3/2}$ (f) $(-8)^{4/3}$ (g) $\left(\dfrac{8}{27}\right)^{1/3}$ (h) $(-1)^{2/3}$

(i) $(0.25)^{3/2}$ (j) $(-1)^{5/3}$ (k) $\left(\dfrac{9}{16}\right)^{-1/2}$ (l) $(36x^2y^4)^{-1.5}$

4. Perform the indicated operations:

(a) $x \cdot x^{1/2}$ (b) $a^{1/2} \cdot a^{3/4}$ (c) $Q^{1/2} \cdot Q^2$ (d) $(\sqrt[3]{P})(P)^{2/3}$

(e) $v \cdot v^{1/2} \cdot v^{1/3}$ (f) $i^2 \cdot i^{-1/2}$ (g) $\left(\sqrt{\dfrac{P}{R}}\right)(V^{3/2})$ (h) $n^{2/3} \div n^{1/3}$

(i) $x^{2/3} \div x^{3/2}$ (j) $a^{0.5} \div a^{1.5}$ (k) $2^{5/2} \div 2^{3/2}$ (l) $Q^{4\frac{1}{2}} \div Q^{2\frac{1}{2}}$

(m) $(a^2)^{3/2}$ (n) $(Q^{3/2})^2$ (o) $(z^4)^{3/8}$ (p) $\sqrt[3]{x^2}$

(q) $\sqrt[4]{x^2y^3}$ (r) $\sqrt{2^6}$ (s) $\dfrac{ax^{1/2}}{\sqrt{4x^4}}$ (t) $\dfrac{\sqrt{\dfrac{v^2}{R}}}{(i^2R)^{1/2}}$

12.3   IMAGINARY NUMBERS

The solution of electric circuit problems will inevitably lead to the use of a special kind of number called an *imaginary number*. In fact, imaginary numbers are encountered in many different subject areas in applied mathematics as well as in science and engineering. Let's see exactly what an imaginary number is.

In the discussion of square roots it is shown that any real, positive number has two square roots. The *principal* root is positive, while the *secondary* root is negative. For example, the square roots of 25 are $+5$ and -5. Notice that the square of either of these two roots yields the positive number 25. That is,

$$(+5)(+5) = 25 \qquad \text{and} \qquad (-5)(-5) = 25$$

Now, one might reasonably ask the question, "Is there any quantity whose square will yield a *negative* number?" Or, one might state the same question by asking, "What is the square root of a *negative* number?"

$$(?)(?) = -25 \qquad \text{or} \qquad \sqrt{-25} = ?$$

From the law of signs it is clear that no real number exists whose square is negative! But many problems in applied mathematics require the evaluation of terms such as $\sqrt{-25}$, $\sqrt{-9a^2}$, or $\sqrt{-4/9}$.

Notice that the rules for operating on radicals can be used to rewrite $\sqrt{-25}$ as $\sqrt{-1}\,\sqrt{25}$. In fact, you can always write

$$\sqrt{-a} = \sqrt{-1} \cdot \sqrt{a}$$

The quantity $\sqrt{-1}$ is defined as the *operator j*.*

OPERATOR j

The imaginary operator j is defined as:

$$j = \sqrt{-1}$$

Thus the square root of any negative number is the product of the operator j and the square root of the number; that is,

$$\sqrt{-a} = \sqrt{-1} \cdot \sqrt{a} = j\sqrt{a}$$

Any real number that is multiplied by the operator j is an imaginary number. For example, $j6$, $-j5$, $j/4$, and $3aj$ are all imaginary numbers. Thus the definition:

IMAGINARY NUMBER

The square root of any negative number a is an imaginary number $j\sqrt{a}$:

$$\sqrt{-a} = j\sqrt{a}$$

Notice that if $j = \sqrt{-1}$, then,

$$j^2 = j \cdot j = \sqrt{-1} \cdot \sqrt{-1} = -1$$

EXAMPLE 12.6.   Evaluate:
(a) $\sqrt{-25}$
(b) $\sqrt{-9a^2}$
(c) $\sqrt{-4/9}$

SOLUTION.
(a) $\sqrt{-25} = \sqrt{-1}\,\sqrt{25} = j\sqrt{25} = j5$
      Check:

$$(j5)(j5) = 25j^2 = 25(-1) = -25$$

(b) $\sqrt{-9a^2} = j\sqrt{9a^2} = 3aj$
      Check:

$$(3aj)(3aj) = 9a^2j^2 = 9a^2(-1) = -9a^2$$

---

*In mathematics $\sqrt{-1}$ is usually defined as i; but to avoid confusion with electric current i, texts dealing with electric circuits usually use the letter j.

(c) $\sqrt{-4/9} = \dfrac{\sqrt{-4}}{\sqrt{9}} = \dfrac{j2}{3}$

Check:

$$\left(\dfrac{j2}{3}\right)\left(\dfrac{j2}{3}\right) = \dfrac{4j^2}{9} = -\dfrac{4}{9}$$

There may be times when it will be necessary to evaluate j raised to a power. This presents no problem, since if

$$j = \sqrt{-1} \quad \text{and} \quad j^2 = j \cdot j = \sqrt{-1} \cdot \sqrt{-1} = -1$$

then

$$j^3 = j(j^2) = j(-1) = -j$$
$$j^4 = (j^2)(j^2) = (-1)(-1) = 1$$

Continuing in this fashion, it is easy to show

$$j = j^5 = j^9 = \cdots = j$$
$$j^2 = j^6 = j^{10} = \cdots = -1$$
$$j^3 = j^7 = j^{11} = \cdots = -j$$
$$j^4 = j^8 = j^{12} = \cdots = +1$$

These four values can then be used to evaluate expressions containing integer powers of the complex operator j.

EXAMPLE 12.7. Evaluate:

(a) $\sqrt{-36b^4}$

(b) $\sqrt{-25x^2/y^3}$

(c) $(j^2)^3$

(d) $(-j^3)^4$

(e) j^{33}

(f) $(j^2)^{-3}$

SOLUTION.

(a) $6b^2 j$

(b) $\dfrac{5xj}{y^{3/2}}$

(c) $(j^2)^3 = (-1)^3 = -1$

(d) $(-j^3)^4 = (-j)^4 = (-1)^4 (j)^4 = 1$

(e) $j^{33} = j^{11} \cdot j^{11} \cdot j^{11} = (-j)(-j)(-j) = (-1)(-j) = j$

(f) $(j^2)^{-3} = \dfrac{1}{(j^2)^3} = \dfrac{1}{(-1)^3} = \dfrac{1}{(-1)} = -1$

Exercises 12.3

Evaluate each of the following:

1. $\sqrt{-9}$

2. $\sqrt{-25}$

3. $\sqrt{-1/9}$

4. $\sqrt{-16/25}$

5. $\dfrac{\sqrt{36}}{\sqrt{-49}}$

6. $\sqrt{-4a^2}$

7. $\sqrt{-8x^2}$ 8. $\sqrt{-15a^2y^4}$ 9. $\dfrac{1}{\sqrt{-6y^3}}$

10. $\sqrt{j^6}$ 11. j^5 12. j^{20}

13. j^{10} 14. j^{100} 15. $(-j)^3$

16. $(-j^2)^4$ 17. $\dfrac{\sqrt{-49}}{j}$ 18. $\dfrac{-11j}{\sqrt{-121}}$

19. $\dfrac{\sqrt{-49x^2}}{j\sqrt{16x}}$ 20. $\dfrac{j4\sqrt{-16}}{-\sqrt{64x^2}}$ 21. $\dfrac{\sqrt{5x^4}}{3j(-6)^{1/2}}$

22. $\dfrac{j(j4)}{\sqrt{-6x}}$ 23. $\left(\dfrac{1}{\sqrt{-7}}\right)\left(\dfrac{j2}{3}\right)$ 24. $(j^2)(3j)(-j^3)\left(\dfrac{1}{j^2}\right)$

13.1 EXPONENTIAL EQUATIONS

From the definitions used with exponents it is known that the algebraic term b^x has a *base* b and an *exponent* x. An exponential term such as b^x can be used to form an algebraic equation such as

$$y = b^x \tag{13.1}$$

In this equation, y and x are the variables, and b is a constant. An algebraic equation in which a variable appears as an exponent is called an *exponential equation*. Exponential equations are encountered frequently in the study of electric circuits (as well as in many other areas of science and mathematics), and we will take the time to become familiar with their characteristics.

In Eq. (13.1), b is a constant. It could be an integer such as $1, 2, 5, 7$, etc., or it could be a mixed number such as $1.5, 7.38, 3.14159\ldots(\pi)$, etc. In any case, it is possible to assign integer values for x, evaluate y, and plot the results. However, in order to evaluate y for values of x that are not integers, it will be necessary to use an electronic calculator or to make use of logarithms.*

EXAMPLE 13.1. Evaluate $y = 2^x$ for the values of x shown in the table. Plot the results.

SOLUTION.

For $x = -3$, $y = 2^{-3} = \dfrac{1}{2^3} = \dfrac{1}{8} = 0.125$

For $x = -2$, $y = 2^{-2} = 2^{1/2} = \dfrac{1}{4} = 0.25$

For $x = -1$, $y = 2^{-1} = \dfrac{1}{2} = 0.5$

For $x = 0$, $y = 2^0 = 1$
For $x = 1$, $y = 2^1 = 2$
For $x = 2$, $y = 2^2 = 4$
For $x = 3$, $y = 2^3 = 8$
For $x = 4$, $y = 2^4 = 16$

x	-3	-2	-1	0	1	2	3	4
y	0.125	0.25	0.5	1	2	4	8	16

The graph is plotted in Figure 13.1.

The exponential curve in Figure 13.1 passes through the point $(0, 1)$; that is, when $x = 0$, $y = 1$. This will be true for any graph of Eq. (13.1), no matter what the value of the base b, since $y = b^0 = 1$.

---

*For example, in order to evaluate $y = 2.6^{1.38}$, use the y^x function on an electronic calculator. Logarithms are covered in a following section; they can also be used to evaluate such expressions.

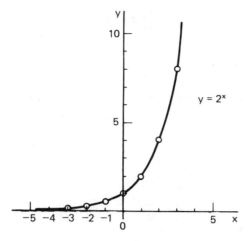

Figure 13.1

For positive values of x, the value of y in Figure 13.1 increases very rapidly. Thus the description "increasing exponentially" simply refers to a very rapidly increasing function.

The curve in Figure 13.1 gets closer to the x-axis for negative values of x; but it never quite touches the x-axis (that is, y is never equal to zero). Any line that a curve approaches but never touches is called an *asymptote*. In this case, the x-axis is an asymptote for the exponential curve $y = 2^x$. One can say, "The curve approaches the x-axis asymptotically."

In the following example, the effect of selecting different values for the base b in Eq. (13.1) is illustrated.

EXAMPLE 13.2. Evaluate y and plot the results for:
(a) $y = 1^x$
(b) $y = 1.5^x$
(c) $y = 3^x$
(d) $y = 0.5^x$

SOLUTION.
(a) $y = 1^x$. For any value of $x, y = 1$. The curve is the straight line shown in Figure 13.2.
(b) $y = 1.5^x$. For $x = -3$,

$$y = 1.5^{-3} = \frac{1}{1.5^3} = \frac{1}{3.375} = 0.296$$

Continuing,

x	-3	-2	-1	0	1	2	3
y	0.296	0.444	0.667	1	1.5	2.25	3.375

(c) $y = 3^x$. For $x = -3$,

$$y = 3^{-3} = \frac{1}{3^3} = \frac{1}{27} = 0.037$$

Continuing,

x	-3	-2	-1	0	1	2	3
y	0.037	0.111	0.333	1	3	9	27

(d) $y = 0.5^x$. For $x = -3$,
$$y = 0.5^{-3} = \frac{1}{0.5^3} = \frac{1}{0.125} = 8$$
Continuing,

x	−3	−2	−1	0	1	2	3
y	8	4	2	1	0.5	0.25	0.125

An explicit solution of the equation $y = b^x$ for x involves the use of logarithms—a subject to be discussed in a later section. However, for a given value of y, it is sometimes possible to find a solution for x by trial and error. For example, if you wish to find x for $8 = 2^x$, you can very quickly determine that $x = 3$ is the solution (since $8 = 2^3$). Similarly, $x = 2$ is clearly a solution for $16 = 4^x$.

If the solution is not easily recognized, you can sometimes resort to a graphical solution. As an example, it is possible to use the curve for $y = 2^x$ plotted in Figure 13.1 to determine a solution for $3 = 2^x$. Estimate from the curve, $x \cong 1.6$ (a formal solution shows $x = 1.5849625$) at the point where $y = 3$.

An electronic calculator can always be used to evaluate exponentials, and you should become proficient in evaluating equations of the form $y = b^x$. The calculator can be used to determine y for given values of b and x directly. It can also be used to estimate x for given values of y and b.

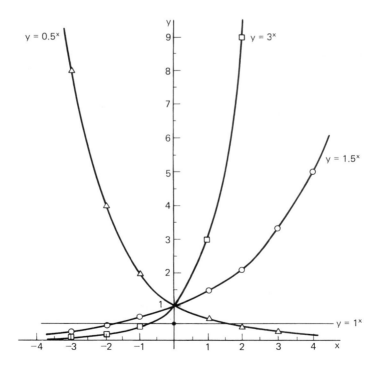

Figure 13.2

Exercises 13.1

1. Evaluate:
    (a) $y = 2^x$ for $x = 0, 1, 2, 5$
    (b) $y = 2^x$ for $x = -1, -3, -5$
    (c) $y = 3^x$ for $x = 1, 3, 6$
    (d) $y = 10^x$ for $x = 0, 1, 2, 3, 4, 5$
    (e) $y = 10^x$ for $x = -1, -6, -12$
    (f) $y = 2.5^x$ for $x = -3, -2, -1, 0, 1, 2, 3$
    (g) $y = 0.3^x$ for $x = -3, -1.0, 1, 3$
    (h) $y = 0.1^x$ for $x = -6, -2, 0, 2, 6$
    (i) $y = 1.2^x$ for $x = -5, -3, -1, 0, 1, 3, 5$
    (j) $y = -2^x$ for $x = -3, -2, -1, 0, 1, 2, 3$
    (k) $y = 2^{-x}$ for $x = -5, -3, -1, 0, 1, 3, 5$
    (l) $y = -1.5^{-x}$ for $x = -5, -3, -1, 0, 1, 3, 5$

2. Calculate enough points to make an accurate plot of:
    (a) $y = 1.5^x$
    (b) $y = 0.2^x$
    (c) $y = -2^x$
    (d) $y = 2^{-x}$
    (e) $y = 1.5^{-x}$
    (f) $y = \left(\dfrac{1}{4}\right)^x$

3. If $y = 1.5^x$, estimate x if y is:
    (a) 2   (b) 3   (c) 1   (d) 0.5

4. If $y = 2^{-x}$, estimate x if y is:
    (a) 1   (b) 2   (c) 0.5   (d) 5

5. Estimate the correct value for x:
    (a) $2.25 = 1.5^x$
    (b) $5.106 = 3^x$
    (c) $2.934 = 4.2^x$
    (d) $0.92 = 0.7^x$
    (e) $2.38 = 2^x$

13.2 e^x AND e^{-x}

The number 2.7182818... is encountered so frequently in mathematics that it is given the special symbol e.* Thus,

$$e = 2.7182818\ldots$$
$$\cong 2.718$$

The number e can be used as the base in an exponential equation, and equations of the form $i = Ae^x$ and $i = Be^{-x}$ can be used to predict the currents (and voltages) in electric circuits. A function of the form $i = Ae^x$ is said to be *exponentially increasing*, and its graph is similar to that shown in Figure 13.1.

The *exponentially decreasing* function $i = Be^{-x}$ is more common in electric circuit problems. In the following Example 13.3, the function $y = e^{-x}$ is evaluated and graphed.

---

*e is the base of natural logarithms, as will be shown in a later section.

EXAMPLE 13.3. Evaluate $y = e^{-x}$ and graph the results in Figure 13.3.

SOLUTION. For $x = -3$,

$$y = e^{-(-3)} = e^3 = 20.09$$

Continuing,

x	-3	-2	-1	0	1	2	3
y	20.09	7.389	2.718	1	0.368	0.135	0.0498

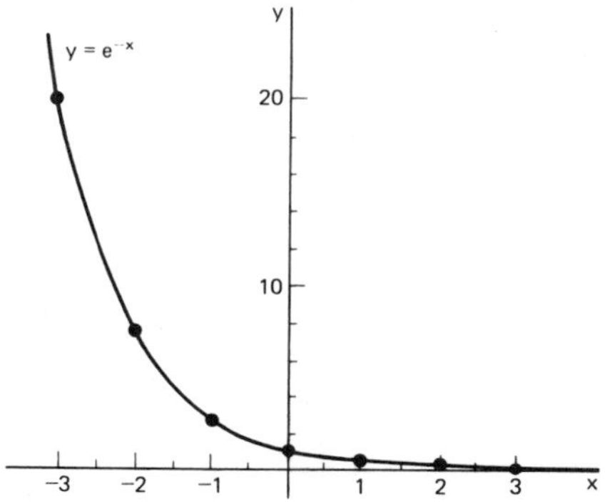

Figure 13.3

The evaluation of the exponentials e^x and e^{-x} is very important in the study of electric circuits. There are at least five different means for evaluating e^x or e^{-x}, even if x is not an integer. These are: (1) tables, (2) the Universal Exponential Curve, (3) a slide rule, (4) an electronic hand calculator, and (5) logarithms.

Tables

In the Appendix at the end of this text there is a table titled "Values of e^x and e^{-x}." The vertical *column* on the left gives the first two numbers in x, while the *row* across the top gives the third number in x. For example, to find the value of $e^{1.23}$, note that $x = 1.23$. Now, find 1.2 in the left column, then move over under 0.03 on the top row, and read the value 3.4212. Thus, $e^{1.23} = 3.4212$.

EXAMPLE 13.4. Use the math tables in the Appendix to evaluate e^x and e^{-x} for the values of x shown.

SOLUTION. For $x = 0$, both e^0 and e^{-0} are equal to 1.0. For $x = 0.62$, find 0.6 in the left-hand column, move over under 0.02 on the top row, and read values $e^x = 1.8589$ and $e^{-x} = 0.5379$. Continuing,

x	0	0.62	1.25	2.37	4.90
e^x	1.0	1.8589	3.4903	10.697	134.29
e^{-x}	1.0	0.5379	0.2865	0.0935	0.0074

Notice that the value of x in the Appendix tables ranges from 0.00 to 5.99. You might wonder how to evaluate e^x for values of x greater than 5.99. For example, how can a value for $e^{6.5}$ be found? The answer is simple—use the Law of Exponents to form exponentials that can be found in the tables. Thus, write

$$e^{6.5} = e^3 \cdot e^{3.5}$$

or

$$e^{6.5} = e^{1.5} \cdot e^5$$

or any other combination desired (recall, $e^a \cdot e^b = e^{(a+b)}$). Then,

$$e^{6.5} = e^3 \cdot e^{3.5} = 20.086 \times 33.115 = 665.14$$

EXAMPLE 13.5. Evaluate e^x and e^{-x} if $x = 10$.

SOLUTION. Write

$$e^{10} = e^2 \cdot e^4 \cdot e^4$$

$$= 7.389 \times 54.6 \times 54.6$$

$$= 2.203 \times 10^4$$

$$e^{-10} = \frac{1}{e^{10}} = \frac{1}{2.203 \times 10^4} = 4.54 \times 10^{-5}$$

Notice that e^{10} could have been evaluated by writing

$$e^{10} = e^5 \cdot e^5 = 148.4 \times 148.4 = 2.203 \times 10^4$$

Any other convenient powers of e^x desired could also be used.

The Universal Exponential Curve shown in Figure 13.4 is essentially a graph of the values of e^{-x} given in the math tables in the Appendix. It is sometimes easier

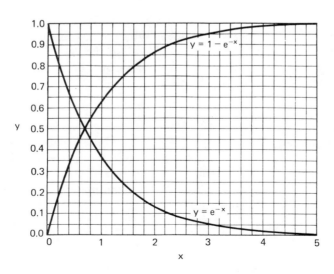

Figure 13.4. Universal exponential curves.

(although usually not as accurate) to evaluate e^x or e^{-x} using this curve. You should select a few values of x, read the corresponding values of e^{-x}, and compare with the table values. Notice that this curve can be used to evaluate x for given values of e^{-x} (or e^x).

EXAMPLE 13.6. Use the Universal Exponential Curve to evaluate: (a) $y = e^{-2}$ (b) $y = e^{1.5}$ (c) x if $0.7 = e^{-x}$ (d) x if $0.21 = e^{-x}$	**SOLUTION.** Examination of the Universal Exponential Curve yields: (a) $y = 0.13$ at the point where $x = 2$ (b) $y = e^{1.5} = \dfrac{1}{e^{-1.5}} = \dfrac{1}{0.22} = 4.5$ (c) $x = 0.36$ at the point where $y = 0.7$ (d) $x = 1.56$ at the point where $y = 0.21$ Check the results by looking in the Appendix math tables

The values of e^x and e^{-x} are quite easily determined using a slide rule or an electronic hand calculator. You should refer to the appropriate instruction manual to learn the proper procedure. The evaluation of these functions using logarithms will be discussed in a later section. For now, use the problems below to become proficient in evaluating e^x or e^{-x} with tables, the Universal Exponential Curve, and either a slide rule or an electronic calculator.

Exercises 13.2

1. Select five different values of x within the limits $0 \leqslant x \leqslant 5.99$. Now use the Appendix tables to evaluate e^x and e^{-x} for each of the five values chosen. Check the results by reading values from the Universal Exponential Curve.

2. Select five different values of x within the limits $0 \leqslant x \leqslant 5.99$. Now, use your slide rule or electronic calculator to evaluate e^x and e^{-x}. Check the results by reading values from the Appendix math tables.

3. Evaluate the following by using the Appendix tables and verify by using a slide rule or electronic calculator:

   (a) $y = e^{4.3}$
   (b) $i = e^{-1.2}$
   (c) $v = e^{-3.7}$
   (d) $v = 4e^{-1.6}$
   (e) $i = 1.2e^{-0.7}$

   (f) $v = e^{10}$
   (g) $i = 3e^{16}$
   (h) $i = 10^{-6} \times e^{8.4}$
   (i) $i = 3 \times 10^{-6} \times e^{11.2}$

4. Evaluate and plot $y = e^x$ for $0 \leqslant x \leqslant 2$. Take x in increments of 0.2 (i.e., $0.0, 0.2, 0.4, 0.6, \ldots$).

5. Evaluate and plot $y = e^{-x}$ for $0 \leqslant x \leqslant 6$. Take x in increments of 0.2.

6. Use the Appendix tables, the Universal Exponential Curve, or a slide rule or electronic calculator to determine x in the following:

(a) $1.3 = e^x$

(b) $0.5 = e^x$

(c) $5.9 \times 10^2 = e^x$

(d) $0.72 = e^x$

(e) $0.11 = e^{-x}$

(f) $21.8 = e^{-x}$

(g) $3.8 = e^{-x}$

(h) $3 \times 10^{-2} = e^{-x}$

(i) $1.6 \times 10^3 = 7e^x$

(j) $15.4 = 1.3e^{-x}$

13.3 RL CIRCUITS

In an electric circuit there is a period of time just after the closing or opening of a switch (or the turning on or off of an electronic device) during which the circuit currents and voltages are settling down to their steady-state values. This period of time is called the *transient period*, and the circuit currents and voltages during this time are referred to as the *transient repsonses*. A complete presentation of transient circuits will not be given here, but some of the more common responses will be discussed since they are exponential in form.

The series circuit shown in Figure 13.5 consists of a resistor (R), an ideal inductor (L), a dc voltage source (V_S), and a switch (SW). When the switch is open, there is no circuit current, and thus no voltages exist across R or L. After the switch is closed, there is a circuit current, and there are voltages across both R and L. Assume that a clock begins counting time just as the switch is closed, and then consider the voltage v_L as a function of time.

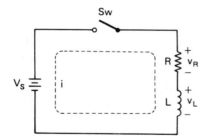

Figure 13.5

The equation for v_L versus time after closing the switch is

$$v_L = V_S e^{-t/\tau^*} \tag{13.2}$$

where

$$V_S = \text{the dc source voltage}$$
$$\tau = \text{the time constant} = L/R$$
$$t = \text{the time measured in seconds}$$

---

*τ is the Greek letter tau. Do not confuse it with the lower-case letter t.

Notice that this is a decaying exponential. When $t=0$ (the switch has just been closed), the value of v_L is

$$v_L(0) = V_S e^{-0/\tau} = V_S$$

since $e^{-0} = 1$. After a long time (say $t = \infty$), v_L has a value of

$$v_L(\infty) = V_S e^{-\infty/\tau} = 0$$

since

$$e^{-\infty} = \frac{1}{e^\infty} = 0$$

Thus the voltage v_L begins at V_S, ends at zero, and is a decaying exponential in between. In Example 13.7, the values of v_L are calculated for various values of time (t), and the resulting data are plotted to provide a curve showing how v_L varies with time t.

EXAMPLE 13.7. For the circuit in Figure 13.5, $R = 10\ \Omega$, $L = 5$ H, and $V_S = 10$ V. Use Eq. (13.2) to evaluate v_L and plot the results.

SOLUTION.

$$\tau = \frac{L}{R} = \frac{5}{10} = 0.5$$

Thus,

$$v_L = V_S e^{-t/\tau}$$

$$= 10 e^{-t/0.5} = 10 e^{-2t}$$

Now, calculate v_L for the values of t chosen. For $t = 0$,

$$v_L = 10 e^{-0} = 10(1) = 10$$

For $t = 0.2$,

$$v_L = 10 e^{-2(0.2)} = 10 e^{-0.4}$$

$$= 10(0.67) = 6.7$$

Continuing,

t	0	0.2	0.4	0.6	0.8	1.0	1.5	2.0	2.5
v_L	10	6.7	4.5	3.0	2.0	1.3	0.5	0.18	0.07

The curve for v_L versus t is plotted in Figure 13.6.

Take another look at the circuit in Figure 13.5. After closing the switch, KVL requires

$$V_S = v_R + v_L$$

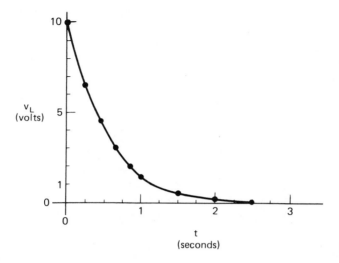

Figure 13.6

Since v_L is known and V_S is a constant, the resistor voltage v_R can be found using Eq. (13.2).

$$v_R = V_S - v_L = V_S - V_S e^{-t/\tau}$$
$$v_R = V_S(1 - e^{-t/\tau}) \tag{13.3}$$

In Example 13.8 a careful plot of v_R is made by using Eq. (13.3).

EXAMPLE 13.8.   Using the same circuit as in Example 13.7, evaluate and plot v_R from Eq. (13.3). Plot the results in Figure 13.7.

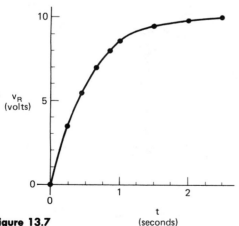

Figure 13.7

SOLUTION.

$$\tau = \frac{L}{R} = \frac{5}{10} = 0.5 \text{ and } V_S = 10$$

Thus, from Eq. (13.3),

$$v_R = V_S(1 - e^{-t/\tau})$$
$$= 10(1 - e^{-2t})$$

For $t = 0$,
$$v_R = 10(1 - e^{-0}) = 10(1 - 1) = 0$$

For $t = \infty$,
$$v_R = 10(1 - e^{-\infty}) = 10(1 - 0) = 10$$

For $t = 0.2$ sec,
$$v_R = 10(1 - e^{-2(0.2)}) = 10(1 - e^{-0.4})$$
$$= 10(1 - 0.67) = 10(0.33) = 3.3$$

Continuing,

t	0	0.2	0.4	0.6	0.8	1.0	1.5	2.0	2.5	∞
v_R	0	3.3	5.5	7.0	8.0	8.7	9.5	9.82	9.93	10

Notice that KVL requires $V_S = v_R + v_L$. Therefore, if the two curves in Figures 13.6 and 13.7 are added point by point, the result should be a constant 10 V.

The circuit current i in Figure 13.5 can easily be found by using Ohm's Law and Eq. (13.3). Thus,

$$i = \frac{v_R}{R} = \frac{V_S}{R}(1 - e^{-t/\tau}) \qquad (13.4)$$

The circuit current curve has exactly the same form as v_R with the appropriate units of current (amperes) on the vertical axis. For the values given in Example 13.7, the current begins at zero amperes and asymptotically approaches $V_S/R = 10/10 = 1$ A for large values of t.

Exercises 13.3

1. Use Eq. (13.2) to evaluate and plot v_L for the circuit in Figure 13.5 for the following element values:

    (a) $R = 5\ \Omega$ 　　　　　(d) $R = 100\ \Omega$
    　　　$L = 5$ H 　　　　　　　　$L = 100$ mH
    　　　$V_S = 5$ V 　　　　　　　$V_S = 28$ V

    (b) $R = 10\ \Omega$ 　　　　　(e) $R = 1$ kΩ
    　　　$L = 1$ H 　　　　　　　　$L = 1$ H
    　　　$V_S = 6$ V 　　　　　　　$V_S = 15$ V

    (c) $R = 100\ \Omega$ 　　　　(f) $R = 2.7$ kΩ
    　　　$L = 2$ H 　　　　　　　　$L = 33$ mH
    　　　$V_S = 12$ V 　　　　　　　$V_S = 5$ V

2. Use Eq. (13.3) to evaluate and plot v_R for the values given in Problem 1(a).
3. Evaluate and plot v_R for the circuit in Problem 1(b).
4. Evaluate and plot v_R for the circuit in Problem 1(c).
5. Evaluate and plot v_R for the circuit in Problem 1(d).
6. Evaluate and plot v_R for the circuit in Problem 1(e).
7. Evaluate and plot v_R for the circuit in Problem 1(f).
8. Using the results from Problems 1(a) and 2, plot v_L and v_R on the same graph and show that $V_S = v_L + v_R$.
9. Using the results from Problems 1(d) and 5, plot v_L and v_R on the same graph and show that $V_S = v_L + v_R$.
10. Use Eq. (13.4) to evaluate and plot i for the element values given in Problem 1(a) through 1(f).

13.4 RC CIRCUITS

The transient responses of the series RC circuit in Figure 13.8 are also exponential in form. When the switch (SW) is open, there is no circuit current and $i=0$. Assuming there is no initial charge on the capacitor, the capacitor voltage v_C is zero before closing the switch. Likewise $v_R=0$ before closing the switch since $i=0$.

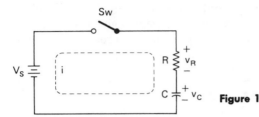

Figure 13.8

After closing the switch, the equation for circuit current in amperes is given by the decaying exponential

$$i = \frac{V_S}{R} e^{-t/\tau} \qquad (13.5)$$

where

$$V_S = \text{the dc source voltage}$$

$$R = \text{the circuit resistance}$$

$$\tau = \text{the time constant} = RC$$

$$t = \text{the time measured in seconds}$$

When $t=0$ (the switch has just been closed), the current is

$$i(0) = \frac{V_S}{R} e^{-0/\tau} = \frac{V_S}{R}$$

since $e^{-0/\tau}=1$.

After a long time $(t=\infty)$, the current is

$$i(\infty) = \frac{V_S}{R} e^{-\infty/\tau} = 0$$

since $e^{-\infty/\tau}=1/e^{\infty/\tau}=0$.

Thus the circuit current begins at V_S/R, ends at zero, and is a decaying exponential in between. In Example 13.9 the values of circuit current i are calculated for various values of time (t), and the resulting data are plotted for the circuit shown in Figure 13.8.

EXAMPLE 13.9. For the circuit in Figure 13.8: $R = 1$ MΩ, $C = 1$ μF, and $V_S = 1000$ V. Use Eq. (13.5) to evaluate i, and then plot i versus time.

SOLUTION.

$$\tau = RC = 10^6 \times 10^{-6} = 1$$

Thus,

$$i = \frac{V_S}{R} e^{-t/\tau}$$

$$= \frac{10^3}{10^6} e^{-t/1}$$

$$= 10^{-3} \times e^{-t} \text{ amperes}$$

$$= e^{-t} \text{ mA}$$

Note carefully that i is now in mA.
For $t = 0$,

$$i = e^{-0} = 1$$

For $t = \infty$,

$$i = e^{-\infty} = 0$$

For $t = 0.5$ sec,

$$i = e^{-0.5} = 0.606$$

For $t = 1.0$ sec,

$$i = e^{-1} = 0.368, \text{ etc.}$$

t	0	0.5	1.0	2.0	3.0	4.0	5.0
i	1	0.606	0.368	0.135	0.05	0.018	0.0067

The curve for i versus time is shown in Figure 13.9.

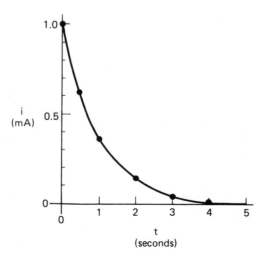

Figure 13.9

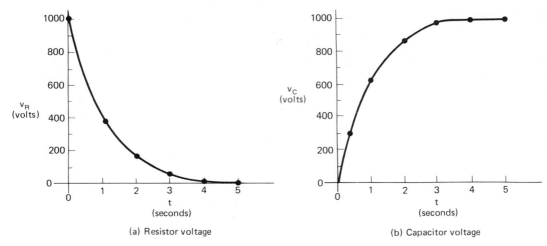

Figure 13.10

Using the equation for circuit current i, Ohm's Law can be used to determine the resistor voltage v_R in the series RC circuit of Figure 13.8. Thus,

$$v_R = iR = \frac{V_S}{R} e^{-t/\tau} \times R$$

$$v_R = V_S e^{-t/\tau} \tag{13.6}$$

The resistor voltage v_R has exactly the same shape as the current i in Figure 13.9, except it begins at V_S and ends at zero. A plot of v_R would therefore appear as shown in Figure 13.10(a).

Returning to Figure 13.8, KVL requires

$$V_S = v_R + v_C$$

Since V_S is a constant, and v_R is known from Eq. (13.6), it is easy to solve for v_C. That is,

$$v_C = V_S - v_R$$
$$= V_S - V_S e^{-t/\tau}$$

Factoring V_S yields

$$v_C = V_S(1 - e^{-t/\tau}) \tag{13.7}$$

In Example 13.10, Eq. (13.7) is used to calculate values of v_C for various values of time in order to plot the curve of v_C as a function of time.

EXAMPLE 13.10. Use Eq. (13.7) to evaluate v_C for the circuit values in Example 13.9. Plot the results.

SOLUTION. From Example 13.9, $\tau = 1$. Thus, using Eq. (13.7),

$$v_C = V_S(1 - e^{-t/\tau})$$

$$= 1000(1 - e^{-t})$$

For $t=0$,

$$v_C = 1000(1 - e^{-0}) = 1000(1 - 1) = 0$$

For $t=\infty$,

$$v_C = 1000(1 - e^{-\infty}) = 1000(1 - 0) = 1000$$

For $t=0.5$ sec,

$$v_C = 1000(1 - e^{-0.5})$$

$$= 1000(1 - 0.606) = 394$$

etc.

t	0	0.5	1.0	2.0	3.0	4.0	5.0
v_C	0	394	632	865	950	982	993

The curve for v_C is plotted in Figure 13.10(b).

Since KVL requires $V_S = v_R + v_C$, the two curves in Figure 13.10 could be added point by point as a check. Clearly, the sum of v_C and v_R at any time t must be equal to 1000 V. For example, at $t=3$ sec, $v_C = 950$ V, $v_R = 50$ V, and thus $v_C + v_R = 1000$ V.

Exercises 13.4

1. Use Eq. (13.5) to evaluate and plot i in Figure 13.8 for the following element values:

   (a) $R = 20\ \Omega$
   $C = 5000\ \mu F$
   $V_S = 20$ V

   (b) $R = 500\ \Omega$
   $C = 1000\ \mu F$
   $V_S = 100$ V

   (c) $R = 2\ k\Omega$
   $C = 400\ \mu F$
   $V_S = 50$ V

   (d) $R = 5\ k\Omega$
   $C = 1\ \mu F$
   $V_S = 25$ V

   (e) $R = 1\ k\Omega$
   $C = 330\ \mu F$
   $V_S = 5$ V

   (f) $R = 47\ k\Omega$
   $C = 33\ \mu F$
   $V_S = 12$ V

2. Use Eq. (13.6) to evaluate and plot v_R for the circuit values given in Problem 1(a) through 1(f).

3. Use Eq. (13.7) to evaluate and plot v_C for the circuit values given in Problem 1(a) through 1(f).

4. Plot both v_R and v_C on the same graph to verify the KVL equation $V_S = v_R + v_C$ for Problem 1(a) through 1(f).

5. It is generally agreed that for all practical purposes the transient period is over after five time constants. That is, the current in Figure 13.8 has decayed to nearly zero when $t = 5\tau$. Can you determine the *percent error* if you assume that $i(5\tau) = 0$? (*Hint:* $\% \text{ ERR} = \dfrac{|\text{ideal-actual}|}{\text{actual}} \times 100$.) Now, ideal assumes that i decays from V_S/R to zero, for a total decay of V_S/R in 5τ. But actually, i decays from V_S/R down to

$$\frac{V_S}{R} e^{-5\tau/\tau}$$

for a total decay of

$$\frac{V_S}{R} - \frac{V_S}{R} e^{-5\tau/\tau}$$

UNIT 14
QUADRATICS

14.1 QUADRATIC EQUATIONS

An equation in one unknown that contains the unknown raised to the second power (squared) is called a *second-degree equation*, or a *quadratic equation*. Some examples:

$P = I^2 R$ is a quadratic equation if I is the unknown and P and R are known.

$P = \dfrac{V^2}{R}$ is a quadratic equation if V is the unknown and P and R are known.

$7i^2 - 28 = 0$ is a quadratic equation where i is the unknown.

$2x^2 + 7x - 3 = 0$ is a quadratic equation where x is the unknown.

The last example, equation $2x^2 + 7x - 3 = 0$, is called a *complete quadratic*, since it contains both the first power and the second power (square) of x. The first three examples are called *incomplete quadratics*, since they do not contain the first power of the unknown.

Finding a solution for an incomplete quadratic is straightforward, as the following examples will show. Notice that there will always be two possible solutions, or two *roots* for a quadratic equation.

EXAMPLE 14.1. Solve the given incomplete quadratics:
(a) $5x^2 = 125$
(b) $7i^2 - 28 = 0$
(c) $P = V^2/R$, solve for V
(d) $\dfrac{1}{x} - x = x$

SOLUTION.

(a) Solving for x:

	$5x^2 = 125$
Divide both sides by 5	$x^2 = 25$
Take the square root	$x = \pm 5$

(b) Solving for i:

	$7i^2 - 28 = 0$
Transposing	$7i^2 = 28$
Divide both sides by 7	$i^2 = 4$
Take the square root	$i = \pm 2$

(c) Solving for V:

	$P = V^2/R$
Multiply both sides by R	$PR = V^2$
Take the square root	$V = \pm \sqrt{PR}$

(d) Solving for x:

	$\dfrac{1}{x} - x = x$
Multiply both sides by x	$1 - x^2 = x^2$
Transposing	$1 = 2x^2$
Divide both sides by 2	$1/2 = x^2$
Take the square root	$x = \pm 1/4$

In order to solve a complete quadratic equation, the following principle is used:

> If the product of two or more factors is zero, then one or more of the factors must equal zero.

According to this principle, if there are two factors, say P and Q, and if $P \cdot Q = 0$, then either $P = 0$, or $Q = 0$, or both $P = 0$ and $Q = 0$.

Now, if a quadratic equation can be factored into two factors, this principle can be used to solve for the two roots of the equation. Here is how it works. Suppose the complete quadratic equation is given as $x^2 + 5x + 4 = 0$. Using previous results, this equation is easily factored into two factors as

$$x^2 + 5x + 4 = 0 \quad \text{(original)}$$
$$(x+1)(x+4) = 0 \quad \text{(factored)}$$

Now, according to the principle given above, if $(x+1)(x+4) = 0$, then either $(x+1) = 0$, or $(x+4) = 0$, or both. This leads to two possible equations:

$$x + 1 = 0 \quad \text{and} \quad x + 4 = 0$$

The solving of these two equations will provide the roots of the given equation. That is:

$$x + 1 = 0 \qquad x + 4 = 0$$
$$x = -1 \qquad x = -4$$

The roots of the given equation are therefore -1 and -4.

The roots should always be substituted back into the original equation to verify the results. Thus:

$$(\text{for } x = -1) \qquad\qquad (\text{for } x = -4)$$

$$x^2 + 5x + 4 = 0 \qquad\qquad x^2 + 5x + 4 = 0$$
$$(-1)^2 + 5(-1) + 4 = 0 \qquad (-4)^2 + 5(-4) + 4 = 0$$
$$1 - 5 + 4 = 0 \qquad\qquad 16 - 20 + 4 = 0$$
$$0 = 0 \qquad\qquad\qquad 0 = 0$$

This principle is now applied to solve three different quadratic equations in the next example.

EXAMPLE 14.2. Use factoring to find roots for the following quadratics:
(a) $x^2 - 3x + 2 = 0$
(b) $i^2 - 9 = 0$
(c) $3x^2 - 3x - 18 = 0$

SOLUTION.
(a) $x^2 - 3x + 2 = 0$

$(x-1)(x-2) = 0$ (factoring)

$(x-1) = 0 \quad (x-2) = 0$ (taking each factor)

$x = 1 \quad \text{or} \quad x = 2$ (solving)

Checking results:

$$\text{(for } x=1) \qquad \text{(for } x=2)$$
$$(1)^2 - 3(1) + 2 = 0 \quad (2)^2 - 3(2) + 2 = 0$$
$$1 - 3 + 2 = 0 \qquad 4 - 6 + 2 = 0$$
$$0 = 0 \qquad\qquad 0 = 0$$

(b) $i^2 - 9 = 0$

$(i+3)(i-3) = 0$ (difference of two squares)

$(i+3) = 0 \qquad (i-3) = 0$ (taking each factor)

$i = -3 \qquad\quad i = 3$ (solving)

Checking results:

$$\text{(for } i = -3) \qquad \text{(for } i = 3)$$
$$(-3)^2 - 9 = 0 \qquad (3)^2 - 9 = 0$$
$$9 - 9 = 0 \qquad\quad 9 - 9 = 0$$
$$0 = 0 \qquad\qquad 0 = 0$$

(c) To facilitate factoring, first divide out 3 from each term:

$$3x^2 - 3x - 18 = 0$$

$x^2 - x - 6 = 0$ (dividing each term by 3)

$(x+2)(x-3) = 0$ (factoring)

$(x+2) = 0 \qquad (x-3) = 0$ (taking each factor)

$x = -2 \qquad\quad x = 3$ (Solving)

Checking results:

$$\text{(for } x = -2) \qquad\qquad \text{(for } x = 3)$$
$$3(-2)^2 - 3(-2) - 18 = 0 \quad 3(3)^2 - 3(3) - 18 = 0$$
$$12 + 6 - 18 = 0 \qquad\qquad 27 - 9 - 18 = 0$$
$$0 = 0 \qquad\qquad\qquad 0 = 0$$

Exercises 14.1

1. Solve the following incomplete quadratics. Check your results by substituting the roots into the original equation.

   (a) $i^2 = 64$ (b) $v^2 = 121$

(c) $4x^2 = 64$

(d) $27 = 3t^2$

(e) $10i^2 = 75$

(f) $\dfrac{V^2}{7} = 3$

(g) $4i^2 - 5 = 15$

(h) $4x = \dfrac{25}{x}$

(i) $\dfrac{V}{2} + \dfrac{4}{V} = V$

(j) $\dfrac{7}{x^2 + 1} = 3$

(k) $\dfrac{1}{x^2 + 1} = \dfrac{1}{2x^2 - 1}$

(l) $(t + 1)(t - 1) = 3t^2 - 33$

(m) Solve for i in $P = i^2 R$

(n) Solve for r in $A = \pi r^2$

(o) Solve for t in $S = \dfrac{1}{2} g t^2$

(p) Solve for d in $A = \dfrac{1}{4} \pi d^2$

(q) Solve for I in $R I^2 = 28$

(r) Solve for V in $P = \dfrac{V^2}{10^4}$

(s) Solve for v in $\dfrac{12 v^2}{R} = 49$

(t) Solve for d in $F = \dfrac{k M_1 M_2}{d^2}$

(u) Solve for x in $\dfrac{1}{x^2} + \dfrac{1}{y^2} = \dfrac{1}{z^2}$

2. Solve the following quadratics by factoring. Check the results by substituting the roots into the original equation.

(a) $x^2 + 5x + 4 = 0$

(b) $x^2 + 3x - 4 = 0$

(c) $x^2 - 7x + 10 = 0$

(d) $2x^2 - 5x - 3 = 0$

(e) $3i^2 + 9i - 54 = 0$

(f) $6v^2 + 12v + 6 = 0$

(g) $y^2 + 7y + 10 = 0$

(h) $a^2 + 7a + 12 = 0$

(i) $p^2 - 7p + 10 = 0$

(j) $Q^2 - 12Q = -35$

(k) $v^2 - 12v + 36 = 0$

(l) $R^2 - 9R + 14 = 0$

(m) $i^2 + 4i - 5 = 0$

(n) $b^2 - 4b - 21 = 0$

(o) $e^2 + 2e - 24 = 0$

(p) $z^2 + 5z - 150 = 0$

(q) $\theta^2 + 2\theta - 48 = 0$

(r) $i^2 + 1.5i + 0.5 = 0$

(s) $v^2 + v + \dfrac{1}{4} = 0$

(t) $8 - 3a + a^2 = 6$

(u) $5s^2 - 13s - 4 = 2$

(v) $i^2 + 4i = 0$

(w) $\dfrac{v^2}{4} = 3v$

(x) $4Q^2 - 1 = 0$

14.2 QUADRATIC FORMULA

A complete quadratic equation can be written in the general form

$$ax^2 + bx + c = 0 \qquad (14.1)$$

where the coefficients a, b, and c can have any numerical values. A solution for this equation can be used to solve any quadratic equation whatever.

To find a solution for this general quadratic equation, a technique known as "completing the square" will be used. Here is how it is done. First, transpose the c,

$$ax^2 + bx = -c$$

Then divide each term by a to obtain

$$x^2 + \frac{b}{a}x = -\frac{c}{a}$$

Now add $(b/2a)^2$ to each side,

$$x^2 + \frac{b}{a}x + \left(\frac{b}{2a}\right)^2 = \left(\frac{b}{2a}\right)^2 - \frac{c}{a}$$

This last step "completes the square," since the left side of the equation is now exactly equal to the square of the term $(x + b/2a)$.... Try it and see. Then rewrite the equation as

$$\left(x + \frac{b}{2a}\right)^2 = \left(\frac{b}{2a}\right)^2 - \frac{c}{a}$$

Simplify the right side of this equation by combining terms,

$$\left(x + \frac{b}{2a}\right)^2 = \frac{b^2}{4a^2} - \frac{c}{a}$$
$$= \frac{b^2 - 4ac}{4a^2}$$

Now take the square root of each side to obtain

$$x + \frac{b}{2a} = \frac{\pm\sqrt{b^2 - 4ac}}{2a}$$

Solving for x yields the *quadratic formula*:

QUADRATIC FORMULA

$$x = \frac{-b \pm \sqrt{b^2 - 4ac}}{2a}$$

The quadratic formula can always be used to find the two roots of any quadratic equation. Naturally, you would not like to go through the process of "completing the square" each time you want to solve a quadratic equation. Therefore you should *memorize* this formula so it can be used at any time you find the need.

The quadratic formula is quite easy to use—simply substitute the appropriate numerical values of a, b, and c and evaluate. Note carefully that there are *two* roots provided by the $\pm$ in front of the square root sign. Here are some examples.

EXAMPLE 14.3. Solve $x^2 + 3x + 2 = 0$ using the quadratic formula.

SOLUTION. The coefficients are: $a = 1$, $b = 3$, $c = 2$. Using the quadratic formula,

$$x = \frac{-3 \pm \sqrt{(3)^2 - 4(1)(2)}}{2(1)}$$

$$= \frac{-3 \pm \sqrt{9 - 8}}{2} = \frac{-3 \pm 1}{2}$$

Thus

$$x = \frac{-3 + 1}{2} = -1$$

or

$$x = \frac{-3 - 1}{2} = -2$$

The two roots are thus -1 and -2. They should be checked by substitution into the original quadratic equation.

Sometimes it will be necessary to rewrite the equation in the standard form like Eq. (14.1), as in the next example.

EXAMPLE 14.4. Solve, using the quadratic formula:

$$\frac{1}{x+1} - \frac{x}{2} = 2$$

SOLUTION. First of all rearrange this equation into the standard form of Eq. (14.1). Combine terms on the left side by multiplying each term by the LCD $2(x + 1)$:

$$2(x+1) \times \frac{1}{x+1} - 2(x+1) \times \frac{x}{2} = 2(x+1)2$$

$$2 - x(x+1) = 4(x+1)$$

Multiply out, and transpose terms,

$$2 - x^2 - x = 4x + 4$$

$$x^2 + 5x + 2 = 0$$

Now, comparing with Eq. (14.1),

$$a = 1 \qquad b = 5 \qquad c = 2$$

Using the quadratic formula,

$$x = \frac{-5 \pm \sqrt{(5)^2 - 4(1)(2)}}{2(1)}$$

$$= \frac{-5 \pm \sqrt{17}}{2} = \frac{-5 \pm 4.123}{2}$$

Thus,

$$x = \frac{-5+4.123}{2} = -0.4385$$

or $\quad x = \dfrac{-5-4.123}{2} = -4.562$

You should check these roots by substitution into the original equation.

Here are two more examples making use of the quadratic formula.

EXAMPLE 14.5.   Solve, using the quadratic formula:
(a) $2x^2 - 1 = 0$
(b) $x^2 + 3x = 0$

SOLUTION.
(a) The coefficients are:

$$a = 2 \qquad b = 0 \qquad c = -1$$

Thus,

$$x = \frac{-0 \pm \sqrt{(0)^2 - 4(2)(-1)}}{2(2)}$$

$$= \frac{\pm\sqrt{8}}{4} = \pm 0.707$$

So, $x = +0.707$
or $x = -0.707$
(b) The coefficients are:

$$a = 1 \qquad b = 3 \qquad c = 0$$

Thus,

$$x = \frac{-3 \pm \sqrt{(3)^2 - 4(1)(0)}}{2(1)} = \frac{-3 \pm 3}{2}$$

So, $\quad x = \dfrac{-3+3}{2} = 0$

or $\quad x = \dfrac{-3-3}{2} = -3$

Again, the solutions should be checked by substitution into the original quadratic equation.

Exercises 14.2

Use the quadratic formula to solve the following equations. Check by substitution. When necessary, round off to the nearest hundredth.

1. $x^2 + 5x + 4 = 0$ 　　　　　　　2. $x^2 - 3x - 4 = 0$

3. $x^2 - 7x + 10 = 0$

4. $2i^2 - 5i - 3 = 0$

5. $3y^2 + 9y - 54 = 0$

6. $6v^2 + 12v + 6 = 0$

7. $s^2 + 4s = 0$

8. $4v^2 - 1 = 0$

9. $e^2 + 2e = 24$

10. $2a^2 + 14a + 24 = 0$

11. $v^2 = 6(2v - 6)$

12. $2\theta = 48 - \theta^2$

13. $\dfrac{R^2}{4} = 3R$

14. $8 - 3x + x^2 = 6$

15. $t^2 + 1.5t + 0.5 = 0$

16. $x^2 + 3x - 1 = 0$

17. $x^2 + 3x + 1 = 0$

18. $2s^2 + 5s + 1 = 0$

19. $2y^2 = 2 - 6y$

20. $\dfrac{x^2}{2} = 7$

21. $i^2 - 5 = 0$

22. $\dfrac{y}{2} + \dfrac{2}{y} = 3$

23. $(x + 1)(x + 2) = 3$

24. $(s + 2) = \dfrac{1}{(s + 3)}$

25. $\dfrac{1}{x^2} - 2 = \dfrac{1}{x}$

26. $\theta^2 + 3\theta = 1$

27. $\dfrac{1}{x + 1} + \dfrac{1}{x - 1} = 1$

14.3   COMPLEX ROOTS

As discussed in a previous section, the operator j is used to evaluate the square root of a negative number. In fact, the square root of a negative number has been defined as an imaginary number. The roots for a quadratic equation may include an imaginary number, since the quadratic formula contains a radical. Consider the equation

$$x^2 + 4 = 0$$

Transposing the 4,

$$x^2 = -4$$

Taking the square root of both sides,

$$x = \pm\sqrt{-4} = \pm j2$$

and the roots are seen to be imaginary numbers. This equation could have been solved using the quadratic formula by noting $a = 1$, $b = 0$, and $c = 4$. Thus,

$$x = \frac{-(0) \pm \sqrt{(0)^2 - 4(1)(4)}}{2(1)} = \frac{\pm\sqrt{-16}}{2} = \pm j2$$

To check on the validity of these two roots, substitute them back into the original equation as follows (recall $j^2 = -1$):

(for $x = +j2$)	(for $x = -j2$)
$x^2 + 4 = 0$	$x^2 + 4 = 0$
$(j2)^2 + 4 = 0$	$(-j2)^2 + 4 = 0$
$j^2 4 + 4 = 0$	$j^2 4 + 4 = 0$
$(-1)4 + 4 = 0$	$(-1)4 + 4 = 0$
$0 = 0$	$0 = 0$

Consider the equation $x^2 - 2x + 2 = 0$. Note that $a = 1$, $b = -2$, $c = 2$, and use the quadratic formula to obtain

$$x = \frac{-(-2) \pm \sqrt{(-2)^2 - 4(1)(2)}}{2(1)} = \frac{2 \pm \sqrt{-4}}{2} = \frac{2 \pm j2}{2} = 1 \pm j$$

The two roots for this equation are thus $(1+j)$ and $(1-j)$. Since each root contains *both* a real number (1) and an imaginary number (j), each root is said to be a *complex number*. In general, any number of the form $a + jb$ is referred to as a *complex number*.*

COMPLEX NUMBER

$$\underbrace{a}_{\substack{\text{Real} \\ \text{part}}} + \underbrace{jb}_{\substack{\text{Imaginary} \\ \text{part}}}$$

Thus the two roots to the equation $x^2 - 2x + 2 = 0$ are the two complex numbers $(1+j)$ and $(1-j)$. To verify the solutions, substitute the two roots back into the original equation to obtain:

(for $x = 1 + j$)	(for $x = 1 - j$)
$x^2 - 2x + 2 = 0$	$x^2 - 2x + 2 = 0$
$(1+j)^2 - 2(1+j) + 2 = 0$	$(1-j)^2 - 2(1-j) + 2 = 0$
$(1 + 2j + j^2) - 2 - 2j + 2 = 0$	$(1 - 2j + j^2) - 2 + 2j + 2 = 0$
$(1 + 2j - 1) - 2 - 2j + 2 = 0$	$(1 - 2j - 1) - 2 + 2j + 2 = 0$
$0 = 0$	$0 = 0$

---

*Complex number operations will be discussed in a following section. Notice that the number $0 + jb = jb$ can be thought of as an imaginary number (jb), or a complex number ($0 + jb$), with its real part equal to zero.

To summarize, the following observations are made as an aid for solving for the roots of a quadratic equation:

QUADRATIC ROOTS

Every quadratic equation has exactly two roots, and these roots are either (1) real and unequal, (2) real and equal, or (3) complex.

QUADRATIC SOLUTIONS

In order to determine the roots for a given quadratic equation, utilize the following techniques:
(1) Solve equations of the form $x^2+b=0$ by transposing the b and taking the square root of both sides.
(2) Try to solve by factoring.
(3) Use the quadratic formula.

The following examples illustrate applications of these techniques.

EXAMPLE 14.6. Solve the quadratic equations:
(a) $x^2-25=0$
(b) $x^2+2x+1=0$
(c) $x^2+2x+2=0$

SOLUTION.
(a) Transposing,

$$x^2-25=0$$
$$x^2=25$$
$$x=\pm 5$$

Check:

$$x=5 \qquad x=-5$$
$$(5)^2-25=0 \quad (-5)^2-25=0$$
$$0=0 \qquad\qquad 0=0$$

(b) Factoring,

$$x^2+2x+1=0$$
$$(x+1)(x+1)=0$$

The two roots are both equal to -1. Check:

$$x=-1$$
$$(-1)^2+2(-1)+1=0$$
$$1-2+1=0$$
$$0=0$$

(c) Using the quadratic formula, $x^2+2x+2=0$,

$$x = \frac{-(2) \pm \sqrt{(2)^2 - 4(1)(2)}}{2(1)}$$

$$= \frac{-2 \pm \sqrt{-4}}{2} = -1 \pm j$$

The roots are $-1+j$ and $-1-j$.
Check

(for $x=-1+j$)	(for $x=-1-j$)
$x^2+2x+2=0$	$x^2+2x+2=0$
$(-1+j)^2+2(-1+j)+2=0$	$(-1-j)^2+2(-1-j)+2=0$
$(1-2j+j^2)-2+2j+2=0$	$(1+2j+j^2)-2-2j+2=0$
$(1-2j-1)-2+2j+2=0$	$(1+2j-1)-2-2j+2=0$
$0=0$	$0=0$

Exercises 14.3

1. Determine by substitution whether the given number is a root of the given equation for each of the following:

   (a) $(-1-j)$ $x^2+2x+2=0$ (d) $(3+2j)$ $x^2=6x-13$

   (b) $(2+j)$ $x^2+4x+4=0$ (e) $(2+3j)$ $x^2=4x-13$

   (c) (2) $x^2-4x+4=0$ (f) $\left(\frac{1}{2}-\frac{1}{2}j\right)$ $x^2=\frac{2x-1}{2}$

2. Solve the following quadratic equations for their complex roots. Verify each solution by substitution into the original equation.

   (a) $x^2+2x+2=0$ (g) $4y^2=-4y-5$
   (b) $x^2-6x+10=0$ (h) $9s^2-6s=-2$
   (c) $x^2-6x+13=0$ (i) $\omega^2-\omega+1=0$
   (d) $x^2-4x+13=0$ (j) $V^2+25=8V$
   (e) $2x^2-2x+1=0$ (k) $I^2=\frac{6I-5}{2}$
   (f) $x^2-2x+17=0$ (l) $1+2P+3P^2=0$

3. Solve each equation by transposing terms, factoring, or using the quadratic equation. Use the method that seems best to you. Express decimal answers to the nearest hundredth. Verify each solution by substitution into the original equation.

   (a) $x^2-49=0$ (c) $4I^2=100$
   (b) $x^2+49=0$ (d) $R^2+6R=-9$

(e) $25 - 16I^2 = 0$

(f) $\dfrac{-V^2}{100} = 4$

(g) $2t^2 = 17t - 21$

(h) $50 = 5P + P^2$

(i) $\dfrac{8 + 2v}{3} = v^2$

(j) $3s^2 + 7s - 1 = 0$

(k) $V^2 + 4V - 4 = 0$

(l) $3 - 8Q = 16Q^2$

(m) $16I = 12I^2 - 3$

(n) $v^2 + v + 2 = 0$

(o) $v^2 - v + 2 = 0$

(p) $\dfrac{v^2}{5} - \dfrac{v}{30} - \dfrac{1}{2} = 0$

(q) $\dfrac{1}{R - 1} - \dfrac{1}{R + 1} = \dfrac{1}{24}$

(r) $1.2x^2 - 1.1x + 2.3 = 0$

4. The power dissipated in a resistance R can be found from the equation $P = I^2 R$. If the power dissipated is 100 W with a current of 5 A, what will be the power dissipation when the current is:

(a) Doubled?   (b) Halved?

5. Two resistors must be chosen such that their sum is 100 Ω, and their product is 1000 Ω. Find the two resistor values.

6. A diffused resistor in an integrated circuit is rectangular in shape. If the perimeter must be 500 mils, and the area must be 3375 square mils, find its dimensions.

7. A 12-volt dc source having an internal resistance of 10 Ω has an unknown resistance R connected across its terminals. Calculate the required value of R if the power dissipation in R is:

(a) 1 W   (b) 2 W   (c) 3 W   (d) 3.6 W.

Hint: Use Ohm's Law to find $I = (V_S)/(R + 10)$. Then the power in R is given by $P = I^2 R$, and by substitution,

$$P = \left(\frac{V_S}{R + 10} \right)^2 R$$

Write this equation in the standard quadratic form with R as the unknown variable.

8. Make a plot of power dissipation in R (on the vertical axis) versus the values of R (on the horizontal axis) for the data points in Problem 7. Note that there are two possible values of R for each value of P calculated. The curve will show that the *maximum* power is dissipated in R when $R = 10$ Ω, and the power is less than this maximum value for all other values of R. This agrees with the well-known Maximum Power Transfer Theorem, which states that

the maximum power is delivered to a load when the load resistance is equal to the source resistance.

9. John can install an electrical panel in 9 hours less time than Bill. Together they can do the installation in 20 hours. How long would it take each one to do the job alone?

10. It takes technician A 3 hours longer to wire a printed circuit board than it takes technician B. Working together, they can wire the board in 8 hours. How long would it take each one to wire the board alone?

TRIGONOMETRY—
APPLICATIONS

UNIT 1
ANGLES AND TRIANGLES

The word *trigonometry* is derived from the three Greek words *tri* (three), *gonia* (angle), and *metron* (measure). Trigonometry is the branch of mathematics dealing with triangles. The six ratios defined as the *trigonometric functions* of an angle are used to specify the relationships that exist among the angles and sides of a triangle. These trigonometric functions not only form the basis for solving problems involving triangles, but they are also used in many diverse areas of mathematics, science, and engineering. A study of trigonometry will provide the basic mathematical tools necessary to solve ac circuit problems.

1.1 ANGLES

An *angle* can be thought of as an "opening" between two lines. Thus in Figure 1.1, the two lines AO and BO are called the sides of the angle, and the point O is its *vertex*.

It is customary to use the mathematical symbol $\angle$ to denote the word *angle*, and the angle formed in Figure 1.1 can be designated in three ways: (1) Use the three letters defining the sides and write them *in order* with the vertex letter in the middle; thus, you can write $\angle AOB$ or $\angle BOA$. (2) Simply use the letter written at the vertex and write $\angle O$. (3) Use the small letter (usually a Greek letter) written inside the angle; in this case, $\angle \alpha$. The α is the Greek letter alpha.

The intersection of two straight lines will generate four angles as shown in Figure 1.2(a). The angles in this case are called α (alpha), β (beta), θ (theta), and ϕ (phi). Notice that $\angle \alpha = \angle \theta$ and $\angle \beta = \angle \phi$. If the two lines are arranged such that their intersection generates four equal angles, the lines are said to be *perpendicular* to one another, and the angles are defined as *right angles*. Thus in Figure 1.2(b), the two lines are perpendicular, the four angles are equal ($\alpha = \beta = \theta = \phi$), and each angle is defined as a right angle. In general, a right angle is denoted in a drawing by means of a small square in the vertex as shown for the right angle α in Figure 1.2(c).

Any angle that is smaller than a right angle is defined as an *acute angle*. For example, α in Figure 1.3(a) is an acute angle.

An *obtuse angle* is an angle greater than a right angle. The angle β in Figure 1.3(b) is an obtuse angle.

Two angles are said to be *complementary* if their sum is equivalent to a right angle. For example, angles α and β in Figure 1.3(c) are complementary. Angle α is the complement of β, or equivalently, β is the complement of α.

Two angles are said to be *supplementary* if their sum is equivalent to a straight line. For example, angles θ and ϕ in Figure 1.3(d) are supplementary angles. Angle θ is the supplement of ϕ, and ϕ is the supplement of θ. Notice that two right angles are supplementary since their sum is equivalent to a straight line as shown in Figure 1.3(e).

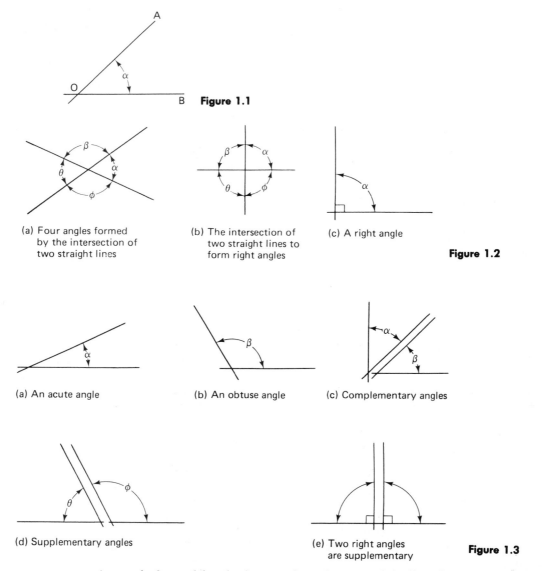

Figure 1.1

(a) Four angles formed by the intersection of two straight lines

(b) The intersection of two straight lines to form right angles

(c) A right angle

Figure 1.2

(a) An acute angle

(b) An obtuse angle

(c) Complementary angles

(d) Supplementary angles

(e) Two right angles are supplementary

Figure 1.3

An angle formed by the intersection of two straight lines is a concept from geometry, but the act of *generating* an angle by rotating a line segment about a point from one position to another is a technique used in trigonometry.

The angle θ in Figure 1.4 was generated by rotating the line segment OA about the point O from its *initial position* on the horizontal line XX' to its *terminal position* as shown. The two sides of θ are called the *initial side* and the *terminal side*, and the point O is called the *vertex*. A *curved arrow* drawn from the initial side to the terminal side shows the direction of rotation used to generate the angle.

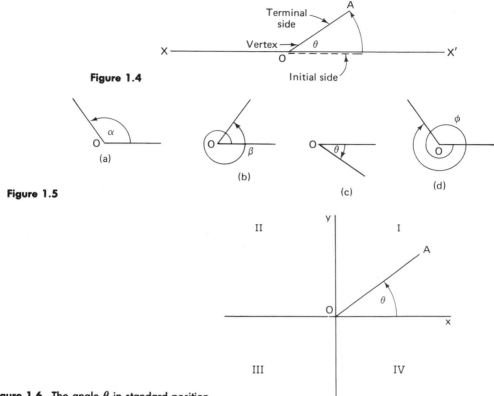

Figure 1.4

Figure 1.5

Figure 1.6. The angle θ in standard position.

Angles generated by rotation in the *counterclockwise* (CCW) direction are defined as *positive*, while rotation in a *clockwise* (CW) direction yields *negative* angles. Thus $\angle\theta$ in Figure 1.4 is a positive angle, and so are α and β in Figure 1.5(a) and (b). But θ and ϕ in Figure 1.5(c) and (d) are negative angles.

A rectangular coordinate system composed of an x-axis and a y-axis as previously discussed is shown in Figure 1.6. The four quadrants are labeled I, II, III, and IV. An angle generated by rotating the line segment OA is said to be in *standard position* on this coordinate system when its vertex is at the origin and its initial side coincides with the positive x-axis. Thus the angle θ in Figure 1.6 is in standard position.

The angle θ in Figure 1.6 is said to be in the first quadrant since its terminal side lies in the first quadrant. In general, an angle is said to be in the quadrant occupied by its terminal side. For example, the angle α in Figure 1.5(a) would be in the second quadrant if placed in standard position. The angle θ in Figure 1.5(c) would be in the fourth quadrant if placed in standard position.

In Figure 1.7(a) the line segment OA has been rotated CCW through exactly one revolution, and the tip of the line describes a circle during rotation. If the circle

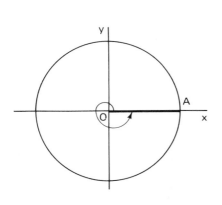

(a) One complete revolution

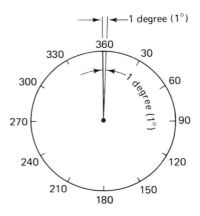

(b) 1° = 1/360 of a circle

Figure 1.7

generated by *OA* is divided into 360 equal angles as shown in Figure 1.7(b), each angle will be exactly equal to 1 *degree* (1°).

The *degree* is the most common unit of angular measure, and it is defined as follows:

> One degree (1°) is the angle formed by a rotation through 1/360 part of a circle.

The angle formed by exactly one revolution in Figure 1.7(a) is equal to 360 degrees (written as 360°). One-half a revolution is equal to 180°. A right angle is exactly 1/4 of a revolution; so a right angle is clearly 360°/4=90°.

EXAMPLE 1.1. Express the following angles in units of degrees:	SOLUTION.
(a) 1/8 of a revolution in a counterclockwise (CCW) direction.	(a) 1/8 of a revolution is equal to 360°/8 = 45°. The angle is positive due to the direction of rotation as shown in Figure 1.8(a).
(b) 1/8 of a revolution in a clockwise (CW) direction.	(b) The angle is again 360°/8=45° in magnitude, but it is written as −45° since the direction of rotation is CW. Notice that it is in the fourth quadrant as shown in Figure 1.8(b).
(c) 3/4 of a revolution CCW.	(c) This angle is $360° \times \frac{3}{4} = 270°$, and it is drawn in Figure 1.8(c).

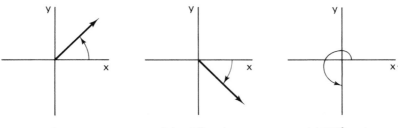

Figure 1.8   (a) 45° angle   (b) − 45° angle   (c) 270° angle

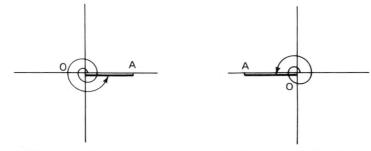

Figure 1.9   (a) Two revolutions of   (b) One and one-half revolutions
                 OA generate a 120° angle   of OA generate a 540° angle

Since a single revolution of *OA* in Figure 1.7(a) generates a 360° angle, two revolutions will generate an angle equal to 360° + 360° = 720° as shown in Figure 1.9(a). The segment *OA* in Figure 1.9(b) has been turned through $1\frac{1}{2}$ revolutions. The first revolution is equivalent to 360°, and the half revolution is equivalent to 180°. The angle generated is thus 360° + 180° = 540°. To determine the angle generated by rotation through more than a single revolution, simply add 360° for each revolution plus the number of degrees generated for any portion of a revolution.

EXAMPLE 1.2. Express the following angles in degrees: (a) 3 revolutions CCW. (b) $2\frac{1}{4}$ revolutions CW. (c) 10.5 revolutions CCW.	SOLUTION. (a) Since each revolution generates 360°, 3 revolutions will generate 360° × 3 = 1080°. The angle is positive due to the CCW direction of rotation. (b) $2\frac{1}{4}$ revolutions will generate 360° × $2\frac{1}{4}$ = 360° + 360° + 90° = 810°. The angle is − 810° since the direction of rotation is CW. (c) 10.5 revolutions will generate an angle equal to 360° × 10.5 = 3780°.

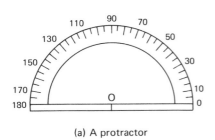

(a) A protractor

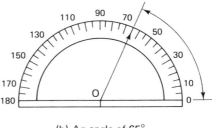

(b) An angle of 65°

Figure 1.10

For more accurate angular measurement, each degree is divided into 60 equal parts called *minutes*, and each minute is further divided into 60 equal parts called *seconds*. Thus an angle might be specified as 48 degrees, 36 minutes, and 24 seconds (written as 48°36′24″).

In electric circuit problems it is common practice to divide the degree *decimally*, instead of using minutes and seconds. Thus the angle 48°36′ would be written as 48.6° (36′ = 36/60 = 0.6°).

The *protractor* shown in Figure 1.10(a) is an instrument for measuring or constructing angles. In practice, the protractor is placed with its center point (labeled 0) placed at the vertex of the angle to be measured (or drawn). The protractor is then rotated until the mark for 0° on its scale is over one side of the angle, and the other side of the angle is on the protractor scale. For example, an angle of 65° is indicated on the protractor in Figure 1.10(b).

The following summary is included as an aid in learning the new terms introduced in this first section on trigonometry.

SUMMARY

ACUTE—An angle smaller than a right angle.

ANGLE—An opening between two lines, or a measure of the rotation of a line segment about a fixed point. The symbol for an angle is ∠ .

CCW—Counterclockwise.

COMPLEMENTARY ANGLES—Two angles are complementary if their sum is equivalent to a right angle (90°).

CW—Clockwise.

DEGREE—A unit of angular measure. One degree (1°) is the angle formed by a rotation through 1/360 part of a circle.

OBTUSE ANGLE—An angle greater than a right angle.

PERPENDICULAR—Two straight lines are perpendicular if their intersection generates four equal angles.

PROTRACTOR—An instrument for measuring or constructing angles.

RIGHT ANGLE—The angle formed by the intersection of two perpendicular straight lines. An angle of exactly 90°.

STANDARD POSITION—An angle located with its vertex at the origin of a rectangular coordinate system and its initial side coincident with the positive *x*-axis.

SUPPLEMENTARY ANGLES—Two angles are supplementary if their sum is equivalent to a straight line (180°).

TRIGONOMETRY—The branch of mathematics dealing with the relationships existing between the sides and angles of triangles.

Exercises 1.1

1. What is the complement of:
   (a) 45°?   (b) 60°?   (c) 90°?   (d) 150°?   (e) −22°?
   (f) 275°?

2. What is the supplement of:
   (a) 30°?   (b) 45°?   (c) 90°?   (d) 180°?   (e) 275°?
   (f) −25°?

3. Express the following angles in degrees:
   (a) 1/6 revolution CCW.
   (b) 1/6 revolution CW.
   (c) 1/12 revolution CCW.
   (d) 5/6 revolution CCW.
   (e) 5/18 revolution CW.
   (f) 2/3 revolution CW.

4. Draw freehand each of the angles in Problem 3 in its standard position and state in which quadrant the angle lies.

5. Express the following angles in degrees:
   (a) 4 revolutions CCW.
   (b) $1\frac{3}{4}$ revolutions CCW.
   (c) $3\frac{1}{6}$ revolutions CCW.
   (d) 5.5 revolutions CW.
   (e) 4.7 revolutions CW.
   (f) 10^3 revolutions CCW.

6. The second hand of a clock will rotate through an angle of how many degrees in:
   (a) 30 sec?   (b) 15 sec?   (c) 20 sec?   (d) 10 sec?   (e) 5 sec?
   (f) 1 sec?

7. Use a protractor to construct the following angles in standard position. Label the angles and draw a curved arrow to show the direction of rotation to generate each angle.

(a) 30°   (b) 75°   (c) 142°   (d) 215°
(e) 345°   (f) −30°   (g) −210°   (h) −300°
(i) 438°   (j) −600°

1.2 RADIAN MEASURE

In the previous section the degree is used as the basic unit for angles. Another system for angular measurement, called *radian measure*, is used extensively in electricity and electronics. The basic unit of angular measure in this system is the *radian* (rad).

The angle *AOB* in Figure 1.11 is equal to one radian. It is generated by rotating the line segment *OA* until the length of the *arc AB* is exactly equal to the length of the line segment *OA*. The radian can be defined as:

RADIAN

An angle with its vertex at the center of a circle is equal to one radian if its sides intercept an arc on the circle having a length exactly equal to the radius of the circle.

In Figure 1.11, the radius of the circle is equal to the length of the line segment *OA*. Since the angle is 1 rad, the length of arc *AB* is exactly equal to the radius. Notice carefully that the arc *AB* is a portion of the circle and is therefore *not* a straight line.

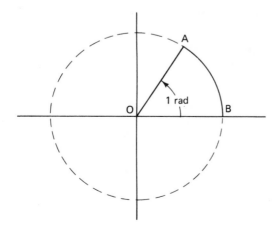

Figure 1.11

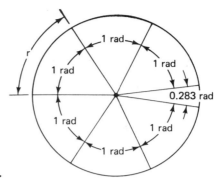

Figure 1.12. 2π (6.283) rad in one revolution.

The equation that expresses the circumference of a circle C as a function of its radius r is given by

$$C = 2\pi r \tag{1.1}$$

This equation states that the circumference of a circle is 2π times longer than the radius. In other words, there are $2\pi(6.283)$ arcs of length r around the circumference as shown in Figure 1.12. Thus the number of radians in the circle must be 2π; that is, one complete revolution of a line segment must generate an angle equal to 2π rad. Since a single revolution of a line segment also generates an angle equal to $360°$,

$$\boxed{\begin{aligned} 2\pi \text{ rad} &= 360° \\ \text{or} \quad \pi \text{ rad} &= 180° \end{aligned}} \tag{1.2}$$

Dividing both sides of the first equation by 2π yields,

$$\boxed{1 \text{ rad} = \frac{360°}{2\pi} = \frac{180°}{\pi}} \tag{1.3}$$

Carrying out the division $180/\pi$,

$$\boxed{1 \text{ rad} = 57.2958° \cong 57.3°} \tag{1.4}$$

Equations (1.2) and (1.4) can be used to convert angular measure from degrees to radians or from radians to degrees. Specifically:

RADIANS TO DEGREES

To convert radians to degrees, multiply the number of radians by $180/\pi$ for absolute accuracy; or multiply by 57.3 for accuracy to three significant figures.

DEGREES TO RADIANS

To convert degrees to radians, divide the number of degrees by $180/\pi$ for absolute accuracy; or divide by 57.3 for accuracy to three significant figures.

EXAMPLE 1.3. Convert the given angles to degrees. As a check, convert them back into radians.
(a) π rad
(b) $\pi/3$ rad
(c) 4.36 rad

SOLUTION.
(a) To convert radians to degrees, multiply by $180/\pi$. Thus,

$$\pi \text{ rad} \times \frac{180}{\pi} = 180°$$

As a check, convert 180° back into radians by dividing by $180/\pi$.

$$\frac{180°}{\dfrac{180}{\pi}} = 180° \times \frac{\pi}{180} = \pi \text{ rad}$$

(b) Multiply by $180/\pi$ to convert radians to degrees. Thus,

$$\frac{\pi}{3} \text{ rad} \times \frac{180}{\pi} = 60°$$

As a check, divide degrees by $180/\pi$ to obtain

$$\frac{60°}{\dfrac{180}{\pi}} = 60° \times \frac{\pi}{180} = \frac{\pi}{3} \text{ rad}$$

(c) For results to three significant figures, use the conversion factor 57.3. Thus,

$$4.36 \text{ rad} \times 57.3 = 250°$$

Suppose the line segment OA in Figure 1.13 is rotating at a constant rate, for instance, like a hand on a clock or like a line drawn on the end of a rotating shaft. If the segment OA rotates through θ angular units (degrees or radians) in a period of time t, then it is said to have an *angular velocity* (δ)* given by the following equation:

$$\text{Angular velocity} = \delta = \frac{\theta}{t} \qquad (1.5)$$

---

*δ is the Greek letter delta.

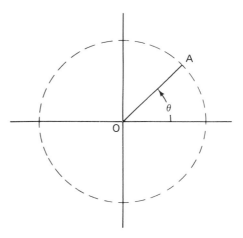

Figure 1.13

If θ is in radians, and t is in seconds, then the units of angular velocity are rad/sec. If θ is measured in degrees and t is in seconds, then δ has units of velocity in degrees/sec.

Frequently the angular velocity of a rotating body is expressed in *revolutions per second* (r/s) or *revolutions per minute* (r/min). Since one revolution is 2π rad or 360°, it is easy to convert from r/s or r/min to degrees/sec or to rad/sec. The following examples deal with angular velocity.

EXAMPLE 1.4. Express the angular velocity of the second hand on a clock in units of rad/sec.	SOLUTION. The second hand rotates at an angular velocity of 1 revolution (2π rad) for each 60 sec. Thus, using Eq. (1.5): $$\delta = \frac{\theta}{t} = \frac{2\pi}{60} = \frac{\pi}{30} \cong 0.105 \text{ rad/sec}$$
EXAMPLE 1.5. The shaft of an electric motor has an angular velocity of 3600 r/min. What is this angular velocity in rad/sec? In degrees/sec?	SOLUTION. First, change r/min to r/s by dividing by 60. Thus, $$\frac{3600 \text{ r/min}}{60} = 60 \text{ r/s}$$ Since one revolution is 2π rad, multiply by 2π to obtain, $$60 \text{ r/s} \times 2\pi = 120\pi \text{ rad/sec}$$ $$\cong 377 \text{ rad/sec}$$ One revolution is also 360°. Thus, multiply by 360 to obtain $$60 \text{ r/s} \times 360 = \frac{21{,}600°}{\text{sec}}$$

Exercises 1.2

1. If one revolution of a line segment generates an angle equal to 2π radians, what is the radian measure of the angle generated by:
   - (a) 2 revolutions?
   - (b) 10 revolutions?
   - (c) 1/4 revolution?
   - (d) 1/8 revolution?
   - (e) 5/6 revolution?
   - (f) $1\frac{2}{3}$ revolutions?

2. Calculate the circumference of a circle having a radius of:
   - (a) 2 cm   (b) 6 in.   (c) 1.72 m   (d) 31 mm

3. Calculate the radius of a circle having a circumference of:
   - (a) 7 in.   (b) 22 cm   (c) 6π ft   (d) 9.63 mm

4. What is the length of the arc AB in Figure 1.11 if the radius of the circle (OA) is 1.73 cm?

5. What must be the radius OA of the circle in Figure 1.11 if the length of the arc AB is 3.71 in.?

6. If the radius of the circle in Figure 1.12 is 10 cm, what must be the length of arc on the circle intercepted by the sides of an angle equal to:
   - (a) 1 rad?   (b) 2 rad?   (c) 6 rad?   (d) 2π rad?   (e) 1.50 rad?
   - (f) 3.75 rad?

7. Use the conversion factor $180/\pi$ to express the following angles in radians. Draw each angle in standard form on a rectangular coordinate system:
   - (a) 30°   (b) 45°   (c) 60°   (d) 90°   (e) 120°
   - (f) 135°   (g) 150°   (h) 180°

8. A right angle contains how many degrees? How many radians?

9. Use the conversion factor $180/\pi$ to express the following angles in degrees. Draw each angle in standard form on a rectangular coordinate system:
   - (a) 2π rad   (b) π rad   (c) $7\pi/6$ (d) $5\pi/3$ (e) $7\pi/4$
   - (f) $\pi/18$ (g) $7\pi/9$ (h) $11\pi/36$

10. Use the conversion factor 57.3 to express the following to three significant figures:
    - (a) 21° in rad   (b) 33.7° in rad   (c) 68.4° in rad
    - (d) 222° in rad   (e) 5° in rad

11. Use the conversion factor 57.3 to express the following angles in degrees to three significant figures:
    - (a) 0.106 rad   (b) 0.288 rad   (c) 1.12 rad
    - (d) 6 rad   (e) 3.51 rad

12. Express the angular velocity of the minute hand of a clock in units of rad/sec.

13. Through how many radians does the second hand of a clock rotate in:
    (a) 15 sec?   (b) 30 sec?   (c) 1 min?   (d) 5 min?

14. A radar antenna rotates at 36 r/min. What is its angular velocity in rad/sec? In degrees/sec? Through what angle will this antenna rotate in 5 sec?

15. A radar antenna rotates with an angular velocity of 1.5 rad/sec. Express this angular velocity in r/s. How long does it take for this antenna to sweep through a 90° angle?

1.3   TRIANGLES

A triangle can be defined as *a closed plane figure having three sides*. Each side is a segment of a straight line, and the sides intersect one another to form three vertices as shown in Figure 1.14. Thus there are three *interior angles* associated with a triangle, and they are labeled α, β, and γ in Figure 1.14. An important result from geometry provides the following relationship concerning the interior angles of a triangle:

INTERIOR ANGLES

The sum of the interior angles of a triangle is equal to 180°.

Applying this relationship to the triangle in Figure 1.14 yields the algebraic expression

$$\alpha + \beta + \gamma = 180° \tag{1.6}$$

From this expression, one angle of a triangle can be calculated provided the other two angles are known.

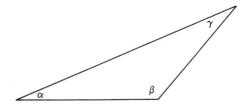

Figure 1.14. A triangle.

EXAMPLE 1.6. Find the unknown interior angle in a triangle if the two known angles are:
(a) $\alpha = 20°$ and $\beta = 120°$

SOLUTION. Use Eq. (1.6) to find the unknown angle. Thus:
(a) $\gamma = 180° - \alpha - \beta$
$\qquad = 180° - 20° - 120° = 40°$

(b) $\beta = 90°$ and $\gamma = 32°$
(c) $\alpha = 45°$ and $\gamma = 45°$

(b) $\alpha = 180° - \beta - \gamma$
$= 180° - 90° - 32° = 58°$
(c) $\beta = 180° - \alpha - \gamma$
$= 180° - 45° - 45° = 90°$

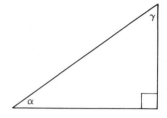

Figure 1.15. A right triangle.

If one of the interior angles of a triangle is a right angle, that triangle is said to be a *right triangle*. The right triangle in Figure 1.15 has its right angle designated with the small square drawn in the vertex, and the other two angles are labeled α and γ.

Since the sum of the interior angles in a right triangle must be 180°, and one of them is known to be 90°, Eq. (1.6) can be applied to the triangle in Figure 1.15 to obtain,

$$\alpha + 90° + \gamma = 180°$$
$$\alpha + \gamma = \quad 90° \text{ (for a right triangle)} \tag{1.7}$$

This last expression leads to the following statements regarding the two variable interior angles of a right triangle:

1. The sum of the two angles must be exactly 90°.

2. Since the sum of the two angles is 90°, each angle must be an acute angle (less than 90°).

EXAMPLE 1.7. Determine the unknown angle in the right triangle shown in Figure 1.15 if:
(a) $\alpha = 27°$
(b) $\gamma = 42°$

SOLUTION. Use Eq. (1.7) to determine the unknown angle:
(a) $\gamma = 90° - \alpha = 90° - 27° = 63°$
(b) $\alpha = 90° - \gamma = 90° - 42° = 48°$

EXAMPLE 1.8. Is it possible to construct a right triangle having the two interior angles $\alpha = 46°$ and $\gamma = 51°$?

SOLUTION. From Eq. (1.7), the sum of the two variable angles in a right angle must be 90°. But, in this case,

$$\alpha + \gamma = 46° + 51° = 97°$$

Therefore, such a right triangle is not possible.

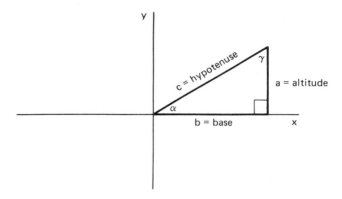

Figure 1.16. A right triangle in standard position.

A right triangle is said to be in standard position if it is drawn on a rectangular coordinate system with one of its acute angles in standard position. The right triangle in Figure 1.16 is drawn in standard position. The side c of the right triangle opposite the right angle is called the *hypotenuse*. The side a opposite the angle drawn in standard position is called the *opposite side*, or the *altitude*. The side b next to the angle in standard position is called the *adjacent side*, or the *base*.

A very important property involving the sides of a right triangle is given by the Pythagorean Theorem:

PYTHAGOREAN THEOREM

The square of the hypotenuse of a right triangle is equal to the sum of the squares of the other two sides. Thus, for the triangle in Figure 1.16,

$$c^2 = a^2 + b^2$$

The Pythagorean Theorem provides the means for determining any one of the sides of a right triangle if the other two sides are known.

EXAMPLE 1.9. For the right triangle in Figure 1.16, the altitude is 6 cm and the base is 8 cm. Determine the hypotenuse.	SOLUTION. Use the Pythagorean Theorem:

EXAMPLE 1.9.   For the right triangle in Figure 1.16, the altitude is 6 cm and the base is 8 cm. Determine the hypotenuse.

SOLUTION.   Use the Pythagorean Theorem:

$$c^2 = a^2 + b^2$$

Taking the square root:

$$c = \sqrt{a^2 + b^2}$$

Substituting values:

$$c = \sqrt{6^2 + 8^2} = \sqrt{36 + 64} = \sqrt{100} = 10 \text{ cm}$$

EXAMPLE 1.10. Find the altitude a in Figure 1.16 if $b=3.17$ in. and $c=5.73$ in.

SOLUTION. Begin with the Pythagorean Theorem:

$$c^2 = a^2 + b^2$$

Transpose: $a^2 = c^2 - b^2$

Square root: $a = \sqrt{c^2 - b^2}$

Substitution:

$$a = \sqrt{5.73^2 - 3.17^2}$$
$$= \sqrt{32.8 - 10.05}$$
$$= \sqrt{22.8}$$
$$= 4.77 \text{ in.}$$

To summarize, the following two important properties of right triangles are stated for the triangle in Figure 1.16:

> For the right triangle in Figure 1.16:
> $\alpha + \gamma = 90°$ and $c^2 = a^2 + b^2$

Exercises 1.3

1. Draw the triangles in Example 1.6.
2. Draw the triangles in Example 1.7.
3. For the triangle in Figure 1.14, solve for the unknown angle, given the two angles:

    (a) $\alpha = 31°$, $\beta = 112°$ (b) $\alpha = 66°$, $\gamma = 21°$ (c) $\gamma = 91°$, $\beta = 3°$

4. A triangle is known to have two equal interior angles of $60°$. What must be the size of the third angle? Draw the triangle.
5. For the right triangle in Figure 1.16, use the given sides and angles to solve for the unknown sides and angles:

    (a) $a=3$, $b=4$, $\alpha = 53.1°$. Find c and γ.
    (b) $b=11$, $c=21$, $\gamma = 31.6°$. Find a and α.
    (c) $a=6.32$, $c=9.11$; $\alpha = 43.9°$ .Find b and γ.
    (d) $a=4.33$, $b=2.50$, $\alpha = 30°$. Find c and γ.
    (e) $a=7$, $b=7$, $\gamma = 45°$. Find c and α.

6. Draw the triangle in Problem 5(a) in standard position with α at the origin.
7. Draw the triangles in Problem 5(d) and (e) in standard position with α at the origin.

8. A TV antenna has a 30 ft tower. The antenna must have a guy wire from a point 5 ft below its top to a point 25 ft from its base, as shown in Figure 1.17. What must be the length of the guy wire? (There must of course be two other guy wires to support the tower.)

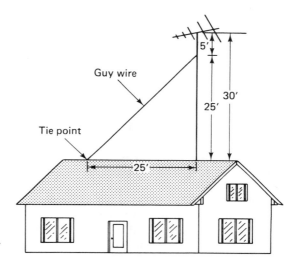

Figure 1.17

9. A second guy wire is to be connected between the midpoint on the TV mast in Figure 1.17 and the same tie point on the roof. What must be the length of this guy wire?

10. A tower 125 ft high is to be used as an antenna support. A guy wire 175 ft long is attached to its top. How far from the base of the tower will the guy point be located? See Figure 1.18.

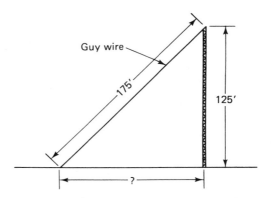

Figure 1.18. Antenna tower.

1.4   POLAR COORDINATES

A unique point on a plane can be determined by specifying the x and y coordinates of the point relative to a rectangular coordinate system. For example, the point P in Figure 1.19 is uniquely determined by specifying the x-coordinate x_1, and the y-coordinate y_1.

Another method that can be used to find the unique location of a single point is shown in Figure 1.20. In this system, a straight line is drawn from the origin O of the coordinate axes (the x-axis and y-axis in this case) to the desired point P. In order to determine the point P uniquely, two quantities must be specified:

1. The *direction* from the origin to the point P. This is determined by specifying the angle θ.

2. The *distance* from the origin to the point P. This is determined by specifying the length of the line OP, which is defined as the *radius vector* ρ (ρ is the Greek letter rho). Since the radius vector is a distance measured from the origin out to a point P, it is *always positive*.

A system that uses distance from the origin (ρ) at a given angle (θ) to uniquely determine a point is referred to as a *polar coordinate system*, and the elements ρ and θ are called the *polar coordinates* of the point. The notation $P(\rho, \theta)$ means, "the point P located at a distance ρ from the origin in a direction given by the angle θ." Polar coordinate graph paper is made specifically for plotting points in polar coordinates (see Figure 1.21).

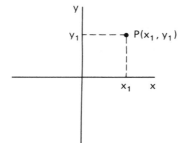

Figure 1.19. Rectangular coordinate system.

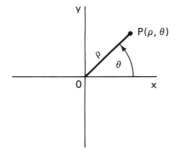

Figure 1.20. Polar coordinate system.

T. Maxeph

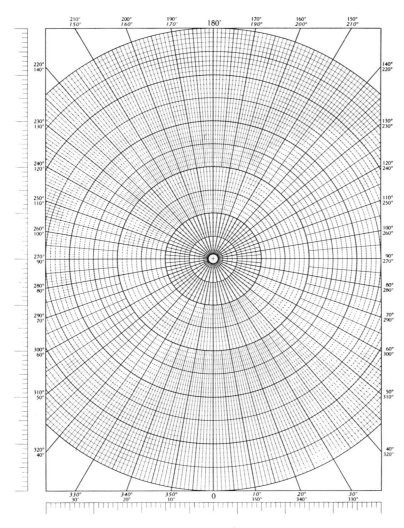

Figure 1.21.

EXAMPLE 1.11. Plot the following points on a polar coordinate system: (a) $(2, 45°)$ (b) $(3, 90°)$ (c) $(4, -135°)$ (d) $(2.5, -30°)$ (e) $(5, 200°)$	**SOLUTION.** The points are plotted in Figure 1.22. You should examine each point carefully to be certain you understand the proper determination of both ρ and θ. Note that the negative angles are measured CW.

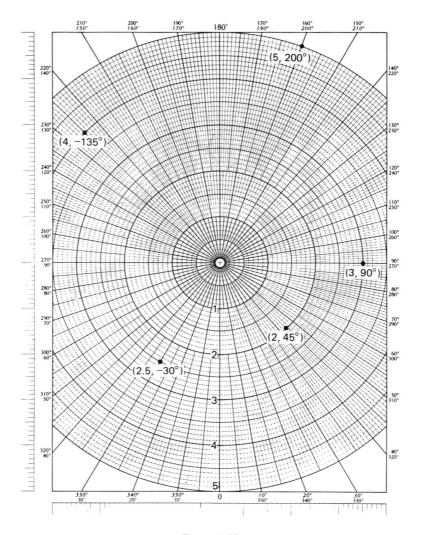

Figure 1.22

The point P in Figure 1.23 has been given *both* rectangular coordinates (x_1, y_1) and polar coordinates (ρ, θ). Notice that a right triangle in standard position is formed if the radius vector ρ is considered to be the hypotenuse, the y-coordinate y_1 to be the altitude, and the x-coordinate x_1 to be the base. The polar coordinate θ is the angle located in standard position with its vertex at the origin.

The Pythagorean Theorem can immediately be used to establish a connection between rectangular and polar coordinates. The relationship given for a right

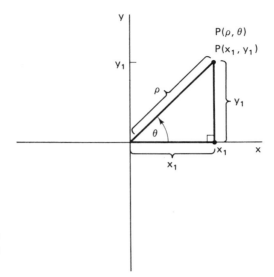

Figure 1.23. A point *P* having rectangular coordinates (x_1, y_1) and polar coordinates (ρ, θ).

triangle, as in Figure 1.16, is

$$c^2 = a^2 + b^2$$

Comparing this to the right triangle in Figure 1.23, and substituting the appropriate coordinates,

$$\rho^2 = y_1^2 + x_1^2 \tag{1.8}$$

This relationship can be used to calculate the radius vector ρ for a given set of rectangular coordinates (x_1, y_1). The relationship between x_1, y_1, and θ will be developed in a following section after the introduction of trigonometric functions.

EXAMPLE 1.12.   Find the radius vector ρ in Figure 1.23 if the point P has the rectangular coordinates:
(a) $(3, 4)$
(b) $(\sqrt{2}, \sqrt{7})$
(c) $(-2, 3)$

SOLUTION.   Using Eq. (1.8):
(a) $\rho^2 = y_1^2 + x_1^2$
  Substitute:
$$\rho^2 = 4^2 + 3^3 = 16 + 9 = 25$$
  Thus, $\rho = 5$
(b) $\rho^2 = y_1^2 + x_1^2$
  Substitute:
$$\rho^2 = (\sqrt{7})^2 + (\sqrt{2})^2 = 7 + 2 = 9$$
  Thus, $\rho = 3$

(c) $\rho^2 = y_1^2 + x_1^2$
Substitute:

$$\rho^2 = (-2)^2 + (3)^2 = 4 + 9 = 13$$

Thus, $\rho = 3.61$
Notice that the point $(-2, 3)$ lies in the second quadrant. Note also that ρ is always positive, so the principal root is always used.

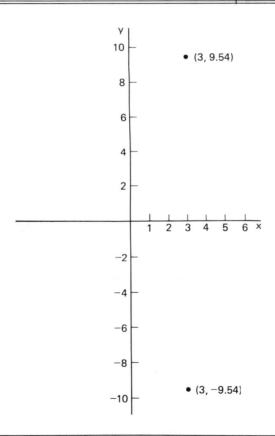

Figure 1.24

EXAMPLE 1.13. Given the radius vector $\rho = 10$, and $x_1 = 3$, find the y-coordinate y_1 in Figure 1.23.

SOLUTION. Use Eq. (1.8). Transposing:

$$y_1^2 = \rho^2 - x_1^2$$

Square root:

$$y_1 = \pm \sqrt{\rho^2 - x_1^2}$$

Substitute:

$$y_1 = \pm \sqrt{10^2 - 3^2}$$

$$= \pm \sqrt{100 - 9} = \sqrt{91}$$

$$= \pm 9.54$$

Take care to note that there are two possible points: $(3, +9.54)$ and $(3, -9.54)$. These two points are plotted in Figure 1.24. A unique solution (a single point) is determined by specifying an angle θ.

Exercises 1.4

1. Plot the given polar coordinates on polar graph paper:

   (a) $(3, 45°)$ (e) $(2.1, -180°)$
   (b) $(1.5, 120°)$ (f) $(2.1, 180°)$
   (c) $(3.75, 250°)$ (g) $(3.2, 340°)$
   (d) $(1.2, -20°)$ (h) $(4.7, -300°)$

2. Plot each of the following points given in rectangular coordinates. In each case, calculate the radius vector ρ:

   (a) $(6, 8)$ (b) $(\sqrt{2}, 3)$ (c) $(-3, 4)$ (d) $(-5, -5)$
   (e) $(1.5, -2.0)$ (f) $(-\sqrt{3}, 5)$ (g) $(-\sqrt{1.5}, -\sqrt{2.3})$
   For each point, draw the right triangle formed by $x_1, y_1,$ and ρ.

3. In each of the following cases, two of the three coordinates $\rho, x_1,$ and y_1 are given. In each case, determine the third coordinate by using Eq. (1.8). Then plot the two possible points on a rectangular coordinate system:

   (a) $x_1 = 4.0, \rho = 8.0; y_1 = ?$
   (b) $y_1 = 2.5, \rho = 5.0; x_1 = ?$
   (c) $x_1 = -7.20, \rho = 10.18; y_1 = ?$
   (d) $y_1 = -\sqrt{2}, \rho = 3.78; x_1 = ?$

4. If the polar coordinate θ is equal to 60°, which of the two possible values for y_1 must be selected in Problem 3(a) above? Plot the correct point.

5. If the polar coordinate θ is equal to 150°, which of the two possible values for x_1 must be selected in Problem 3(b) above? Plot the correct point.

6. The polar coordinates for a point are given as $P(10, 120°)$. If the rectangular x-coordinate is $x_1 = -5$, what must be the y-coordinate y_1?

7. The polar coordinates for a point are given as $P(3.75, 27°)$. If the rectangular y-coordinate is $y_1 = 1.70$, what must be the x-coordinate x_1?

UNIT 2
THE TRIGONOMETRIC FUNCTIONS

2.1 TRIGONOMETRIC FUNCTIONS OF A GENERAL ANGLE

The six *trigonometric functions* form the foundation on which the subject of trigonometry is based. The names of these six functions along with their abbreviations are:

sine (sin)   cosine (cos)

tangent (tan)   cotangent (cot)

secant (sec)   cosecant (csc)

The trigonometric functions are defined in terms of some angle θ. Thus, it is correct to say "the sine of the angle θ," abbreviated as $\sin\theta$. Similarly, cos means "the cosine of the angle θ," and $\tan\theta$ means "the tangent of the angle θ," and so on. Notice that the abbreviations sin, cos, etc., written by themselves have no mathematical significance.

In the previous section it was shown how a point P is uniquely located on a plane by specifying either its rectangular coordinates $P(x,y)$ or its polar coordinates $P(\rho,\theta)$. Of course, every point on a plane has both rectangular and polar coordinates as shown in Figure 1.23. The coordinates x, y, and ρ are used to define the trigonometric functions of the angle θ (the fourth coordinate) in the following way:

1. Draw the angle θ in standard position as shown in Figure 2.1(a).

2. Select any convenient point $P(x,y)$ on the terminal side of θ, and draw a line through P perpendicular to the x-axis. This forms a right triangle of reference for θ as shown in Figure 2.1(b). The three sides of the reference triangle are the base x, the altitude y, and the hypotenuse ρ.

3. The point P in Figure 2.1(b) has the three coordinates x, y, and ρ, and they are used in the definitions of the trigonometric functions of the angle θ. Thus,

---

TRIGONOMETRIC FUNCTIONS

The six trigonometric functions of the angle in Figure 2.1(b) are:

$$\sin\theta = \frac{y}{\rho} \qquad \cos\theta = \frac{x}{\rho}$$

$$\tan\theta = \frac{y}{x} \qquad \cot\theta = \frac{x}{y}$$

$$\sec\theta = \frac{\rho}{x} \qquad \csc\theta = \frac{\rho}{y}$$

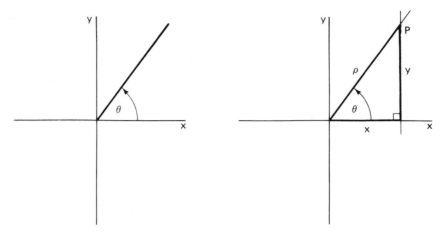

Figure 2.1   (a) θ is drawn in standard position   (b) Reference triangle for the angle θ

The coordinates x, y, and ρ are real numbers (either positive or negative). As a result, each trigonometric function is itself a positive or negative number, since each function is simply the quotient of two coordinates (or the quotient of two sides of the reference triangle).

EXAMPLE 2.1. The point P in Figure 2.1(b) has the rectangular coordinates $(3,4)$. Calculate the six trigonometric functions of the angle θ.

SOLUTION. First, use the Pythagorean Theorem to determine the radius vector ρ:
$$\rho^2 = x^2 + y^2$$
Substitute:
$$\rho^2 = 3^2 + 4^2 = 9 + 16 = 25$$
Square root: $\rho = +5$
The reference triangle has the sides: $x = 3$, $y = 4$, $\rho = 5$. Using the definitions for the functions:

$$\sin\theta = \frac{y}{\rho} = \frac{4}{5} = 0.8$$

$$\cos\theta = \frac{x}{\rho} = \frac{3}{5} = 0.6$$

$$\tan\theta = \frac{y}{x} = \frac{4}{3} = 1.33\ldots$$

$$\cot\theta = \frac{x}{y} = \frac{3}{4} = 0.75$$

$$\sec\theta = \frac{\rho}{x} = \frac{5}{3} = 1.66\ldots$$

$$\csc\theta = \frac{\rho}{y} = \frac{5}{4} = 1.25$$

EXAMPLE 2.2. Calculate the six trigonometric functions of the angle θ in Figure 2.2.

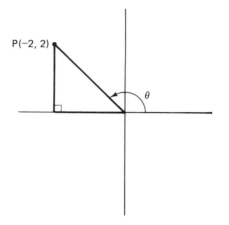

P(−2, 2)

θ

Figure 2.2

SOLUTION. From Figure 2.2, the base of the reference triangle is $x = -2$, and the altitude is $y = +2$. The radius vector ρ is found using the Pythagorean Theorem:

$$\rho^2 = x^2 + y^2$$

Substitute:

$$\rho^2 = (-2)^2 + 2^2 = 4 + 4 = 8$$

Square root: $\rho = \sqrt{8} = 2\sqrt{2} = 2.828$

The six functions are thus:

$$\sin\theta = \frac{y}{\rho} = \frac{2}{2\sqrt{2}} = \frac{1}{\sqrt{2}} = 0.707$$

$$\cos\theta = \frac{x}{\rho} = \frac{(-2)}{2\sqrt{2}} = \frac{-1}{\sqrt{2}} = -0.707$$

$$\tan\theta = \frac{y}{x} = \frac{2}{(-2)} = -1$$

$$\cot\theta = \frac{x}{y} = \frac{(-2)}{2} = -1$$

$$\sec\theta = \frac{\rho}{x} = \frac{2\sqrt{2}}{(-2)} = -\sqrt{2} = -1.414$$

$$\csc\theta = \frac{\rho}{y} = \frac{2\sqrt{2}}{2} = \sqrt{2} = 1.414$$

It is important to realize that the·trigonometric functions of an angle θ depend only on the size of the angle θ, and they do not in any way depend on the size of the reference triangle in Figure 2.1(b). An angle $\theta = 45°$ is drawn in standard position in Figure 2.3(a). A reference triangle drawn using the point $P_1(1,1)$ has a base $x = 1$ and an altitude $y = 1$. The tangent of the 45° angle can then be calculated using $\tan\theta = y/x$. Thus,

$$\tan 45° = \frac{y}{x} = \frac{1}{1} = 1.0$$

If the point $P_2(2,2)$ were chosen instead to draw the reference triangle, the sides of the reference triangle would be $x = 2$ and $y = 2$. Then,

$$\tan 45° = \frac{y}{x} = \frac{2}{2} = 1.0$$

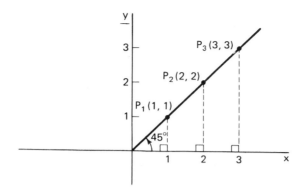

Figure 2.3. A 45°angle.

Now it should be apparent that the tangent of a 45° angle will *always* be 1.0, no matter what point is chosen to draw the reference triangle.

To prove that the tangent of an angle θ is indeed independent of the point chosen to draw the reference triangle, consider the angle θ drawn in standard position in Figure 2.4(a). The two points P_1 and P_2 lead to the two reference triangles in Figure 2.4(b).

By definition, two triangles are said to be *similar* if their corresponding angles are equal. In other words, *similar triangles* have exactly the same *shape* but are not necessarily the same size. The two reference triangles in Figure 2.4(b) are similar since their corresponding angles are equal.

An important property of similar triangles is given as:

SIMILAR TRIANGLES

The ratio of any two sides of a triangle is exactly equal to the ratio of the corresponding two sides of a *similar* triangle.

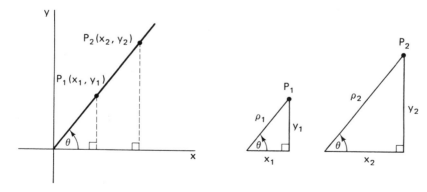

Figure 2.4   (a) Angle θ in standard position   (b) Two similar reference triangles

If this property is applied to the two reference triangles in Figure 2.4(b), the ratio of altitude to base for each yields:

$$\frac{y_1}{x_1} = \frac{y_2}{x_2}$$

But these two ratios are exactly equal to $\tan\theta$. Therefore, since P_1 and P_2 can be any two points in general, it is proved that $\tan\theta$ is indeed independent of the reference point P.

Notice that if θ is changed, the value of y_1/x_1 (and thus $\tan\theta$) is changed. Since the value of $\tan\theta$ depends on the value of θ, and is independent of the choice of P, it can be said that $\tan\theta$ is a *function* of θ (recall the definition of a function).

This same argument can be applied to the other five trigonometric functions with the following result:

DEPENDENCE OF FUNCTIONS

The trigonometric functions of an angle θ depend solely on the value of θ and not on the size of the reference triangle.

EXAMPLE 2.3. Calculate the sine of a 45° angle using two different reference triangles.

SOLUTION. Use the 45° angle drawn in Figure 2.3. For the reference triangle drawn using point $P_1(1, 1)$, the radius vector is

$$\rho_1 = \sqrt{x_1^2 + y_1^2} = \sqrt{1^2 + 1^2} = \sqrt{2}$$

The sine of 45° is then,

$$\sin 45° = \frac{y_1}{\rho_1} = \frac{1}{\sqrt{2}} = 0.707$$

For the reference triangle drawn using point $P_2(2, 2)$, the radius vector is

$$\rho_2 = \sqrt{x_2^2 + y_2^2} = \sqrt{2^2 + 2^2} = \sqrt{8} = 2\sqrt{2}$$

The sine of 45° is then,

$$\sin 45° = \frac{y_2}{\rho_2} = \frac{2}{2\sqrt{2}} = \frac{1}{\sqrt{2}} = 0.707$$

Clearly, for either point P_1 or P_2 chosen, the result will be

$$\sin 45° = 0.707$$

EXAMPLE 2.4.   Use Figure 2.4 to prove that the sine of an angle θ is independent of the point P chosen for the reference triangle.

SOLUTION.   For the smaller triangle, by definition,

$$\sin\theta = \frac{y_1}{\rho_1}$$

For the larger triangle, by definition,

$$\sin\theta = \frac{y_2}{\rho_2}$$

But, for *similar* triangles,

$$\frac{y_1}{\rho_1} = \frac{y_2}{\rho_2}$$

Therefore the value of $\sin\theta$ is the same for both triangles.

The key ideas regarding the trigonometric functions are: (1) the fact that each function is a ratio of two sides of a triangle and the size of the triangle is of absolutely no importance; and (2) the fact that the value of each function is determined solely by the size of the angle θ.

Exercises 2.1

1. Each of the given points lies on the terminal side of an angle θ in standard position. Draw θ in standard position and use a curved arrow to indicate θ. Draw and label the sides of the reference triangle using the given point. Calculate the six trigonometric functions of the angle θ to three decimal places in each of the following:

    (a) $(4, 3)$ (g) $(8, -5)$
    (b) $(-1, 1)$ (h) $(1, 3)$
    (c) $(-1, -1)$ (i) $(-1, 3)$
    (d) $(2, -2)$ (j) $(\sqrt{3}, -1)$
    (e) $(-3, -2)$ (k) $(-\sqrt{2}, -2\sqrt{2})$
    (f) $(-5, 6)$

2. Which of the following are not possible?

    (a) $\sin\theta = 36$ (c) $\cos\theta = -0.71$
    (b) $\tan\theta = 54$ (d) $\sin\theta = 0$

3. The angle θ is in the first quadrant. What can you say about its approximate size if:

    (a) $\sin\theta = 0.002$?   (d) $\tan\theta = 0.005$?
    (b) $\cos\theta = 0.002$?   (e) $\sin\theta = 0.999$?
    (c) $\tan\theta = 1000$?   (f) $\cos\theta = 0.999$?

4. The terminal side of the angle θ in Figure 2.4 can be thought of as a segment of a straight line passing through the origin. Recall that the equation of a straight line is given by

$$y = mx + b$$

where b is the y-intercept and m is the slope. What is the relationship between the slope of the line forming the terminal side of θ and $\tan \theta$ in Figure 2.4?

5. Prove that the cosine of an angle θ shown in Figure 2.4 is independent of the point P chosen for the reference triangle.

2.2   COMPUTATION OF TRIGONOMETRIC FUNCTIONS

It has been proved in the previous section that each of the six ratios defined as the trigonometric functions is solely a function of the angle θ. In other words, for any given angle θ, each trigonometric function has a unique value. For instance, the sine of a 30° angle will *always* be equal to $1/2$ ($\sin 30° = 0.5$).

This suggests the idea of calculating the trigonometric functions for a number of different angles and tabulating them for future reference. It should not be surprising to learn that the trigonometric functions have already been calculated and tabulated, and are readily available in a Table of Trigonometric Functions. This table is included in the Appendix of this text. Furthermore, operations with trigonometric functions are included with nearly all electronic hand calculators, and thus the values of the trigonometric functions are quite literally "at your fingertips."

In this section a graphical technique will be used to calculate the trigonometric functions of a number of different angles in order to demonstrate some of their important characteristics. The use of the trigonometric tables will then be discussed in the following section.

First of all, take another look at the basic definitions of the trigonometric functions.

$$\sin \theta = \frac{y}{\rho} \qquad\qquad \csc \theta = \frac{\rho}{y}$$

$$\cos \theta = \frac{x}{\rho} \qquad\qquad \sec \theta = \frac{\rho}{x}$$

$$\tan \theta = \frac{y}{x} \qquad\qquad \cot \theta = \frac{x}{y}$$

Notice that the sine of the angle θ is given by the ratio y/ρ, and the cosecant of θ is given by the ratio ρ/y. But,

$$\frac{y}{\rho} = \frac{1}{\dfrac{\rho}{y}}$$

That is, these two ratios are reciprocals. Therefore $\sin\theta$ must be the reciprocal of $\csc\theta$. For instance, the sine of 30° is $1/2$ ($\sin 30° = 1/2$), and thus the cosecant of 30° must be $1/(1/2)=2$ ($\csc 30° = 2$). Therefore,

$$\sin\theta = \frac{1}{\csc\theta} \qquad (2.1)$$

Similarly, since $\cos\theta = x/\rho$ and $\sec\theta = \rho/x$, these ratios are also reciprocals. That is,

$$\cos\theta = \frac{1}{\sec\theta} \qquad (2.2)$$

Also, $\tan\theta = y/x$ and $\cot\theta = x/y$, and clearly these too are reciprocals. Thus,

$$\tan\theta = \frac{1}{\cot\theta} \qquad (2.3)$$

These reciprocals are tabulated as follows:

RECIPROCALS

$\sin\theta$ and $\csc\theta$
$\cos\theta$ and $\sec\theta$
$\tan\theta$ and $\cot\theta$

Clearly a knowledge of the values of the three functions in either the left-hand column or the right-hand column (in the display above) is sufficient, since the unknown functions can be calculated as reciprocals. For this reason, the three functions $\sin\theta$, $\cos\theta$, and $\tan\theta$ are usually the only ones given in a table of trigonometric functions, or included as the operations on an electronic hand calculator.

EXAMPLE 2.5. Given $\sin 30° = 0.500$, $\cos 30° = 0.866$, and $\tan 30° = 0.577$, use the reciprocal relationships to calculate $\sec 30°$, $\csc 30°$, and $\cot 30°$.

SOLUTION.

$$\csc 30° = \frac{1}{\sin 30°} = \frac{1}{0.500} = 2.000$$

$$\sec 30° = \frac{1}{\cos 30°} = \frac{1}{0.866} = 1.155$$

$$\cot 30° = \frac{1}{\tan 30°} = \frac{1}{0.577} = 1.733$$

The graph in Figure 2.5 has 36 radius vectors drawn with 10° angles between them. The outer circle further divides each 10° angle into 1° segments. This graph can be used to calculate the trigonometric functions of an angle. With care, it is

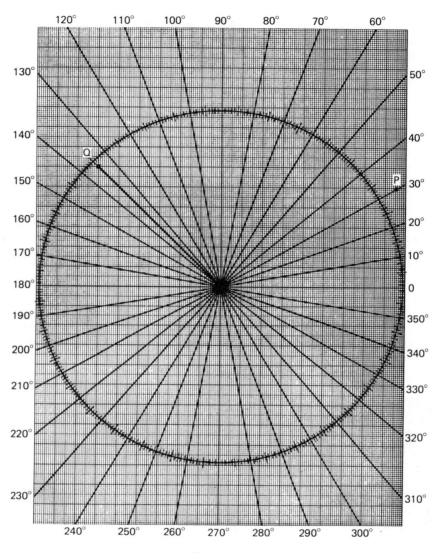

Figure 2.5

possible to locate the angle to the nearest degree, and the functions can then be calculated to perhaps two significant figures. For example, the point P located on the 30° radius vector has the coordinates (8.6, 5.0)—you should find this point. Using these coordinates, the tangent of 30° can be calculated as

$$\tan 30° = \frac{y}{x} = \frac{5.0}{8.6} = 0.58$$

EXAMPLE 2.6.   Use the point P in Figure 2.5 to calculate $\sin 30°$ and $\cos 30°$.

SOLUTION.   The point P is used to form a reference triangle having a base $x = 8.6$ and an altitude $y = 5.0$. The hypotenuse ρ is found using the Pythagorean Theorem:

$$\rho^2 = x^2 + y^2$$

Substitute:

$$\rho^2 = 8.6^2 + 5.0^2 = 74 + 25 = 99$$

Square root: $\rho \cong 10$
Then,

$$\sin 30° = \frac{y}{\rho} = \frac{5.0}{10} = 0.50$$

$$\cos 30° = \frac{x}{\rho} = \frac{8.6}{10} = 0.86$$

In general, Figure 2.5 can be used to calculate the trigonometric functions of an angle θ as follows:

1. Draw the angle θ in standard position. Attempt accuracy to at least the nearest degree.

2. Locate any convenient point on the terminal side of θ and determine the coordinates (x,y) of the selected point to two significant figures.

3. Use the Pythagorean Theorem to calculate the radius vector to the selected point.

4. Use x, y, and ρ to calculate the trigonometric functions to two significant figures.

EXAMPLE 2.7.   Calculate the trigonometric functions of the angle $\theta = 135°$.

SOLUTION.   The radius vector for a 135° angle is drawn in Figure 2.5, and the point Q is chosen for reference. The coordinates of the point Q are $(-6.0, 6.0)$. Notice carefully the signs: $x = -6.0$, $y = 6.0$. The Pythagorean Theorem is used to calculate:

$$\rho^2 = x^2 + y^2$$

Substitute:

$$\rho^2 = (-6.0)^2 + 6.0^2$$

$$= 36 + 36 = 72$$

Square root: $\rho = 8.5$

The trigonometric functions are then calculated as:

$$\sin 135° = \frac{y}{\rho} = \frac{6.0}{8.5} = 0.71$$

$$\csc 135° = \frac{1}{\sin 135°} = \frac{1}{0.71} = 1.42$$

$$\cos 135° = \frac{x}{\rho} = \frac{-6.0}{8.5} = -0.71$$

$$\sec 135° = \frac{1}{\cos 135°} = \frac{1}{-0.71} = -1.42$$

$$\tan 135° = \frac{y}{x} = \frac{6.0}{-6.0} = -1.00$$

$$\cot 135° = \frac{1}{\tan 135°} = \frac{1}{-1.00} = -1.00$$

In the previous examples it is apparent that the values of the trigonometric functions are sometimes positive and sometimes negative. Let's examine them more carefully to determine exactly when each function is positive and when it is negative.

The trigonometric functions are determined as the various ratios formed using the three sides of a reference triangle x, y, and ρ. The ρ is the hypotenuse of the triangle, or the radius vector to the chosen reference point $P(x,y)$. Thus, ρ is *always* positive. However, the coordinates x and y can have either positive or negative values depending on the quadrant in which $P(x,y)$ is located.

The sine of the angle θ is defined as

$$\sin\theta = \frac{y}{\rho}$$

Since ρ is always positive; the sign of the function $\sin\theta$ is determined by the sign of the coordinate y. Positive values of y occur only above the x-axis, and thus $\sin\theta$ is positive whenever θ is in quadrant I or II as shown in Figure 2.6(a). Clearly, $\sin\theta$ is negative whenever θ is in quadrant III or IV.

Since $\csc\theta = 1/\sin\theta$, the sign of this function is exactly the same as that of $\sin\theta$.

The cosine of the angle θ is defined as

$$\cos\theta = \frac{x}{\rho}$$

Since ρ is always positive, and x is positive only to the right of the y-axis, it is clear that $\cos\theta$ is positive in quadrants I and IV. $\cos\theta$ is negative in quadrants II and III as shown in Figure 2.6(b). Furthermore, $\sec\theta$ has exactly the same sign as $\cos\theta$ since they are reciprocals.

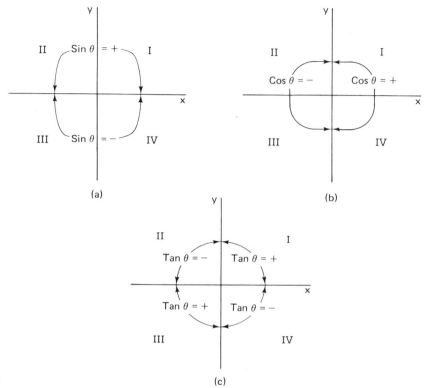

Figure 2.6

(a)   (b)   (c)

Finally, $\tan\theta$ is defined as

$$\tan\theta = \frac{y}{x}$$

Thus, $\tan\theta$ will be positive whenever x and y have the *same* signs. This occurs in quadrant I (x and y both positive) and in quadrant III (x and y both negative). In quadrants II and IV, x and y have opposite signs, and $\tan\theta$ is therefore negative. The sign for $\tan\theta$ in each quadrant is shown graphically in Figure 2.6(c). Obviously $\cot\theta$ has the same sign as $\tan\theta$ since they both are reciprocals.

The signs of each of the functions in the various quadrants are summarized in Table 2.1 below.

Table 2.1

QUADRANT	SIN θ	CSC θ	COS θ	SEC θ	TAN θ	COT θ
I	+	+	+	+	+	+
II	+	+	−	−	−	−
III	−	−	−	−	+	+
IV	−	−	+	+	−	−

EXAMPLE 2.8. In which quadrant is θ if:
(a) $\sin\theta$ is $+$?
(b) $\cos\theta$ is $-$?
(c) $\sin\theta$ is $+$ and $\tan\theta$ is $+$?
(d) $\cos\theta$ is $+$ and $\csc\theta$ is $-$?

SOLUTION. Use Table 2.1 to determine the proper quadrant(s):
(a) $\sin\theta$ is positive in both quadrants I and II. Therefore, θ could be in either quadrant I or quadrant II.
(b) $\cos\theta$ is negative in both quadrants II and III. Therefore, θ could be in either quadrant II or quadrant III.
(c) The only quadrant in which *both* $\sin\theta$ and $\tan\theta$ are positive is quadrant I.
(d) Quadrant IV is the only quadrant where $\cos\theta$ is $+$ while $\csc\theta$ is $-$.

Calculating the trigonometric functions of $0°$, $90°$, $180°$, and $270°$ leads to a special problem. The reference triangle for any of these angles disappears since either the value of x or the value of y is zero. The problem of division by zero is thus encountered. For example, the angle $\theta = 90°$ is shown in standard position in Figure 2.7. Clearly for any reference point $P(x,y)$ on the terminal side of θ, $x = 0$. Thus, an attempt to evaluate the tangent yields

$$\tan 90° = \frac{y}{x} = \frac{y}{0} = ?$$

But division by zero is undefined.

The following argument can be used to resolve this dilemma:

1. As θ increases from some acute angle and approaches closer and closer to $90°$, the value of x becomes smaller and smaller.

2. If the value of x gets smaller and smaller, then $\tan\theta = y/x$ gets larger and larger, and approaches infinity (∞).

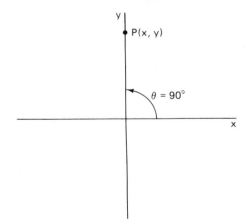

Figure 2.7

3. To avoid division by zero, one says, "The value (limit) of $\tan\theta = y/x$ approaches infinity as x approaches zero." In mathematical notation, this is written as

$$\lim_{x\to 0} \tan\theta = \infty$$

Even though it is not mathematically correct (since division by zero is undefined), the quotient of any number A divided by zero is commonly given the value infinity; that is, $A/0 = \infty$. Notice that the reciprocal of infinity is zero; that is,

$$1/\infty = \frac{1}{\dfrac{A}{0}} = \frac{0}{A} = 0$$

EXAMPLE 2.9.   Calculate the trigonometric functions of the 90° angle in Figure 2.7.

SOLUTION.   The coordinates of the reference point P are $x=0$, $y=y_1$. Clearly, $\rho = y_1$
Then,

$$\sin 90° = \frac{y}{\rho} = \frac{y_1}{y_1} = 1.0$$

$$\csc 90° = \frac{1}{\sin 90°} = \frac{1}{1.0} = 1.0$$

$$\cos 90° = \frac{x}{\rho} = \frac{0}{y_1} = 0$$

$$\sec 90° = \frac{1}{\cos 90°} = \frac{1}{0} = \infty$$

$$\tan 90° = \frac{y}{x} = \frac{y_1}{0} = \infty$$

$$\cot 90° = \frac{1}{\tan 90°} = \frac{1}{\infty} = 0$$

Exercises 2.2

1. Use Figure 2.5 to determine the trigonometric functions of the following angles. Obtain results to two decimal places.

   (a) 45°   (b) 60°   (c) 23°   (d) 152°   (e) 247°
   (f) 300°   (g) 351°   (h) 198°   (i) 105°   (j) 333°

2. In which quadrant is θ if:

   (a) tan is +?
   (b) tan is −?

(c)   csc is + and sec is −?
(d)   tan is + and sin is −?
(e)   cot is − and sin is −?

3. Calculate the trigonometric functions of 0°.

4. Calculate the trigonometric functions of 180°.

5. Calculate the trigonometric functions of 270°.

6. Given the following information, locate a reference triangle in standard position on Figure 2.5. Determine the angle θ, and calculate the remaining trigonometric functions of θ.

(a) $\sin \theta = 1/4$, θ is in quadrant II
(b) $\sin \theta = -1/4$, θ not in quadrant IV
(c) $\tan \theta = 3/4$, θ not in quadrant III
(d) $\tan \theta = 3/4$, θ not in quadrant I

2.3   TABLE OF TRIGONOMETRIC FUNCTIONS

In the Appendix there is a table of trigonometric functions titled "Natural Sines, Cosines, and Tangents." The term *natural* is used to distinguish these functions from other tables that give the logarithms of the functions. Only the sine, cosine, and tangent are included since the other three functions can be found using the reciprocal relationships.

The value of the sine, cosine, and tangent of any angle from 0° to 90°, in 0.1° increments, is given to four decimal places. This table, therefore, gives the values of the trigonometric functions of *acute* angles. There are two different problems to consider:

1. Given an angle θ, find one of its trigonometric functions.

2. Given a trigonometric function of some angle θ, determine the angle.

Consider first the problem of finding the trigonometric function of a given angle. Refer to the first page in the table of "Natural Sines, Cosines, and Tangents." The *column* on the left with "Degs." at its head gives the angle θ to the nearest degree. The first page of the table begins at 0 and continues up to 14. The *row* across the top of the table gives the nearest tenth degree of θ. It begins at 0.0°, 0.1°, 0.2°, and so on, and ends at 0.9°.

Now, the trigonometric functions of an angle to the nearest 1/10° are found by locating the intersection of:

1. The horizontal row giving degrees, and

2. The vertical column giving tenths of a degree.

For instance, the trigonometric functions for $\theta = 8.7°$ are found by moving *down* the left column to 8°, and then moving to the *right* to a position under 0.7°. At this position there are three entries:

$$0.1513$$
$$0.9885$$
$$0.1530$$

The top number is $\sin\theta$, the middle number is $\cos\theta$, and the bottom number is $\tan\theta$; as seen by the notation $\begin{smallmatrix}\sin\\\cos\\\tan\end{smallmatrix}$ next to the left-hand column. Thus,

$$\sin 8.7° = 0.1513 \qquad \cos 8.7° = 0.9885 \qquad \tan 8.7° = 0.1530$$

At this point, you should determine how to find trigonometric functions on your own hand calculator.

EXAMPLE 2.10. Use the tables to determine sine, cosine, and tangent of:
(a) $\theta = 51.3°$
(b) $\theta = 21.0°$
Find these same functions by using an electronic hand calculator. This will serve as a check on your results.

SOLUTION.
(a) Moving down the left column to 51° and then across under 0.3°, read the following entries:

sin 0.7804
cos 0.6252
tan 1.2482

(b) Move down the left column to 21° and then across under 0.0°. The entries are:

sin 0.3584
cos 0.9336
tan 0.3839

The problem of determining an angle corresponding to a given trigonometric function is simply a question of searching for the function in the table, and then reading the appropriate angle.

For instance, given $\tan\theta = 1.0951$, search under the $\tan\theta$ entries in the table until the number 1.0951 is found. It happens to occur corresponding to the angle $\theta = 47.6°$—you should verify that this is so.

It is possible that the exact value of a given function will not appear in the table. For instance, $\sin\theta = 0.2610$ does not appear in the table. But it is possible to find

$$\sin 15.1° = 0.2605$$
$$\sin 15.2° = 0.2622$$

Since 0.2610 is closer to 0.2605 than to 0.2622, choose $\theta = 15.1$ to obtain an answer to the nearest $1/10°$.

To facilitate the solution of this type of problem, the following "mathematical shorthand" is used:

$\sin^{-1} x$ or $\arcsin x$ means "the angle whose sine is x"

$\cos^{-1} x$ or $\arccos x$ means "the angle whose cosine is x"

$\tan^{-1} x$ or $\arctan x$ means "the angle whose tangent is x"

Therefore, the problem is to find θ, given $\sin\theta = 0.2605$. The problem is stated as:

$$\theta = \sin^{-1} 0.2605$$

or $$\theta = \arcsin 0.2605$$

which means "θ is the angle whose sine is 0.2605." Then, finding 0.2605 in the trigonometric tables provides the value for θ.

$$\theta = 15.1°$$

Now, use the booklet with your hand calculator to learn the method for determining $\sin^{-1}$, $\cos^{-1}$, and $\tan^{-1}$ (or arcsin, arccos, and arctan).

EXAMPLE 2.11. Use the table to determine θ if:

(a) $\theta = \sin^{-1} 0.7230$

(b) $\theta = \cos^{-1} 0.9977$

(c) $\theta = \tan^{-1} 13.00$

In each case, determine the same angle using an electronic hand calculator. This will serve as a check.

SOLUTION.

(a) The entry 0.7230 is found at the junction of 46° and 0.3° under the sine. Thus, $\theta = 46.3°$

(b) The entry 0.9977 under cosine is found at the intersection of 3° and 0.9°. Thus, $\theta = 3.9°$

(c) The entry 13.00 under tangent is found at the intersection of 85° and 0.6°. Thus, $\theta = 85.6°$

Exercises 2.3

1. Use the trigonometric table to find the values of the following trigonometric functions to four decimal places. Use an electronic hand calculator to verify each value.

(a) $\sin 21.3°$

(b) $\cos 14.8°$

(c) $\tan 11.6°$

(d) $\cos 51.6°$

(e) $\tan 84.2°$

(f) $\sin 57.1°$

(g) $\tan 59.0°$

(h) $\sin 1.5°$

(i) $\cos 88.4°$

(j) $\sin 89.1°$

(k) $\cos 2.3°$

(l) $\tan 1.8°$

2.   Use the trigonometric table to find θ to the nearest degree corresponding to the given function. Use an electronic hand calculator to verify each solution.

(a) $\theta = \sin^{-1} 0.3681$ (f) $\theta = \sin^{-1} 0.8996$

(b) $\theta = \cos^{-1} 0.9298$ (g) $\theta = \tan^{-1} 3.3980$

(c) $\theta = \tan^{-1} 0.3959$ (h) $\theta = \sin^{-1} 0.1800$

(d) $\theta = \cos^{-1} 0.2538$ (i) $\theta = \cos^{-1} 0.8360$

(e) $\theta = \tan^{-1} 1.1918$

3.   Use the trigonometric tables and the reciprocal relationships to determine:

(a) $\csc 21.3°$ (b) $\sec 14.8°$ (c) $\cot 11.6°$

(d) $\sec 51.6°$ (e) $\cot 84.2°$ (f) $\csc 57.1°$

UNIT 3
RIGHT TRIANGLES

3.1 STANDARD NOTATION

One of the most widely used geometrical figures is the right triangle. Numerous mathematical problems are formulated in terms of a right triangle, and the solutions to these problems are obtained by utilizing the trigonometric functions—problems in surveying, navigation, drafting, architecture, civil engineering, mechanical engineering, and electrical engineering, to name a few. Since this figure is so widely used, there is a standard notation and terminology for a right triangle.

If an *acute* angle θ is drawn in standard position and a point $P(x,y)$ is selected on its terminal side, then the reference triangle formed is a right triangle as shown in Figure 3.1(a). The radius vector ρ is known as the *hypotenuse* (hyp) of this reference triangle. The y-coordinate is opposite the angle θ and is known as the *opposite side* (opp). The x-coordinate is adjacent to the reference angle θ and is known as the *adjacent side* (adj).

Using this terminology, a right triangle can be drawn in any random position (not in standard position), as shown in Figure 3.1(b). The acute angle θ selected as reference has a hypotenuse, an opposite side, and an adjacent side relative to that angle as shown.

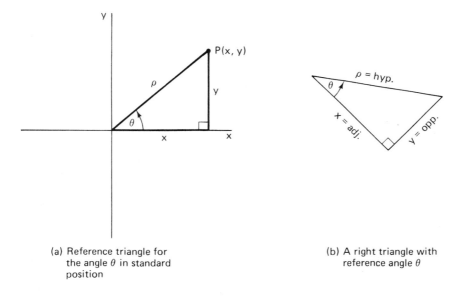

(a) Reference triangle for the angle θ in standard position

(b) A right triangle with reference angle θ

Figure 3.1

EXAMPLE 3.1.   For the right triangle in Figure 3.1(b), relabel the sides if the other acute angle (not θ) is selected as the reference angle. Label this other angle ϕ.

SOLUTION.   The triangle is shown in Figure 3.2. For the angle ϕ, the hyp remains the same. But the side opposite ϕ corresponds to the adj side of θ. The side adjacent to ϕ corresponds to the opp side of θ.

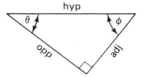

Figure 3.2. The hypotenuse, opposite, and adjacent sides of the triangle relative to the angle θ.

To summarize, the trigonometric functions of an acute angle θ in a right triangle are frequently defined in terms of hyp, opp, and adj sides. Thus, for the triangle in Figure 3.1(b),

---

TRIGONOMETRIC FUNCTIONS

For an acute angle θ in a right triangle—Figure 3.1(b):

$$\sin\theta = \frac{\text{opp}}{\text{hyp}}$$

$$\cos\theta = \frac{\text{adj}}{\text{hyp}}$$

$$\tan\theta = \frac{\text{opp}}{\text{adj}}$$

The other three trigonometric functions can, of course, be determined using the reciprocal relationships.

EXAMPLE 3.2.   Determine the sine, cosine, and tangent of the angle α in Figure 3.3.

SOLUTION.   For the angle α:

hyp $= 11$ cm

opp $= 6.6$ cm

adj $= 8.8$ cm

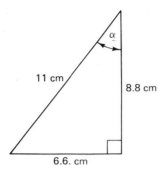

Figure 3.3

Thus,

$$\sin \alpha = \frac{\text{opp}}{\text{hyp}} = \frac{6.6}{11} = 0.60$$

$$\cos \theta = \frac{\text{adj}}{\text{hyp}} = \frac{8.8}{11} = 0.80$$

$$\tan \theta = \frac{\text{opp}}{\text{adj}} = \frac{6.6}{8.8} = 0.75$$

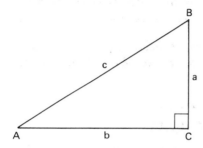

Figure 3.4. Standard notation for a right triangle.

It is common practice to label the sides and angles of a right triangle as shown in Figure 3.4. The capital letter C is reserved for the right angle, and the lower-case letter c is used for the hyp. One of the acute angles is designated with the capital letter A, and the side opposite $\angle A$ is labeled with the lower-case letter a. The other acute angle is designated with the capital letter B, and the side opposite $\angle B$ is labeled with the lower-case letter b.

Now, the sine of $\angle A$ in Figure 3.4 can be written as

$$\sin A = \frac{\text{opp}}{\text{hyp}} = \frac{a}{c}$$

Similarly, the cosine of angle B in Figure 3.4 is

$$\cos B = \frac{\text{adj}}{\text{hyp}} = \frac{a}{c}$$

Clearly, $\sin A = \cos B$. In a similar way, it can be shown

$$\cos A = \sin B$$
$$\tan A = \cot B$$
$$\sec A = \csc B$$
$$\csc A = \sec B$$
$$\cot A = \tan B$$

By definition, the sine and the cosine are said to be *cofunctions*; that is, the sine is the *cofunction* of the cosine, and the cosine is the *cofunction* of the sine.

Similarly, the tangent, and cotangent are *cofunctions*, and the secant and cosecant are *cofunctions*.

Using these definitions and the above results:

COFUNCTIONS

A trigonometric function of any acute angle is equal to the cofunction of its complementary angle.

For example: $\sin 30° = \cos 60°$, $\tan 50° = \cot 40°$, $\sec 63° = \csc 27°$, and so on.

EXAMPLE 3.3. For the triangle in Figure 3.4, show that $\cos A = \sin B$.

SOLUTION.

$$\cos A = \frac{\text{adj}}{\text{hyp}} = \frac{b}{c}$$

$$\sin B = \frac{\text{opp}}{\text{hyp}} = \frac{b}{c}$$

Thus $\cos A = \sin B$

EXAMPLE 3.4. Suppose the unlabeled acute angle in Figure 3.3 is called β. Clearly β is the complement of α since this is a right triangle. Calculate $\sin \beta$ and $\cos \beta$. Use the results from Example 3.2 to show that

$$\sin \alpha = \cos \beta$$

and

$$\cos \alpha = \sin \beta$$

SOLUTION.

$$\sin \beta = \frac{\text{opp}}{\text{hyp}} = \frac{8.8}{11} = 0.80$$

$$\cos \beta = \frac{\text{adj}}{\text{hyp}} = \frac{6.6}{11} = 0.60$$

From Example 3.2,

$$\sin \alpha = 0.60$$

and

$$\cos \alpha = 0.80$$

Clearly,

$$\sin \alpha = \cos \beta = 0.60$$

and

$$\cos \alpha = \sin \beta = 0.80$$

Exercises 3.1

1. Calculate the sine, cosine, and tangent of the angle θ in Figure 3.1(b) to three-digit accuracy if:

   (a) hyp $= \sqrt{5}$, adj $= 1$, opp $= 2$
   (b) hyp $= \sqrt{13}$, adj $= 3$, opp $= 2$
   (c) hyp $= 4$, adj $= 3$, opp $= \sqrt{7}$
   (d) hyp $= \dfrac{\sqrt{13}}{6}$, adj $= \dfrac{1}{2}$, opp $= \dfrac{1}{3}$

   (e) adj $= 3$, opp $= 6$
   (f) adj $= 1.5$, opp $= 2.5$
   (g) hyp $= 4.8$, adj $= 3.6$
   (h) hyp $= 11.2$, opp $= 5.9$

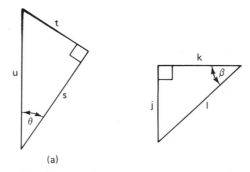

(a)

Figure 3.5

2.   Calculate the sine, cosine, and tangent of the angle θ in Figure 3.5(a).

3.   Calculate the sine, cosine, and tangent of the angle β in Figure 3.5(b).

4.   Using the triangle in Figure 3.4, prove:

   (a) $\tan A = \cot B$
   (b) $\sec A = \csc B$
   (c) $\csc A = \sec B$
   (d) $\cot A = \tan B$

5.   If the unlabeled angle in Figure 3.3 is called β, show that

   (a) $\tan \alpha = \cot \beta$
   (b) $\cot \alpha = \tan \beta$

3.2   RIGHT TRIANGLES FOR REFERENCE

There are a number of different right triangles that are commonly used for reference purposes. Three of the most common ones are: (1) the 30°–60°–90° and (2) the 45°–90° (their names reflect the sizes of their angles), and (3) the 3–4–5 (its name refers to the lengths of its sides).

The triangle drawn in Figure 3.6(a) has three equal sides (of length 2) and three equal angles (each angle is 60°). Such a figure is called an *equilateral* triangle. The equilateral triangle can be divided into two identical triangles by drawing a *bisector* (dashed line) from the vertex of one angle to the midpoint of the side opposite that angle. Such a *bisector* divides the 60° angle into two 30° angles, and it is perpendicular to the opposite side. Bisection of this equilateral triangle forms the 30°–60° right triangle shown in Figure 3.6(b). By comparison with the original, it is easy to see that the length of the hypotenuse is 2, and the length of the bisected side is 1. The length of the *bisector* is found using the Pythagorean Theorem. Thus,

$$\text{Length of bisector} = \sqrt{2^2 - 1^2} = \sqrt{4 - 1} = \sqrt{3}$$

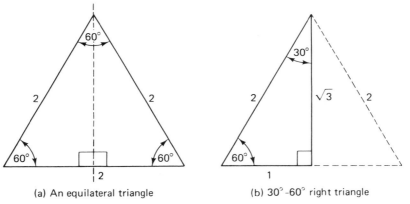

Figure 3.6 (a) An equilateral triangle (b) 30°-60° right triangle

The sine, cosine, and tangent of a 30° angle and a 60° angle are now easily calculated by referring to the lengths of the sides shown in Figure 3.6(b).

(for 30° angle)

$$\sin 30° = \frac{\text{opp}}{\text{hyp}} = \frac{1}{2}$$

$$\cos 30° = \frac{\text{adj}}{\text{hyp}} = \frac{\sqrt{3}}{2}$$

$$\tan 30° = \frac{\text{opp}}{\text{adj}} = \frac{1}{\sqrt{3}}$$

(for 60° angle)

$$\sin 60° = \frac{\text{opp}}{\text{hyp}} = \frac{\sqrt{3}}{2}$$

$$\cos 60° = \frac{\text{adj}}{\text{hyp}} = \frac{1}{2}$$

$$\tan 60° = \frac{\text{opp}}{\text{adj}} = \frac{\sqrt{3}}{1}$$

The results are checked by noting that each trigonometric function of 30° is equal to the cofunction of 60°. The reciprocal relationships can be used to determine the other three functions of each angle.

EXAMPLE 3.5. The line-of-sight distance between two microwave repeater stations is known to be 4.3 mi. Station B is at an elevation 11,350 ft higher than station A (as shown in Figure 3.7). At what angle of depression should the antenna at station B be sighted toward station A?

SOLUTION. Changing the line-of-sight distance from mi to ft yields:

$$4.3 \times 5280 \cong 22{,}700 \text{ ft}$$

Then,

$$\sin \alpha = \frac{\text{opp}}{\text{hyp}} = \frac{11{,}350}{22{,}700} = 0.50$$

From a standard 30°–60° right triangle, it is known that $\sin 30° = 1/2$. Thus,

$$\alpha = \sin^{-1} 1/2 = 30°$$

Incidentally, notice that antenna A must have an angle of *elevation* exactly equal to α.

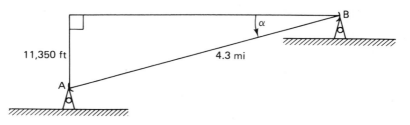

Figure 3.7

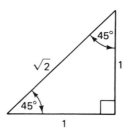

Figure 3.8. A 45° right triangle.

A 45° right triangle is shown in Figure 3.8. Each acute angle is equal to 45°, and the side opposite each acute angle is given a length of 1. The length of the hypotenuse is found using the Pythagorean Theorem. Thus,

$$\text{Hypotenuse} = \sqrt{1^2 + 1^2} = \sqrt{2}$$

The sine, cosine, and tangent of a 45° angle can then be calculated using the lengths of the sides shown in Figure 3.8.

$$\sin 45° = \frac{\text{opp}}{\text{hyp}} = \frac{1}{\sqrt{2}}$$

$$\cos 45° = \frac{\text{adj}}{\text{hyp}} = \frac{1}{\sqrt{2}}$$

$$\tan 45° = \frac{\text{opp}}{\text{adj}} = \frac{1}{1} = 1$$

The other three trigonometric functions of a 45° angle can be found by using the reciprocal relationships.

EXAMPLE 3.6. Calculate the secant, cosecant, and cotangent of a 45° angle using the above results.

SOLUTION. Using the reciprocal relationships:

$$\sec 45° = \frac{1}{\cos 45°} = \frac{1}{1/\sqrt{2}} = \sqrt{2}$$

$$\csc 45° = \frac{1}{\sin 45°} = \frac{1}{1/\sqrt{2}} = \sqrt{2}$$

$$\cot 45° = \frac{1}{\tan 45°} = \frac{1}{1} = 1$$

The sine, cosine, and tangent of 0°, 30°, 45°, 60°, and 90° are summarized in Table 3.1. The trigonometric functions of these angles are encountered often enough that you might consider the value of memorizing them.*

Table 3.1

	0°	30°	45°	60°	90°
sine	0	$1/2$	$1/\sqrt{2}$	$\sqrt{3}/2$	1
cosine	1	$\sqrt{3}/2$	$1/\sqrt{2}$	$1/2$	0
tangent	0	$1/\sqrt{3}$	1	$\sqrt{3}$	∞

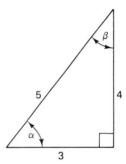

Figure 3.9. A 3-4-5 right triangle. $\alpha = 53.1°$ $\beta = 36.9°$

The 3–4–5 right triangle is a convenient configuration because of the lengths of its sides as shown in Figure 3.9. The two acute angles can be determined by using the trigonometric functions. For instance, the larger acute angle α is found by noting

$$\alpha = \tan^{-1}\frac{\text{opp}}{\text{adj}} = \tan^{-1}\frac{4}{3} = \tan^{-1}1.333 = 53.1°$$

The smaller acute angle can be found as the complement of α:

$$\beta = 90° - \alpha = 90° - 53.1° = 36.9°$$

Alternatively, and as a check,

$$\beta = \tan^{-1}\frac{\text{opp}}{\text{adj}} = \tan^{-1}\frac{3}{4} = \tan^{-1}0.75 = 36.9°$$

---

*The entries in this table form an interesting pattern. Going from left to right, the values of the sine are $\sqrt{0}/2$, $\sqrt{1}/2$, $\sqrt{2}/2$, $\sqrt{3}/2$, and $\sqrt{4}/2$. The cosine entries are the same when going from right to left in the table.

EXAMPLE 3.7. The slope of a roof is given as 3/4, and it is shown by the small triangle drawn on the rafter in Figure 3.10(a). If the roof ridge is 6 ft above the attic floor, how wide is the attic floor? How long should each rafter be if a 2 ft overhang is required? What is the angle α between the rafters and the attic floor?

SOLUTION. The rafter and the attic floor form a triangle similar to a 3–4–5 right triangle, but twice the size, as shown in Figure 3.10(b). Clearly, the sides are 6, 8, and 10 ft long. The attic floor is therefore 16 ft wide. Each rafter must be 10 ft long plus 2 ft for overhang, to total 12 ft in length. The angle α corresponds to the smaller acute angle of a 3–4–5 right triangle. Thus,

$$\alpha = 36.9°$$

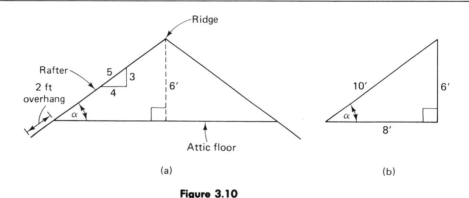

Figure 3.10

Exercises 3.2

1. Calculate the value of each expression to three significant figures. Take care to note that $\sin^2\theta = (\sin\theta)^2$.

    (a) $2\sin 30°$
    (b) $1.8\cos 45°$
    (c) $11\tan 60°$
    (d) $10\cos 30°$
    (e) $5\sin 45°$
    (f) $3\tan 45°$
    (g) $4\tan 30°$
    (h) $6\sin 60°$

    (i) $7\cos 60°$
    (j) $3\cos 90° + 2\sin 30°$
    (k) $5\sin 0° - 2\tan 0°$
    (l) $\sin^2 30° + \cos^2 30°$
    (m) $\sin^2 60° - \cos^2 60°$
    (n) $y = 3\tan^2 45° + 1$
    (o) $v = \sqrt{2}\,\cos 45° + \sqrt{3}\,\sin 60°$

2. A guy wire from a power pole to the ground makes an angle of 60° with the ground at a point 22 ft from the base of the pole. What must be the length of the guy wire?

3. How high above the ground is the guy wire attached to the pole in Problem 2?

4. The angle of depression α in Figure 3.7 is 30°, and the difference in elevation between the two microwave stations is known to be 4000 ft. What must be the line-of-sight distance between the two stations?

5. If the angle α in Figure 3.7 is 45°, and the difference in elevation between the two stations is 2000 ft., what must be the line-of-sight distance between them?

6. Calculate $\sin\alpha$, $\cos\alpha$, $\sin\beta$, and $\cos\beta$ for the triangle in Figure 3.9 using the lengths of the sides. Verify that the functions do indeed correspond to 53.1° and 36.9°.

7. Use the Pythagorean Theorem to verify that the length of the hypotenuse of a 3–4–5 right triangle is indeed 5 if the other two sides have lengths of 3 and 4. In other words, solve for ρ: $\rho=\sqrt{x^2+y^2}$, if $x=3$ and $y=4$.

3.3   SOLVING RIGHT TRIANGLES

To solve a right triangle means to determine the values of the unknown sides and angles in terms of the known sides and angles.

The right triangle in Figure 3.11 has three sides (a, b, and c) and three angles (A, B, and C). Of course $\angle C$ is known to be 90°. In order to find a solution for a right triangle, a certain minimum amount of information must be known in order to completely specify the size and shape of the triangle. When the size and shape are uniquely specified, the triangle is said to be *determined*.

In addition to knowing that $\angle C=90°$, the right triangle in Figure 3.11 is *determined* by:

1. Any two sides, or
2. One side and an acute angle.

In other words:

In order to solve a right triangle, you must know either
(1) The lengths of two sides, or
(2) The length of one side and the size of one acute angle.

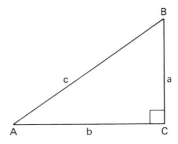

Figure 3.11. A right triangle with standard notation.

EXAMPLE 3.8. The two acute angles of a right triangle are known to be 30° and 60°. Does this determine the triangle?

SOLUTION. Specifying the three angles determines the *shape* of the triangle. But since no mention is made of the length of any side, the triangle can be of any size. In fact, there are an infinite number of *similar* 30°-60°-90° triangles, as suggested in Figure 3.12. Therefore specifying the two acute angles does not determine a right triangle.

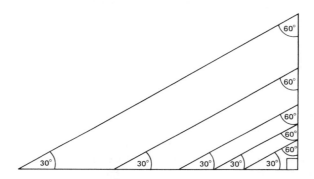

Figure 3.12

Once a right triangle is determined, it can be solved by using the given data in conjunction with the Pythagorean Theorem, the complementary angle formula, and the trigonometric functions. To summarize, the relationships available for solving a right triangle are:

SOLVING RIGHT TRIANGLES

For the right triangle in Figure 3.11:

Pythagorean Theorem: $c^2 = a^2 + b^2$

Complementary angles: $90° = A + B$

Trigonometric functions of an acute angle θ:

$$\sin\theta = \frac{\text{opp}}{\text{hyp}}, \quad \cos\theta = \frac{\text{adj}}{\text{hyp}}, \quad \tan\theta = \frac{\text{opp}}{\text{adj}}$$

As in solving any problem, the solution for a right triangle should proceed in a logical, orderly fashion. The following steps are suggested:

1. Make a careful sketch of the triangle. Label clearly all known quantities. List the unknowns to be determined.

2. To solve for an unknown, select an equation that contains the unknown and two known quantities. Solve for the unknown by substitution. Do calculations to three significant figures and the nearest 0.1. This will provide sufficiently accurate results for most problems.

3. Always compare results with the original sketch as a check on the "reasonableness" of the solution. As an aid, note that the length of one of the sides (*a* or *b* in Figure 3.11) can *never* exceed the length of the hypotenuse (*c*).

The remainder of this section will be devoted to the solutions of a variety of right-triangle problems. The first two examples deal with right triangles determined by the lengths of two sides.

EXAMPLE 3.9. Find the unknown side and the acute angles for the triangle in Figure 3.13.

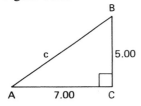

Figure 3.13

SOLUTION. Since *two sides* are specified, the triangle is determined and a solution is possible. The Pythagorean Theorem is used to find the hypotenuse *c*:

$$c = \sqrt{a^2 + b^2}$$

Substitute:

$$c = \sqrt{5^2 + 7^2} = \sqrt{74} = 8.60$$

Then the size of angle *A* can be found using its tangent. Thus,

$$A = \tan^{-1} \frac{\text{opp}}{\text{adj}} = \tan^{-1} \frac{5}{7} = 35.5°$$

The complementary angle formula can next be used to find angle *B* as

$$B = 90° - A = 90° - 35.5° = 54.5°$$

As a check,

$$B = \tan^{-1} \frac{\text{opp}}{\text{adj}} = \tan^{-1} \frac{7}{5} = 54.5°$$

EXAMPLE 3.10. Find the unknown side and the acute angles for the triangle in Figure 3.14.

SOLUTION. This triangle is determined by specifying the lengths of *two sides*, one of which is the hypotenuse. The side *x* is determined using the Pythagorean Theorem:

$$x = \sqrt{Z^2 - R^2} = \sqrt{10^2 - 7^2}$$

$$= \sqrt{100 - 49} = \sqrt{51} = 7.14$$

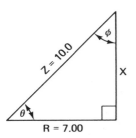

Figure 3.14

The angle θ is found using its cosine:

$$\theta = \cos^{-1} \frac{\text{adj}}{\text{hyp}} = \cos^{-1} \frac{7}{10} = 45.6°$$

The angle ϕ is found using its sine:

$$\phi = \sin^{-1} \frac{\text{opp}}{\text{hyp}} = \sin^{-1} \frac{7}{10} = 44.4°$$

As a check,

$$\theta + \phi = 90°$$

$$45.6° + 44.4° = 90°$$

The next two examples deal with triangles in which the length of one side and the size of one acute angle are known.

EXAMPLE 3.11. Solve the power triangle in Figure 3.15 if $P_A = 50$ and $\theta = 35°$.

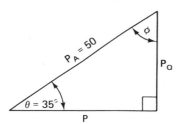

Figure 3.15

SOLUTION. This triangle is determined by *one side* (the hypotenuse) and *an acute angle*. The angle ϕ can be determined as

$$\phi = 90° - \theta = 90° - 35° = 55°$$

To solve for the side P, use the relationship for the $\cos 35°$. Thus,

$$\cos 35° = \frac{\text{adj}}{\text{hyp}} = \frac{P}{50}$$

Solving for P,

$$P = 50 \cos 35° = 50 \times 0.819$$

$$= 41.0$$

To solve for P_Q, use the relationship for the $\sin 35°$. Thus,

$$\sin 35° = \frac{\text{opp}}{\text{hyp}} = \frac{P_Q}{50}$$

Solving for P_Q,

$$P_Q = 50 \sin 35° = 50 \times 0.574$$

$$= 28.7$$

The results of Example 3.11 can be checked by using the Pythagorean Theorem; that is,

$$P_A^2 = P^2 + P_Q^2$$

By substitution,

$$50^2 = 41.0^2 + 28.7^2$$
$$2500 \cong 1681 + 824$$
$$\cong 2505$$

and the result is accurate to three significant figures.

EXAMPLE 3.12. A guy wire attached to the top of a 10 ft TV mast makes an angle of 40° with the mast. Find the length of the guy wire and the distance from the base of the mast to the tie point. A sketch of the mast is shown in Figure 3.16.

SOLUTION. The right triangle to be solved is specified by *one side* (the mast) and a 40° *angle*. The guy wire is the hyp—label it c. The third side of the triangle is the distance from the base of the mast to the tie point—label it b. Use the $\cos 40°$ to determine the guy wire length c. Thus,

$$\cos 40° = \frac{adj}{hyp} = \frac{10}{c}$$

Solving for c,

$$c = \frac{10}{\cos 40°} = \frac{10}{0.766} = 13.1 \text{ ft}$$

Use the $\tan 40°$ to determine the side b. Thus,

$$\tan 40° = \frac{opp}{adj} = \frac{b}{10}$$

Solving for b,

$$b = 10 \tan 40° = 10 \times 0.839 = 8.39 \text{ ft}$$

Clearly, the other acute angle is equal to

$$(90° - 40°) = 50°$$

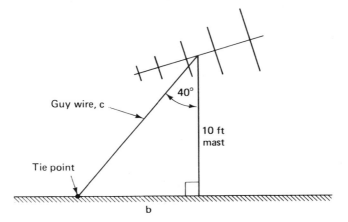

Figure 3.16

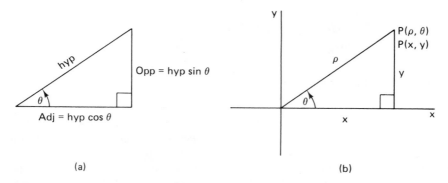

Again, the Pythagorean Theorem can be used as a check on Example 3.12; that is:

$$c^2 = b^2 + 10^2$$
$$13.1^2 \cong 8.39^2 + 10^2$$
$$171.6 \cong 70.4 + 100$$

Example 3.11 illustrates two very useful relationships. Whenever an acute angle θ and the hyp of a right triangle are known, the opp and adj are easily found —see Figure 3.17(a):

$$\text{opp} = \text{hyp} \sin \theta \qquad (3.1)$$
$$\text{adj} = \text{hyp} \cos \theta \qquad (3.2)$$

If θ is drawn in standard position as in Figure 3.17(b), then the hyp and the acute angle θ are seen to be the polar coordinates ρ and θ of a point $P(\rho, \theta)$. Equations (3.1) and (3.2) can then be used to determine the rectangular coordinates x and y of a point $P(x,y)$ using known values of ρ and θ. Thus,

$$y = \rho \sin \theta \qquad (3.3)$$
$$x = \rho \cos \theta \qquad (3.4)$$

The polar coordinates ρ and θ can also be determined from the known values of the rectangular coordinates. First of all, recall from Section 1.4 the equation (1.8), $\rho^2 = x^2 + y^2$. The radius vector is thus,

$$\rho = \sqrt{x^2 + y^2} \qquad (3.5)$$

Also, referring to Figure 3.17(b), it is clear from the definition of the tangent that

$$\theta = \tan^{-1} \frac{y}{x} \qquad (3.6)$$

EXAMPLE 3.13. Find the rectangular coordinates of the point $P(\rho, \theta) = 5 \angle 30°$. As a check, convert the calculated rectangular coordinates back into polar coordinates.

SOLUTION. Polar to rectangular using Eqs. (3.3) and (3.4):

$$y = \rho \sin \theta = 5 \sin 30° = 2.50$$
$$x = \rho \cos \theta = 5 \cos 30° = 4.33$$

Rectangular to polar using Eqs. (3.5) and (3.6):

$$\rho = \sqrt{x^2 + y^2} = \sqrt{2.50^2 + 4.33^2} = 5.00$$

$$\theta = \tan^{-1}\frac{y}{x} = \tan^{-1}\frac{2.50}{4.33} = 30°$$

Note: If your calculator provides direct polar ↔ rectangular conversion, you should learn how to solve such problems at this point.

Exercises 3.3

The triangle shown in Figure 3.18 is known as an *impedance* triangle and it is used in many ac circuit problems. The side Z is called the *impedance*, the side X is the *reactance*, and the side R is the *resistance*. All three sides (Z, X, and R) are measured in units of ohms (Ω). The angle θ is known as the *phase angle*. The angle θ is unnamed and is simply labeled for convenience. Use this impedance triangle in Problems 1 through 25, and solve for the unknowns.

Z = impedance
X = reactance
R = resistance
θ = phase angle

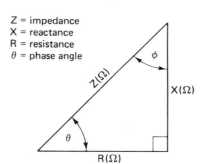

Figure 3.18. An impedance triangle.

Triangles determined by two sides:

1. $Z = 10\ \Omega,\ R = 5\ \Omega$
2. $Z = 150\ \Omega,\ X = 90\ \Omega$
3. $R = 50\ \Omega,\ X = 50\ \Omega$
4. $Z = 2\ k\Omega,\ X = 1\ k\Omega$
5. $Z = 5\ k\Omega,\ R = 3\ k\Omega$

6. $X = 86\ \Omega,\ R = 43\ \Omega$
7. $R = 0\ \Omega,\ X = 2.7\ k\Omega$
8. $R = 100\ \Omega,\ X = 0\ \Omega$
9. $Z = 500\ \Omega,\ X = 0\ \Omega$

Triangles determined by the length of one side and an acute angle:

10. $Z = 500\ \Omega,\ \theta = 30°$
11. $Z = 3.3\ k\Omega,\ \theta = 60°$

12. $Z = 53\ \Omega,\ \theta = 45°$
13. $R = 250\ \Omega,\ \theta = 40°$

14. $R = 75 \ \Omega, \ \theta = 50°$ 18. $R = 3.3 \ \text{k}\Omega, \ \theta = 0°$

15. $X = 10 \ \Omega, \ \theta = 10°$ 19. $X = 800 \ \Omega, \ \theta = 90°$

16. $X = 2.5 \ \text{k}\Omega, \ \theta = 78°$ 20. $Z = 750 \ \Omega, \ \theta = 0°$

17. $X = 950 \ \Omega, \ \theta = 28°$

Use Eqs. (3.1) and (3.2) to determine R and X if:

21. $Z = 75 \ \Omega, \ \theta = 30°$ 24. $Z = 100 \ \Omega, \ \theta = 0°$

22. $Z = 1.5 \ \text{k}\Omega, \ \theta = 45°$ 25. $Z = 5 \ \text{k}\Omega, \ \theta = 90°$

23. $Z = 300 \ \Omega, \ \theta = 60°$

The triangle shown in Figure 3.19 is known as a power triangle, and it is used in many ac circuit problems. The side P_A is the *apparent power* measured in voltamperes (VA), the side P is the *real power* measured in watts (W), and the side P_Q is the *reactive power* measured in vars (var). The angle θ is the *phase angle*. Angle ϕ is unnamed and is simply labeled for convenience. Use this power triangle in Problems 26 through 43, and solve for the unknowns.

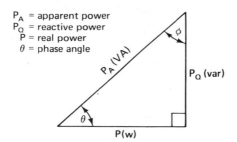

P_A = apparent power
P_Q = reactive power
 P = real power
 θ = phase angle

Figure 3.19. A power triangle.

Triangles determined by two sides:

26. $P_A = 50 \ \text{VA}, \ P = 30 \ \text{W}$ 29. $P = 75 \ \text{mW}, \ P_Q = 16 \ \text{mvar}$

27. $P_A = 1.5 \ \text{VA}, \ P_Q = 0.7 \ \text{var}$ 30. $P_A = 250 \ \text{mVA}, \ P = 250 \ \text{mW}$

28. $P_Q = 5.1 \ \text{var}, \ P = 23 \ \text{W}$ 31. $P_Q = 6.75 \ \text{var}, \ P = 6.75 \ \text{W}$

Triangles determined by the length of one side and an acute angle:

32. $P_A = 65 \ \text{VA}, \ \theta = 30°$ 36. $P = 73.5 \ \text{mW}, \ \theta = 11°$

33. $P_A = 85.2 \ \text{mVA}, \ \theta = 45°$ 37. $P_Q = 3.91 \ \text{var}, \ \theta = 87°$

34. $P_A = 1.73 \ \text{VA}, \ \theta = 60°$ 38. $P_Q = 17.2 \ \text{mvar}, \ \theta = 5.3°$

35. $P = 33.5 \ \text{W}, \ \theta = 18.7°$ 39. $P = 1.73 \ \text{W}, \ \theta = 86.3°$

Use Eqs. (3.1) and (3.2) to determine P and P_Q if:

40. $P_A = 11 \ \text{VA}, \ \theta = 6.5°$ 42. $P_A = 3.72 \ \text{VA}, \ \theta = 90°$

41. $P_A = 159 \ \text{mVA}, \ \theta = 0°$ 43. $P_A = 16.7 \ \text{mVA}, \ \theta = 21.3°$

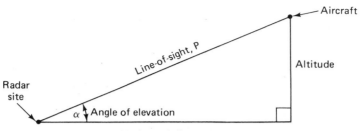

Figure 3.20

44. A radar antenna measures angle of elevation α and line-of-sight distance ρ to an aircraft as shown in Figure 3.20. Determine the altitude of the aircraft and its horizontal distance from the radar site if:

    (a) $\alpha = 30°$, $\rho = 500$ m   (c) $\alpha = 18.3°$, $\rho = 1.73$ km
    (b) $\alpha = 21.2°$, $\rho = 12{,}400$ m   (d) $\alpha = 41.3°$, $\rho = 3{,}820$ m

45. Two microwave repeater stations are located as shown in Figure 3.21. Determine the angle of elevation α and the angle of depression β if:

    (a) $d = 3.7$ km, $h = 1.72$ km
    (b) $d = 14.3$ km, $h = 251$ m
    (c) $d = 5.13$ mi, $h = 410$ ft

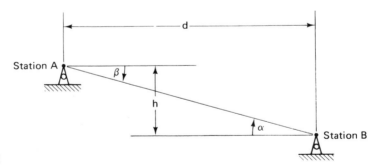

Figure 3.21

46. Determine the line-of-sight distances between stations A and B in each case in Problem 45 above.

47. Use Eqs. (3.3) and (3.4) to determine the rectangular coordinates x and y of a point if the polar coordinates ρ and θ are:

    (a) $\rho = 11$, $\theta = 15°$ (c) $\rho = 34.7$, $\theta = 46.3°$
    (b) $\rho = 1.75$, $\theta = 67.2°$ (d) $\rho = 3.7$, $\theta = 36.4°$

48. Determine the polar coordinates ρ and θ of the point $P(\rho,\theta)$ in Figure 3.17 if its rectangular coordinates are:

    (a) $x = 3$, $y = 7$ (b) $x = 6.5$, $y = 1.32$

    (c) $x = 21.4, y = 25.6$ (f) $x = 0, y = 21$

    (d) $x = 0.72, y = 0.56$ (g) $x = 73.4, y = 0$

    (e) $x = 1\text{-}3/8, y = 2\text{-}1/5$ (h) $x = 1\text{-}7/8, y = 6.92$

49. A piece of conduit is bent such that it has a rise of 34 in. in a distance of 60 in. as shown in Figure 3.22. Determine the angles α and β.

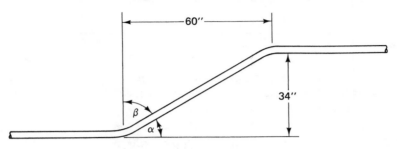

Figure 3.22

UNIT 4
TRIGONOMETRIC FUNCTIONS
OF ANY ANGLE

4.1 RELATED ANGLES

The Trigonometric Table of Natural Sines, Cosines, and Tangents in the Appendix contains values of the trigonometric functions of acute angles. However, the problem of determining a trigonometric function of an angle greater than 90° is readily reduced to finding the same function of an acute angle. In order to determine the functions of angles greater than 90°, the *Related Angle Theorem* is used.

RELATED ANGLE THEOREM

The *related angle* θ_r of any given angle θ is the positive acute angle between the x-axis and the terminal side of θ.

First Quadrant (I)

If the angle θ is in the first quadrant as shown in Figure 4.1(a), then the related angle θ_r is exactly the same as θ. Thus, in quadrant I,

$$\theta_r = \theta$$

Second Quadrant (II)

If the angle θ is in quadrant II as shown in Figure 4.1(b), then θ_r is equal to $180° - \theta$. Thus, in quadrant II,

$$\theta_r = 180° - \theta$$

Third Quadrant (III)

If the angle θ is in quadrant III as shown in Figure 4.1(c), then θ_r is equal to $\theta - 180°$. Thus, in quadrant III,

$$\theta_r = \theta - 180°$$

Fourth Quadrant (IV)

If the angle θ is in quadrant IV as shown in Figure 4.1(d), then θ_r is equal to $360° - \theta$. Thus, in quadrant IV,

$$\theta_r = 360° - \theta$$

To summarize, the related angle θ_r of a given angle θ is shown in Table 4.1 for all four quadrants.

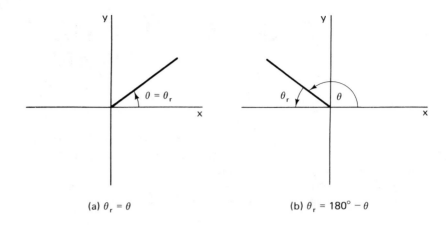

(a) $\theta_r = \theta$ (b) $\theta_r = 180° - \theta$

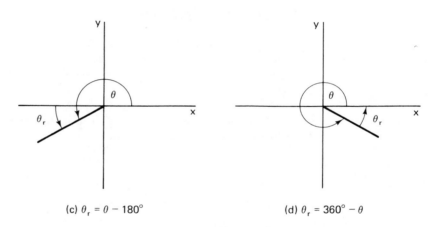

(c) $\theta_r = \theta - 180°$ (d) $\theta_r = 360° - \theta$

Figure 4.1. Related angles.

Table 4.1

Quadrant	Related Angle θ_r
I	$\theta_r = \theta$
II	$180° - \theta$
III	$\theta - 180°$
IV	$360° - \theta$

EXAMPLE 4.1. Find the related angles for θ if:
(a) in Figure 4.1(a), $\theta = 35°$
(b) in Figure 4.1(b), $\theta = 160°$
(c) in Figure 4.1(c), $\theta = 215°$
(d) in Figure 4.1(d), $\theta = 320°$

SOLUTION. Using Table 4.1:
(a) $\theta_r = \theta = 35°$
(b) $\theta_r = 180° - \theta = 180° - 160° = 20°$
(c) $\theta_r = \theta - 180° = 215° - 180° = 35°$
(d) $\theta_r = 360° - \theta = 360° - 320° = 40°$

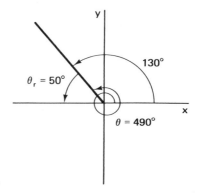

Figure 4.2. Related angle of 490°.

The related angle of an angle greater than 360° (more than a single revolution of the radius vector) is found by first subtracting 360° for each complete revolution, and then using Table 4.1. For example, the related angle of 490° is the same as the related angle of $(490° - 360°) = 130°$. Thus, for $\theta = 490°$,

$$\theta_r = 180° - 130° = 50°$$

as seen in Figure 4.2.

Now, here is how the concept of a related angle can be used to determine the trigonometric functions of an angle greater than 90°. An acute angle θ is drawn in standard position in Figure 4.3(a). The lengths of the sides of the reference triangle are ρ, x, and y. The sine, cosine, and tangent of θ are:

$$\sin\theta = \frac{y}{\rho} \qquad \cos\theta = \frac{x}{\rho} \qquad \tan\theta = \frac{y}{x}$$

An angle in the second quadrant θ_2 is drawn in Figure 4.3(b), and its related angle θ_r is exactly equal to the angle θ in Figure 4.3(a). The radius vector ρ has been chosen the same length for both angles, and thus the two reference triangles are identical. The coordinates for the point P_2 in Figure 4.3(b) are thus $P_2(-x, y)$. The sine, cosine, and tangent for θ_2 are then:

$$\sin\theta_2 = \frac{y}{\rho} \qquad \cos\theta_2 = \frac{-x}{\rho} \qquad \tan\theta_2 = \frac{y}{-x}$$

By comparison with the functions for the angle θ,

$$\sin\theta_2 = \frac{y}{\rho} = \sin\theta \qquad \cos\theta_2 = \frac{-x}{\rho} = -\cos\theta \qquad \tan\theta_2 = \frac{y}{-x} = -\tan\theta$$

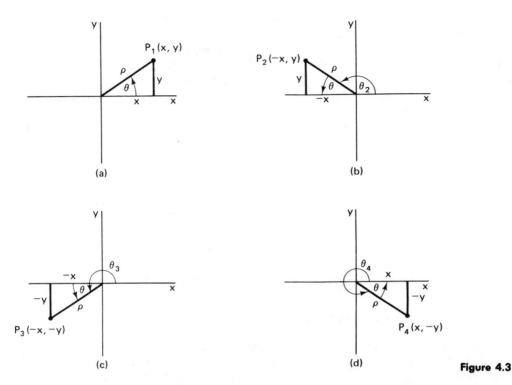

Figure 4.3

But since θ is the related angle of θ_2 in Figure 4.3(b), it is easy to see that the sine of any angle in the second quadrant is exactly equal to the sine of its related angle. The cosine and tangent of any angle in the second quadrant have exactly the same *magnitude* as the related angle, but they are negative in *sign*.

An angle θ_3 in the third quadrant is drawn in Figure 4.3(c) such that its related angle is exactly equal to θ in Figure 4.3(a). The radius vector still has length ρ, and the coordinates of the reference point are $P_3(-x, -y)$. The sine, cosine, and tangent of θ_3 are then:

$$\sin\theta_3 = \frac{-y}{\rho} \qquad \cos\theta_3 = \frac{-x}{\rho} \qquad \tan\theta_3 = \frac{-y}{-x} = \frac{y}{x}$$

Comparing with the functions for the angle θ,

$$\sin\theta_3 = -\sin\theta \qquad \cos\theta_3 = -\cos\theta \qquad \tan\theta_3 = \tan\theta$$

Again, since θ is the related angle of θ_3, the sine and cosine of any third quadrant angle are simply the negative of the sine and cosine of its related angle, while the tangent is exactly the same.

An angle θ_4 is drawn in the fourth quadrant in Figure 4.3(d), and its related angle is also equal to θ in Fig. 4.3(a). By a similar argument,

$$\sin\theta_4 = \frac{-y}{\rho} = -\sin\theta \qquad \cos\theta_4 = \frac{x}{\rho} = \cos\theta \qquad \tan\theta_4 = \frac{-y}{x} = -\tan\theta$$

Thus, the trigonometric functions of any angle in the fourth quadrant are seen to be the same as the functions of its related angle, except for signs.

A similar argument holds for the secant, cosecant, and cotangent since they are simply the reciprocals of the sine, cosine, and tangent. Thus the following theorem:

RELATED ANGLE THEOREM

The trigonometric function of any angle has exactly the same *magnitude* as the same function of its related angle. The appropriate *sign* is determined according to Table 4.2.

Table 4.2

QUADRANT	$\sin\theta/\csc\theta$	$\cos\theta/\sec\theta$	$\tan\theta/\cot\theta$
I	+	+	+
II	+	−	−
III	−	−	+
IV	−	+	−

EXAMPLE 4.2. Determine the sine and cosine of a 150° angle using related angles.

SOLUTION. The related angle of 150° is

$$\theta_r = 180° - \theta = 180° - 150° = 30°$$

Since 150° is in quadrant II, the sine is *positive* and the cosine is *negative*. Thus,

$$\sin 150° = +\sin 30° = 0.500$$
$$\cos 150° = -\cos 30° = -0.866$$

EXAMPLE 4.3. Find the tangent of 150° using related angles.

SOLUTION. The related angle of 150° is

$$\theta_r = 180° - 150° = 30°$$

Since 150° is in quadrant II, the tangent is *negative*, and thus,

$$\tan 150° = -\tan 30° = -0.577$$

EXAMPLE 4.4. Use related angles to determine:
(a) $\sin 250°$
(b) $\cos 310°$
(c) $\tan 300°$

SOLUTION.
(a) The related angle of 250° is

$$\theta_r = 250° - 180° = 70°$$

(d) $\cot 210°$

250° is in quadrant III, and the sine is *negative*. Thus,

$$\sin 250° = -\sin 70° = -0.940$$

(b) The related angle is

$$\theta_r = 360° - 310° = 50°$$

The cosine is *positive* in quadrant IV, and thus,

$$\cos 310° = +\cos 50° = 0.643$$

(c) The related angle is

$$\theta_r = 360° - 300° = 60°$$

The tangent is *negative* in quadrant IV, and thus,

$$\tan 300° = -\tan 60° = -1.732$$

(d) First determine $\tan 210°$. The related angle is

$$\theta_r = 210° - 180° = 30°$$

In quadrant III the tangent is *positive*. Thus,

$$\tan 210° = \tan 30° = 0.577$$

Using the reciprocal relationship,

$$\cot 210° = \frac{1}{\tan 210°} = \frac{1}{0.577} = 1.732$$

Exercises 4.1

1. Find the related angle θ_r for each given angle. In each case draw the given angle in standard position and show θ_r:

    (a) 60°   (b) 160°   (c) 135°
    (d) 241°   (e) 319°   (f) 185°
    (g) 291°   (h) 252°   (i) 340°
    (j) 420°   (k) 862°   (l) 1000°

2. Find the sine and cosine of each given angle by using related angles:

    (a) 60°   (b) 160°   (c) 135°
    (d) 241°   (e) 319°   (f) 185°
    (g) 291°   (h) 252°   (i) 340°

3. Evaluate, using related angles:

   (a) $\tan 195°$
   (b) $\cos 283°$
   (c) $\cot 310°$
   (d) $1 - \sin 193°$
   (e) $15 \cos 211°$
   (f) $7.6 \sin 129°$

   (g) $167 \cos 315°$
   (h) $21 - 15 \sin 300°$
   (i) $1.95 \cos 99°$
   (j) $\sin^2 411° + \cos^2 411°$
   (k) $31.5 \sin 743°$
   (l) $9.3 \cos 1100°$

4. Can you find two angles in two different quadrants such that:

   (a) $|\sin \theta| = 0.500?$
   (b) $|\cos \theta| = 0.500?$
   (c) $|\tan \theta| = 1.000?$

4.2 TRIGONOMETRIC FUNCTIONS OF A NEGATIVE ANGLE

Let θ be an angle generated by rotation of a radius vector ρ in a CCW direction as shown in Figure 4.4. Then a *negative* angle $(-\theta)$ is generated by rotation of the radius vector through the same angle, but in a CW direction. The reference triangle for θ is exactly the same size as the reference triangle for $-\theta$. The reference point P_+ for θ has coordinates $P_+(x,y)$, and the reference point P_- for $-\theta$ has coordinates $P_-(x, -y)$.

The sine, cosine, and tangent for the two angles are:

(for θ)   (for $-\theta$)

$\sin\theta = \dfrac{y}{\rho}$ $\sin(-\theta) = \dfrac{-y}{\rho}$

$\cos\theta = \dfrac{x}{\rho}$ $\cos(-\theta) = \dfrac{x}{\rho}$

$\tan\theta = \dfrac{y}{x}$ $\tan(-\theta) = \dfrac{-y}{x}$

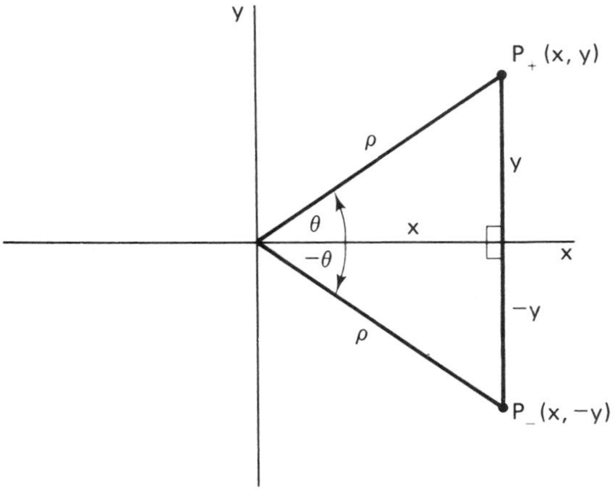

Figure 4.4

By direct comparison,

$$\sin(-\theta) = \frac{-y}{\rho} = -\sin\theta$$

$$\cos(-\theta) = \frac{x}{\rho} = \cos\theta$$

$$\tan(-\theta) = \frac{-y}{x} = -\tan\theta$$

If this argument is repeated for an angle θ drawn in each of the other three quadrants, the same results will be obtained. Thus, the following theorem:

NEGATIVE ANGLE THEOREM

For any angle θ,

$$\sin(-\theta) = -\sin\theta$$
$$\cos(-\theta) = \cos\theta$$
$$\tan(-\theta) = -\tan\theta$$

The secant, cosecant, and cotangent of a negative angle are determined by using the reciprocal relationships.

EXAMPLE 4.5. Show the validity of the Negative Angle Theorem for an angle θ drawn in quadrant II.

SOLUTION. An angle θ and an angle $-\theta$ are shown in Figure 4.5. The related angle for θ is θ_r, and the related angle for $-\theta$ is $-\theta_r$ as shown. The reference triangles for each angle are identical, and the two reference points have coordinates $P_+(-x,y)$ and $P_-(-x,-y)$. The sine, cosine, and tangent for each angle are found using the related angles.

$$\begin{array}{cc} \text{(for } \theta) & \text{(for } -\theta) \\[4pt] \sin\theta = \sin\theta_r = \dfrac{y}{\rho} & \sin(-\theta) = \sin(-\theta_r) = \dfrac{-y}{\rho} \\[12pt] \cos\theta = \cos\theta_r = \dfrac{-x}{\rho} & \cos(-\theta) = \cos(-\theta_r) = \dfrac{-x}{\rho} \\[12pt] \tan\theta = \tan\theta_r = \dfrac{y}{-x} & \tan(-\theta) = \tan(-\theta_r) = \dfrac{-y}{-x} \end{array}$$

By direct comparison,

$$\sin(-\theta) = -\sin\theta$$

$$\cos(-\theta) = \cos\theta$$

$$\tan(-\theta) = -\tan\theta$$

and the theorem is verified for second quadrant angle.

EXAMPLE 4.6. Determine the sine, cosine, and tangent of $-65°$.

SOLUTION. Using the Negative Angle Theorem:

$$\sin(-65°) = -\sin 65° = -0.906$$

$$\cos(-65°) = \cos 65° = 0.423$$

$$\tan(-65°) = -\tan 65° = -2.15$$

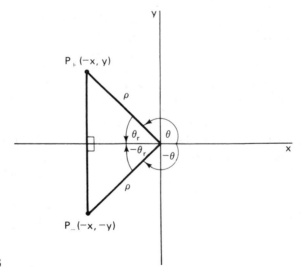

Figure 4.5

Exercises 4.2

1. Use the Negative Angle Theorem to determine the sine, cosine, and tangent of the given angles:

   (a) $-11°$ (b) $-74°$ (c) $-121°$

   (d) $-310°$ (e) $-241°$ (f) $-322°$

   (g) $-198°$ (h) $-95°$ (i) $-45°$

2. Evaluate the following:

(a) $11\cos(-50°)$ (b) $3.8\sin(-30°)$ (c) $6\sin(-40°)+2$

(d) $\dfrac{5\cos(-15°)}{3\sin(-15°)}$ (e) $\dfrac{1.5\sin(-60°)}{\cos(-60°)}$ (f) $\dfrac{2.2\sin(-30°)}{\cos(40°)}$

4.3   TRIGONOMETRIC FUNCTIONS OF ANGLES MEASURED IN RADIANS

There may be occasions where it is desired to determine the trigonometric functions of an angle expressed in radians. Such problems can be resolved by changing the given angle from radians to degrees and then determining the functions. Or the functions can be determined directly from the Table of Trigonometric Functions in Radian Measure, as given in the Appendix. Furthermore, most electronic hand calculators provide the sine, cosine, and tangent of angles entered in radian measure.

The Table of Trigonometric Functions in Radian Measure contains the values of functions for acute angles. The column on the left gives the value of the angle in radians to two decimal places. The first entry is 0.00 radians (corresponding to 0°). Values are then given for angles up to 1.57 radians (1.57 rad ≅ 90°) in steps of 0.01 radians. In other words, it is possible to find the value of a function for an angle between 0.00 rad and 1.57 rad (0° to 90°) to the nearest 0.01 rad.

EXAMPLE 4.7. Use the Table of Trigonometric Functions in Radian Measure to evaluate the following functions. The angles are given in radians. (a) $\sin 0.15$ (b) $\cos 0.62$ (c) $\tan 1.45$ Verify the results by using an electronic hand calculator.	SOLUTION. Use the Table of Trigonometric Functions in Radian Measure: (a) Find 0.15 in the left-hand column under *rad*. Move to the right and read "0.1494" uner *sin* in the first column. Thus, $$\sin 0.15 = 0.1494$$ (b) Locate 0.62 in the left-hand column. Move to the right under *cos* and read "0.8139." Thus, $$\cos 0.62 = 0.8139$$ (c) Locate 1.45 under *rad* in the left-hand column. To the right of this under *tan*, read "8.238." Thus, $$\tan 1.45 = 8.238$$

The Table of Trigonometric Functions in Radian Measure in the Appendix is for acute angles only. Fortunately, the techniques in both Sections 4.1 and 4.2 can be applied equally to angles measured in either degrees or radians. Thus to find the trigonometric functions of an angle greater than 1.57 rad (90°), the Related Angle

Theorem is used. Of course, the angular measure is now in radians rather than degrees, and the related angle is determined by using π rad instead of 180°, and 2π rad instead of 360°. To summarize, the related angle θ_r of a given angle θ measured in radians is shown in Table 4.3 below for all four quadrants.

Table 4.3

QUADRANT	RELATED ANGLE θ_r
I	$\theta_r = \theta$
II	$\pi - \theta$
III	$\theta - \pi$
IV	$2\pi - \theta$

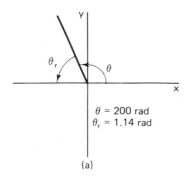

θ = 200 rad
θ_r = 1.14 rad

(a)

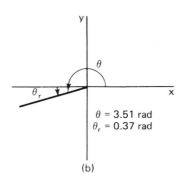

θ = 3.51 rad
θ_r = 0.37 rad

(b)

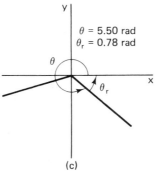

θ = 5.50 rad
θ_r = 0.78 rad

(c)

Figure 4.6

EXAMPLE 4.8. Find the related angle θ_r if:
(a) $\theta = 2.00$ rad
(b) $\theta = 3.51$ rad
(c) $\theta = 5.50$ rad

SOLUTION. Using Table 4.3:
(a) 2.00 rad is an angle in quadrant II as shown in Figure 4.6(a). Its related angle θ_r is:

$$\theta_r = \pi - \theta = 3.1416 - 2.00 = 1.1412 \text{ rad}$$

(b) 3.51 rad is an angle in quadrant III as shown in Figure 4.6(b). Its related angle θ_r is:

$$\theta_r = \theta - \pi = 3.51 - 3.1416 = 0.368 \text{ rad}$$

(c) 5.50 rad is an angle in quadrant IV as shown in Figure 4.6(c). Its related angle θ_r is:

$$\theta_r = 2\pi - \theta = 6.2832 - 5.50 = 0.783 \text{ rad}$$

EXAMPLE 4.9. Use the results of Example 4.8 above and the trigonometric tables in the Appendix to determine:
(a) $\sin 2.00$
(b) $\cos 3.51$ rad
(c) $\tan 5.5$ rad

SOLUTION. Using the Related Angle Theorem:
(a) The sine is positive in quadrant II. The related angle $\theta_r \cong 1.14$ rad is found in the Appendix tables. Thus,

$$\sin 2.00 = +\sin 1.14 = 0.9086$$

You should verify this solution by determining $\sin 2.00$ on an electronic hand calculator.

(b) The cosine is negative in quadrant III. The related angle $\theta_r \cong 0.37$ is found in the tables, and

$$\cos 3.51 = -\cos 0.37 = -0.9323$$

(c) The tangent is negative in quadrant IV. The related angle is $\theta_r \cong 0.78$ rad, and

$$\tan 5.5 = -\tan 0.78 = -0.9893$$

You should verify $\cos 3.51$ and $\tan 5.5$ using an electronic hand calculator.

Exercises 4.3

1. Use the trigonometric tables in the Appendix to evaluate the following functions of angles given in radians. Verify results using an electronic hand calculator.

    (a) $\sin 0.81$ (b) $\cos 0.12$ (c) $\tan 1.23$
    (d) $\cos 1.50$ (e) $\tan 0.38$ (f) $\sin 1.11$
    (g) $\tan 0.74$ (h) $\sin 0.33$ (i) $\cos 0.48$

2. Determine the related angle θ_r in radians for each given angle θ:

    (a) 1.63 rad   (b) 4.88 rad   (c) 4.11 rad
    (d) 6.00 rad   (e) 3.00 rad   (f) 4.00 rad

3. Determine the sine, cosine, and tangent of each given angle by using the related angle theorem. Verify results by direct determination of each function using an electronic hand calculator.

    (a) 1.63 rad   (b) 4.88 rad   (c) 4.11 rad
    (d) 6.00 rad   (e) 3.00 rad   (f) 4.00 rad

UNIT 5
GRAPHS OF TWO
TRIGONOMETRIC FUNCTIONS*

5.1 $y = \rho \sin \theta$

The Sine Function

A radius vector of length ρ, drawn at an angle θ, is shown on a standard x-y coordinate system in Figure 5.1. The point P is uniquely determined, either by specifying its polar coordinates ρ and θ, or by using its rectangular coordinates x and y. There are of course definite relationships between the polar and the rectangular coordinates. In fact, it has been shown previously that the x and y coordinates can be found in terms ρ and θ by using the following equations:

$$y = \rho \sin \theta \qquad (5.1)$$
$$x = \rho \cos \theta \qquad (5.2)$$

For instance, if $\rho = 2.00$ and $\theta = 30°$, then

$$y = \rho \sin \theta = 2.00 \sin 30° = 2.00 \times 0.500 = 1.00$$
$$x = \rho \cos \theta = 2.00 \cos 30° = 2.00 \times 0.866 = 1.73$$

A radius vector having a length $\rho = 1.00$ is known as a *unit vector*. In Figure 5.2(a), a unit vector is drawn with a zero angle—that is, $\theta_a = 0°$. This vector is then rotated CCW, and Figures 5.2(b) through (h) show the vector at various angles of rotation, namely, $\theta_b, \theta_c, \theta_d, \ldots, \theta_h$. For each of these angles of rotation, there are unique values for y and x. Equation (5.1) can be used to calculate each value of y, and since $\rho = 1.00$, Eq. (5.1) becomes

$$y = \sin \theta$$

Now, visualize the value of y as the vertical line drawn from the x-axis to the tip of the radius vector for each angular position in Figure 5.2.

Figure 5.2(a): The value of y is clearly zero. Note that $\theta_a = 0°$, and $y_a = \sin \theta_a = \sin 0° = 0.00$.

Figure 5.2(b): The value of y is a positive number less than 1.00. This is true since the sine of an angle in the first quadrant is between 0.00 and 1.00, and $y_b = \sin \theta_b$.

Figure 5.2(c): $\theta_c = 90°$, and thus $y_c = \sin \theta_c = \sin 90° = 1.00$.

Figure 5.2(d): y_d is a positive number less than 1.00 since $y_d = \sin \theta_d$, and θ_d is a positive number between 0.00 and 1.00 in the second quadrant.

---

*There are of course graphs for each of the six trigonometric functions, but only the sine and the cosine will be considered here.

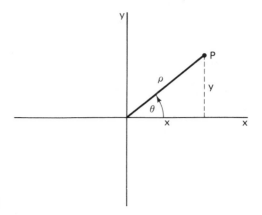

Figure 5.1

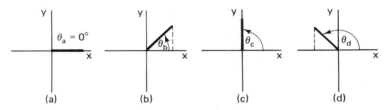

Figure 5.2   (e)   (f)   (g)   (h)

Figure 5.2(e): $\theta_e = 180°$; thus $\sin\theta_e = \sin 180° = 0.00$ and $y_e = \sin\theta_e = 0.00$.

Figure 5.2(f): θ_f is in the third quadrant, and $\sin\theta_f$ has a negative value with a magnitude between 0.00 and 1.00. Thus $y_f = \sin\theta_f$ is a negative number with magnitude less than 1.00.

Figure 5.2(g): $\theta_g = 270°$, and $\sin\theta_g = \sin 270° = -1.00$. Thus, $y_g = \sin\theta_g = -1.00$.

Figure 5.2(h): The value of y_h is again a negative number with magnitude less than 1.00. $y_h = \sin\theta_h$, and $\sin\theta_h$ is negative with a magnitude between 0.00 and 1.00 for values of θ_h in the fourth quadrant.

If the unit vector is rotated to 360°, the value of $\sin 360°$ is the same as $\sin 0°$ given in (a) above. Rotation beyond 360° (one revolution) will simply result in a repeat of the eight positions analyzed above. Notice that for each value θ, there is a unique value of y, and thus y is said to be a function of θ. The equation $y = \sin\theta$ is

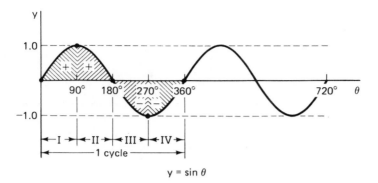

y = sin θ

Figure 5.3

called the *sine function*, and it can be plotted as a smooth curve on a rectangular coordinate system.

By choosing a rectangular coordinate system having y on the vertical axis and θ on the horizontal axis, as shown in Figure 5.3, the information obtained in analyzing Figure 5.2 can be graphed to get an idea of the shape of the sine function. First, $y = \sin\theta$ is equal to zero for $\theta = 0°$, 180°, and 360°. These three points are plotted on Figure 5.3. Next, $y = \sin\theta$ is equal to 1.00 for $\theta = 90°$, and -1.00 for $\theta = 270°$. These points are also plotted on Figure 5.3. Now, $y = \sin\theta$ is positive in quadrants I and II, and negative in quadrants III and IV. When all these points are connected with a smooth continuous curve, the sine function will appear as shown in Figure 5.3. This curve is generally referred to as a *sine wave*. A very accurate curve can of course be plotted by calculating $y = \sin\theta$ for numerous values of θ. The sine wave has its very own special *shape*. It is said to be *sinusoidal*. Note that it is *not* simply *circular* or *elliptical*.

Notice in Figure 5.3 that the sine wave will simply repeat as θ varies from 360° to 720°, and indeed for any 360° segment selected along the horizontal axis. One *cycle* of the sine wave is generated by one single rotation of the radius vector through an angle of 360° (beginning in this case at 0°). Thus, three cycles of a sine wave would appear as drawn in Figure 5.4. There is of course no reason why $y = \sin\theta$ could not be plotted for negative values of θ; the wave form would simply be extended to the left of the y-axis, as shown in Figure 5.4.

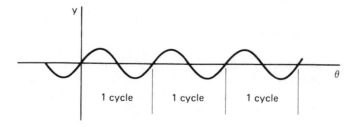

Figure 5.4

EXAMPLE 5.1. Calculate values for $y = \sin\theta$ as θ varies from 0° to 360° in 30° increments. Plot the resulting sine wave.

SOLUTION. Make a table of values and write in the given values of θ. Then calculate the appropriate values of y using a hand calculator (or tables).

$$\theta = 0°, y = \sin\theta = \sin 0° = 0$$

$$\theta = 30°, y = \sin\theta = \sin 30° = 0.500$$

⋮

θ	y
0°	0
30°	0.500
60°	0.866
90°	1.000
120°	0.866
150°	0.500
180°	0
210°	−0.500
240°	−0.866
270°	−1.000
300°	−0.866
330°	−0.500
360°	0

The curve $y = \sin\theta$ is plotted in Figure 5.5 using these data.

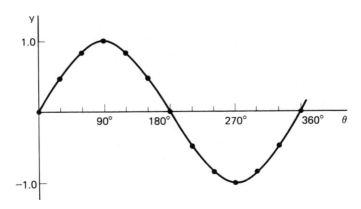

Figure 5.5

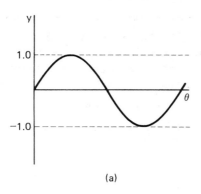

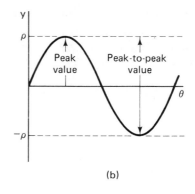

(a)   (b)

Figure 5.6

Amplitude

The previous discussion has developed the concept of a sinusoidal wave form plotted as $y = \sin \theta$. In this particular discussion, the value of ρ is chosen to be 1.00, and thus the value of y is always within the limit $-1 \leqslant y \leqslant 1$, as shown in Figure 5.6(a).

If ρ is allowed to have a positive value other than 1.00, then the values of y are calculated according to $y = \rho \sin \theta$. This is another sine wave, having the same sinusoidal shape, but the values of y are now confined to the limits $-\rho \leqslant y \leqslant \rho$, as seen in Figure 5.6(b). Thus, positive values of $\rho > 1.0$ will *amplify* the basic wave form (increase its limits), while positive values of $\rho < 1.0$ will *attenuate* the basic wave form (reduce its limits).

For instance, if $\rho = 7.5$, then at $\theta = 90°$

$$y = \rho \sin \theta = 7.5 \sin 90° = 7.5 \times 1.0 = 7.5$$

Note y is 7.5 times greater than when $\rho = 1.0$. The basic wave form has been amplified by 7.5 times.

Or, if $\rho = 0.3$, then at $\theta = 90°$

$$y = \rho \sin \theta = 0.3 \sin 90° = 0.3 \times 1.0 = 0.3$$

The basic wave form has been attenuated to 0.3 times its original value.

The *peak value* or *amplitude* of a sinusoidal wave form is simply the maximum positive value of the wave taken when $\theta = 90°$. This is shown in Figure 5.6(b), and clearly the peak value is equal to ρ.

Another important quantity associated with a sine wave is its *peak-to-peak value*. This is simply twice the peak value as shown in Figure 5.2(b), and is clearly equal to 2ρ.

| EXAMPLE 5.2. What is the peak value and the peak-to-peak value of $y = 13 \sin \theta$? | SOLUTION. The peak value y_p is simply the coefficient of $\sin \theta$, or 13. The peak-to-peak value y_{p-p} is twice the peak value, or,

 $$y_{p-p} = 2y_p = 26$$ |

EXAMPLE 5.3.   Calculate values for $y = 3\sin\theta$ as θ varies from 0° to 360° in 30° increments.

SOLUTION.   The table of values will be exactly as in Example 5.1, except the y values will be multiplied by 3. That is:

$$\theta = 0°, y = 3\sin 0° = 3 \times 0 = 0$$

$$\theta = 30°, y = 3\sin 30° = 3 \times 0.500 = 1.50$$

$$\vdots$$

The wave form will be exactly as in Figure 5.5 if the y coordinates are multiplied by 3.

Radian Measure

In some applications of sinusoidal wave forms involving calculus, it is considerably more convenient to use a coordinate system where 1.0 on the vertical axis is also equal to 1.0 on the horizontal axis. As a result, you may very well encounter sinusoidal wave forms plotted with θ measured in radians rather than degrees. In this case, the number 1.0 on the vertical scale represents 1.0 radians on the horizontal scale.

It is really no problem to plot $y = \rho\sin\theta$ where θ is measured in radians. Simply proceed as before, but determine values for $\sin\theta$ using a calculator (or tables) with θ entered in radians.

For convenience, the radian measure of θ is sometimes given as a fraction. Some of the more common values are given in Table 5.1.

Table 5.1

θ (degrees)	0°	30°	45°	60°	90°	120°	135°	150°	180°
θ (radians)	0	$\dfrac{\pi}{6}$	$\dfrac{\pi}{4}$	$\dfrac{\pi}{3}$	$\dfrac{\pi}{2}$	$\dfrac{2\pi}{3}$	$\dfrac{3\pi}{4}$	$\dfrac{5\pi}{6}$	π

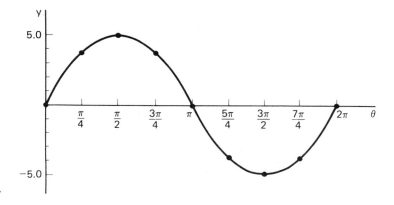

Figure 5.7

EXAMPLE 5.4.   Calculate values for $y = 5 \sin \theta$ as θ varies from 0 to 2π radians in steps of $\pi/4$ radians. Plot the results.

SOLUTION.   Make a table of values containing the given values of θ. Then calculate y for each θ.

$$\theta = 0, y = 5 \sin \theta = 5 \times 0 = 0$$

$$\theta = \frac{\pi}{4}, y = 5 \sin \frac{\pi}{4} = 5 \times 0.707 = 3.54$$

$$\theta = \frac{\pi}{2}, y = 5 \sin \frac{\pi}{2} = 5 \times 1.00 = 5.00$$

$$\vdots$$

θ (rad)	0	$\pi/4$	$\pi/2$	$3\pi/4$	π	$5\pi/4$	$3\pi/2$	$7\pi/4$	2π
y	0	3.54	5.00	3.54	0	−3.54	−5.00	−3.54	0

The results are plotted in Figure 5.7.

Exercises 5.1

1. Calculate values for $y = \sin \theta$ as θ varies from 0° to 360° in 40° increments. Include the points $\theta = 0°$, 90°, 180°, and 270°. Plot the results.

2. Calculate values for $y = \sin \theta$ as θ varies from 0° to 360° in 15° increments. Plot the results.

3. Calculate values for $y = -\sin \theta$ as θ varies from 0° to 360° in 30° increments.

4. Evaluate $y = \rho \sin \theta$ for the following:
    (a) $\rho = 2.00, \theta = 75°$
    (b) $\rho = 1.88, \theta = 178°$
    (c) $\rho = 0.61, \theta = 47°$
    (d) $\rho = 0.18, \theta = 315°$
    (e) $\rho = 3 \times 10^{-3}, \theta = 217°$
    (f) $\rho = -21.6, \theta = 117°$
    (g) $\rho = -17.1, \theta = 270°$
    (h) $\rho = 2 \times 10^{2}, \theta = \frac{\pi}{3}$ rad
    (i) $\rho = 6.15, \theta = 1.17$ rad

5. What is the amplitude (peak value) and the peak-to-peak value of:
    (a) $y = 1.8 \sin \theta$?
    (b) $y = 11.6 \sin \theta$?
    (c) $y = 0.18 \sin \theta$?
    (d) $v = 17 \sin \theta$?
    (e) $i = 31 \times 10^{-3} \sin \theta$?
    (f) $v = 167 \sin \theta$?

6. Plot one cycle of the sine wave $y = 1.8 \sin \theta$.

7. Plot one cycle of the sine wave $v = 21 \sin \theta$.

8. Plot one cycle of the sine wave $i = 41 \times 10^{-3} \sin \theta$.

9. Plot two cycles of the sine wave $v = 167 \sin \theta$.

10. Plot two cycles of the sine wave $i = 0.8 \sin \theta$.

11. Write the equation for a sine wave having:
    (a) A peak value of 6.1
    (b) A peak-to-peak value of 5.8
    . (c) A peak value of 0.198
    (d) A peak-to-peak value of 3.4×10^{-3}

12. An equation for the ac voltage supplied by a local power company is $v = 167 \sin \theta$ (volts). Plot one cycle of this sinusoidal voltage.

13. The current in an ac circuit is given as $i = V/R \sin \theta$ (amp). Plot three cycles of i if:
    (a) $V = 10$ V, $R = 12$ Ω
    (b) $V = 167$ V, $R = 85$ Ω
    (c) $V = 16$ V, $R = 1.5$ kΩ
    (d) $V = 2.1$ V, $R = 4.7$ kΩ

14. The ac voltage across the terminals of a resistor is given as $v = IR \sin \theta$ (volts). Plot two cycles of v if:
    (a) $I = 1.7$ A, $R = 55$ Ω
    (b) $I = 3 \times 10^{-3}$ A, $R = 2.2$ kΩ
    (c) $I = 38$ mA, $R = 6.8$ kΩ

15. Calculate values for $y = \sin \theta$ as θ varies from 0 rad to 2π rad in increments of $\pi/6$ rad. Plot the results.

16. Calculate values for $y = 1.8 \sin \theta$ as θ varies from 0 rad to π rad in increments of 0.2 rad. Plot the results.

5.2 $y = \rho \cos \theta$

The Cosine Function

In the previous section, a radius vector is drawn on an x-y-coordinate system in Figure 5.1. The x-coordinate of the point P in Figure 5.1 can be found using Eq. (5.2) (as given previously):

$$x = \rho \cos \theta \tag{5.2}$$

If $\rho = 1.0$, then $x = \rho \cos \theta$ reduces to $x = \cos \theta$. In this equation, for each assigned value of θ, x has a unique value. Therefore, x is said to be a function of θ and the equation $x = \cos \theta$ is called the *cosine function*. The cosine function can be plotted as a smooth curve on a rectangular coordinate system just as was done for the sine function.

A general idea of the shape of the cosine function can be determined by using a unit vector as shown in Figure 5.8. This unit vector is rotated CCW from 0° to 360°, and Figures 5.8(a) through (h) show the vector in various angular positions. The value of x in each figure can be visualized as the horizontal line drawn from the y-axis to the tip of the unit vector.

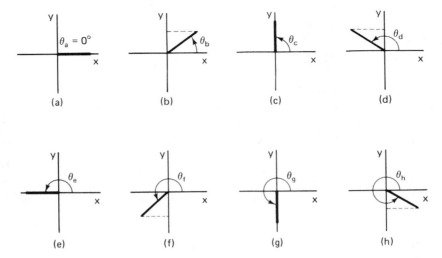

Figure 5.8

1. Figure 5.8(a): Clearly the value of x is 1.0. Note that $\theta_a = 0°$, and $x_a = \cos\theta_a = \cos 0° = 1.0$.

2. Figure 5.8(b): The value of x is a positive number less than 1.0, since $x_b = \cos\theta_b$, and the cosine of an angle in quadrant I is between 1.0 and 0.0.

3. Figure 5.8(c): The value of x is zero. Note that $\theta_c = 90°$ and $x_c = \cos\theta_c = \cos 90° = 0$.

4. Figure 5.8(d): θ_d is in quadrant II and the cosine of θ_d must be negative with magnitude less than 1.0. Thus $x_d = \cos\theta_d$ must be negative with magnitude less than 1.0.

5. Figure 5.8(e): Since $\theta_e = 180°$ and $\cos 180° = -1.0$, the value of x_e must be $x_e = \cos\theta_e = -1.0$.

6. Figure 5.8(f): In quadrant III, the cosine is negative with magnitude less than 1.0, and thus $x_f = \cos\theta_f$ must be negative with magnitude less than 1.0.

7. Figure 5.8(g): $\theta_g = 270°$, and since $\cos 270° = 0$, the value of x_g must be $x_g = \cos\theta_g = 0$.

8. Figure 5.8(h): In quadrant IV, the cosine is positive with magnitude less than 1.0, and thus $x_h = \cos\theta_h$ must be positive with magnitude less than 1.0.

If the unit vector is rotated to 360°, the value of $\cos 360°$ is the same as $\cos 0°$ given in (a) above. Rotation beyond 360° (one revolution) will simply result in a repeat of the eight positions analyzed above.

The information obtained in analyzing Figure 5.8 can be used to sketch a cosine function on a rectangular coordinate system as shown in Figure 5.9; the vertical axis is labeled x.

First, since $\cos\theta$ is equal to 1.0 when $\theta = 0°$ and 360°, plot these two points in Figure 5.9. Then, $\cos\theta = 0$ when $\theta = 90°$ and 270°—locate these two points. $\cos\theta = -1.0$ when $\theta = 180°$; locate this point. Finally, the cosine function is positive in quadrants I and IV, and negative in quadrants II and III. When these points are connected with a smooth continuous curve, the cosine function will appear as shown in Figure 5.9. This curve is generally referred to as a *cosine wave*.

A very accurate cosine wave can of course be plotted by calculating $x = \cos\theta$ for numerous values of θ. The cosine wave has exactly the same shape as the sine wave, and it is therefore *sinusoidal*. In fact, you can visualize a cosine wave as a sine wave that has been shifted to the left a distance of 90° along the θ-axis.

Notice that the cosine wave will simply repeat itself as θ varies from 360° to 720°, and indeed for any 360° segment selected along the horizontal axis. One *cycle* of the cosine wave is generated by one single rotation of the radius vector through an angle of 360° (beginning at 0° in this case).

EXAMPLE 5.5. Calculate values for $x = \cos\theta$ as θ varies from 0° to 360° in 30° increments. Plot the resulting cosine wave.

SOLUTION. Make a table of values and write in the given values of θ. Then, calculate the appropriate values of x using a hand calculator (or tables). Thus,

$$\theta = 0°, \ x = \cos\theta = \cos 0° = 1.0$$

$$\theta = 30°, \ x = \cos\theta = \cos 30° = 0.866$$

$$\vdots$$

θ	x
0°	1.00
30°	0.866
60°	0.500
90°	0
120°	−0.500
150°	−0.866
180°	−1.0
210°	−0.866
240°	−0.500
270°	0
300°	0.500
330°	0.866
360°	1.000

The curve $x = \cos\theta$ is plotted in Figure 5.10 using these data.

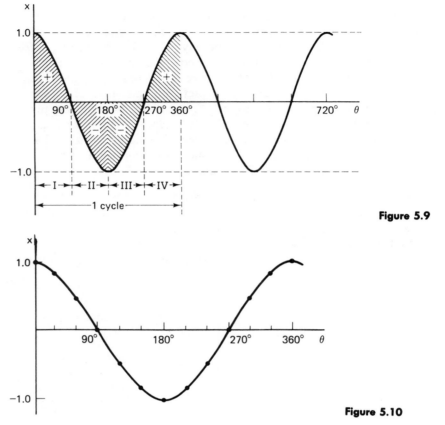

Figure 5.9

Figure 5.10

Amplitude

The general expression for a cosine wave is $x = \rho\cos\theta$, and the coefficient ρ controls the *amplitude* of the wave form just as it did with the sine wave in the previous section. Thus, in Figure 5.11, it is clear that the amplitude or *peak value* of $x = \rho\cos\theta$ is equal to ρ. Also, the peak-to-peak value is twice the peak value as shown.

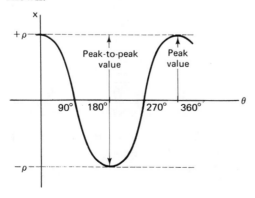

Figure 5.11

EXAMPLE 5.6. What is the peak value, and the peak-to-peak value of $x = 7.5\cos\theta$?	SOLUTION. The peak value x_p is simply the coefficient of $\cos\theta$, or 7.5. The peak-to-peak value x_{p-p} is twice the peak value. So, $$x_{p-p} = 2x_p = 2 \times 7.5 = 15$$
EXAMPLE 5.7. Calculate the values for $x = 4\cos\theta$ as θ varies from 0° to 360° in 30° increments.	SOLUTION. The table of values will be exactly as in Example 5.5, except the x values will each be multiplied by 4. That is: $\theta = 0°$, $x = 4\cos\theta = 4\cos 0° = 4 \times 1 = 4.0$ $\theta = 30°$, $x = 4\cos 30° = 4 \times 0.866 = 3.46$ $\vdots$ The wave form will be exactly as shown in Figure 5.10 if the x coordinates are multiplied by 4.

Radian Measure

It is possible to graph the cosine function $x = \rho\cos\theta$ on a rectangular coordinate system where 1.0 on the vertical x-axis is also equal to 1.0 on the horizontal θ-axis. In order to do this, it is necessary to measure θ in radians, and then 1.0 on the θ-axis corresponds to 1.0 radians. For convenience, the radian measure of θ is sometimes given as a fraction, as shown previously in Table 5.1.

EXAMPLE 5.8 Calculate values for $x = 4\cos\theta$ as θ varies from 0 to 2π radians in steps of $\pi/4$ radians. Plot the resulting wave form.	SOLUTION. Make a table of values containing the given values of θ. Then calculate x for each value of θ. $\theta = 0$, $x = 4\cos\theta = 4\cos 0 = 4 \times 1 = 4.0$ $\theta = \dfrac{\pi}{4}$, $x = 4\cos\dfrac{\pi}{4} = 4 \times 0.707 = 2.83$ $\theta = \dfrac{\pi}{2}$, $x = 4\cos\dfrac{\pi}{2} = 4 \times 0 = 0$ $\vdots$

θ(rad)	0	$\pi/4$	$\pi/2$	$3\pi/4$	π	$5\pi/4$	$3\pi/2$	$7\pi/4$	2π
x	4	2.83	0	-2.83	-4	-2.83	0	2.83	4.0

The results are plotted in Figure 5.12.

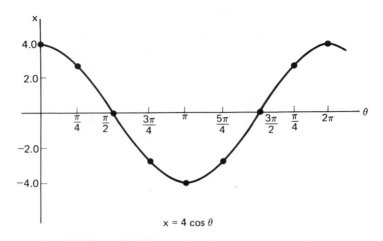

$x = 4 \cos \theta$

Figure 5.12 $x = 4 \cos \theta$.

Exercises 5.2

1. Calculate values for $x = \cos\theta$ as θ varies from $0°$ to $360°$ in $40°$ increments. Include the points $\theta = 0°$, $90°$, $180°$, $270°$, and $360°$.

2. Calculate values for $x = \cos\theta$ as θ varies from $0°$ to $360°$ in $15°$ increments. Plot the results.

3. Calculate values for $x = -\cos\theta$ as θ varies from $0°$ to $360°$ in $30°$ increments.

4. Evaluate $x = \rho\cos\theta$ for each of the following:

   (a) $\rho = 2.00$, $\theta = 75°$ (f) $\rho = -21.6$, $\theta = 117°$

   (b) $\rho = 1.88$, $\theta = 178°$ (g) $\rho = 17.1$, $\theta = 270°$

   (c) $\rho = 0.61$, $\theta = 47°$ (h) $\rho = 2 \times 10^2$, $\theta = \dfrac{\pi}{3}$ rad

   (d) $\rho = 0.18$, $\theta = 315°$ (i) $\rho = 6.15$, $\theta = 1.17$ rad

   (e) $\rho = 3 \times 10^{-3}$, $\theta = 217°$

5. What is the amplitude (peak value) and the peak-to-peak value of:

   (a) $x = 1.8\cos\theta$?   (d) $v = 17\cos\theta$?

   (b) $x = 11.6\cos\theta$?   (e) $i = 31 \times 10^{-3}\cos\theta$?

   (c) $x = 0.18\cos\theta$?   (f) $v = 167\cos\theta$?

6. Plot one cycle of the cosine wave $x = 1.8\cos\theta$.

7. Plot one cycle of the cosine wave $v = 21\cos\theta$.

8. Plot one cycle of the cosine wave $i = 41 \times 10^{-3}\cos\theta$.

9. Plot two cycles of the cosine wave $v = 167\cos\theta$.

10. Plot two cycles of the cosine wave $i = 0.8\cos\theta$.

11. Write the equation for a cosine wave having:

    (a) A peak value at 6.1

    (b) A peak-to-peak value of 5.8

   (c) A peak value of 0.198
   (d) A peak-to-peak value of 3.4×10^{-3}

12. An equation for the ac voltage supplied by a local power company is $v = 167 \cos\theta$ (volts). Plot one cycle of this voltage wave form.

13. The current in an ac circuit is given as $i = V/R \cos\theta$ (amperes). Plot three cycles of i if:

   (a) $V = 10$ V, $R = 12$ Ω
   (b) $V = 167$ V, $R = 85$ Ω
   (c) $V = 16$ V, $R = 1.5$ kΩ
   (d) $V = 2.1$ V, $R = 4.7$ kΩ

14. The ac voltage across the terminals of a resistor is given as $v = IR \cos\theta$ (volts). Plot two cycles of v if:

   (a) $I = 1.7$ A, $R = 55$ Ω
   (b) $I = 3 \times 10^{-3}$ A, $R = 2.2$ kΩ
   (c) $I = 38$ mA, $R = 6.8$ kΩ

15. Calculate values for $x = \cos\theta$ as θ varies from 0 rad to 2π rad in increments of $\pi/6$ rad. Plot the results.

16. Calculate values for $x = 1.8 \cos\theta$ as θ varies from 0 rad to 2π rad in increments of 0.2 rad. Plot the results.

UNIT 6
SINUSOIDAL WAVEFORMS

6.1 PERIOD AND FREQUENCY

Many natural phenomena are periodic in nature—the tides, sound waves, light waves, radio waves, and machine motion, to name a few. The analysis of each of these phenomena is aided by considering *periodic wave forms* that help describe their behavior. In particular, alternating currents and voltages can be described using either sine waves or cosine waves, and a clear understanding of sinusoidal wave forms is essential when dealing with ac circuit problems.

Periodic Functions

By definition, a *periodic function* is any function that repeats itself over a given interval or time period. The sinusoidal wave form plotted in Figure 5.3 (or Figure 5.4) can be thought of as a number of identical one-cycle-segments connected end to end. Since this sinusoidal wave form repeats itself in exactly the same order at regular intervals, it is called a *periodic function*. A similar argument applies to the cosine wave plotted in Figure 5.9, and it is also seen to be a periodic function.

Period

It has been shown previously that if an angle θ is increased by $360°$ (or 2π rad), the value of the sine or cosine of the angle is unchanged. For instance, for any angle θ,

$$\sin\theta = \sin(\theta + 360°) \quad \text{and} \quad \cos\theta = \cos(\theta + 360°)$$

Clearly the values of the sine and cosine of an angle recur periodically, the period of recurrence being $360°$. This is illustrated graphically by noting that the value of $\sin\theta$ is exactly the same at points P and Q in Figure 6.1. The horizontal distance between point P and point Q covers one complete cycle of the wave form, and it is defined as the period.

A function of θ, which for any value of θ is unchanged when θ is increased by an amount τ, is said to be a periodic function of θ with a period τ. Clearly the functions $y = \sin\theta$ and $x = \cos\theta$ are periodic functions, and the period τ of each is equal to $360°$ (or 2π rad).

Angular Velocity

In Section 5.1 it was shown how a sinusoidal wave form is related to a rotating radius vector ρ. Specifically, Eqs. (5.1) and (5.2) are the equations for a sine wave and a cosine wave, respectively.

$$y = \rho\sin\theta \tag{5.1}$$
$$x = \rho\cos\theta \tag{5.2}$$

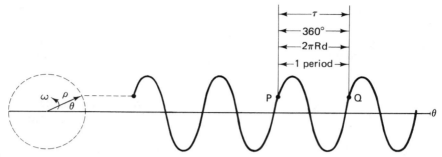

Figure 6.1

It is generally agreed that the radius vector ρ, as shown in Figure 6.1, rotates CCW with a constant angular velocity ω.* The units of angular velocity are typically revolutions per second, revolutions per minute, degrees per second, radians per second, etc. In any case, since the angular velocity is constant, it will require a certain fixed amount of time for the radius vector to complete exactly one revolution. Since one revolution of the radius vector generates exactly one cycle of the sinusoidal wave form, the time of one revolution or the time of one cycle is known as the *time period*.

You will recall that rate (or velocity) is equal to distance divided by time. For example,

$$\text{Velocity(mi/h)} = \frac{\text{distance(miles)}}{\text{time(hours)}}$$

A similar expression is true for angular velocity. Namely,

$$\text{Angular velocity(rad/s)} = \frac{\text{angle(rad)}}{\text{time(s)}}$$

This relationship can be expressed in equation form:

$$\omega = \frac{\theta}{t} \qquad (6.1)$$

where,

$\omega = $ angular velocity, in radians/second

$\theta = $ angle of rotation, in radians

$t = $ time of rotation, in seconds

It should be noted that units other than radians and seconds could be used for ω, θ, and t, provided there is consistency. For example, measuring θ in degrees and t in seconds requires that ω be measured in degrees/second.

---

*Angular velocity has been discussed in Section 1.2, Radian Measure.

EXAMPLE 6.1. The shaft of an electric generator rotates at an angular velocity of 3600 r/min. Find its angular velocity in rad/s.

SOLUTION. 3600 r/min is equivalent to $3600/60 = 60$ revolutions per second. Since each revolution is equal to an angle of 2π rad, the shaft rotates through an angle of $\theta = 60 \times 2\pi = 120\pi$ rad in each second. Therefore, the angular velocity is

$$\omega = \frac{\theta}{t} = \frac{120\pi \text{ rad}}{1 \text{ s}} = 120\pi \text{ rad/s}$$

Equation (6.1) can of course be rearranged to solve for either θ or for t as follows:

$$\theta = \omega t \tag{6.2}$$

$$t = \frac{\theta}{\omega} \tag{6.3}$$

EXAMPLE 6.2. Through what angle θ will the shaft in Example 6.1 above rotate in a time of 1 ms?

SOLUTION. The angular velocity is known to be $\omega = 120\pi$ rad/s, and $t = 10^{-3}$ s. Using Eq. (6.2),

$$\theta = \omega t = 120\pi \times 10^{-3} = 0.377 \text{ rad}$$

or,

$$\theta = 21.6°$$

EXAMPLE 6.3. What is the time required for the shaft in Example 6.1 to complete one revolution?

SOLUTION. The angular velocity is known to be $\omega = 120\pi$ rad/s. One revolution is equivalent to an angle of rotation of 2π rad. Using Eq. (6.3),

$$t = \frac{\theta}{\omega} = \frac{2\pi}{120\pi} = \frac{1}{60} = 16.7 \text{ ms}$$

Frequency

In Example 6.1, the generator shaft has an angular velocity of 3600 revolutions per minute, or 60 revolutions per second. Since each revolution generates exactly one cycle of the sine wave, there will be exactly 60 cycles of the wave form in any one-second interval. The *frequency* of a sinusoidal wave form can be defined as the number of cycles occurring in a one-second time period. In this particular example, the frequency is said to be 60 cycles per second. It is customary to refer to one cycle per second as *one hertz* (Hz), and thus the frequency in this case is 60 Hz.

The frequency f of a sinusoidal wave form is exactly equal to the reciprocal of its time period τ. Thus,

$$f = \frac{1}{\tau} \tag{6.4}$$

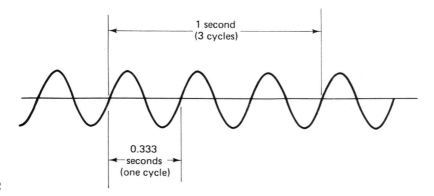

Figure 6.2

and

$$\tau = \frac{1}{f} \tag{6.5}$$

where f is the frequency in hertz (Hz) and τ is the time period in seconds.

EXAMPLE 6.4. A sinusoidal wave form has a frequency of 3 Hz. Calculate its time period.

SOLUTION. Using Eq. (6.5),

$$\tau = \frac{1}{f} = \frac{1}{3} = 0.333 \text{ s}$$

The wave form is shown in Figure 6.2.

EXAMPLE 6.5. A sinusoidal wave form representing an ac voltage is displayed on an oscilloscope as shown in Figure 6.3. Calculate the frequency of the wave form.

SOLUTION. The horizontal scale is set at 5 milliseconds per centimeter. Since one cycle occupies 4 cm, the period of the wave form is

$$\tau = 4 \text{ cm} \times 5 \frac{\text{ms}}{\text{cm}} = 20 \text{ ms}$$

Then using Eq. (6.4), the frequency is found to be

$$f = \frac{1}{\tau} = \frac{1}{20 \text{ ms}} = 50 \text{ Hz}$$

An important relationship between angular velocity and frequency can be developed by applying Eq. (6.1) to one single cycle of a wave form. For a single cycle, the time t is equal to τ, and the angle θ is equal to 2π rad. Thus,

$$\omega = \frac{\theta}{t} = \frac{2\pi}{\tau}$$

But, from Eq. (6.5), $\tau = 1/f$. By substitution,

$$\omega = \frac{2\pi}{\tau} = \frac{2\pi}{\dfrac{1}{f}}$$

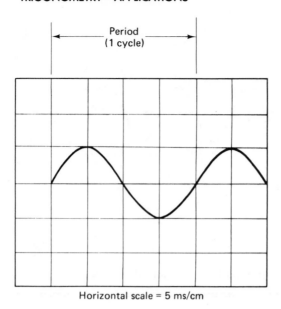

Horizontal scale = 5 ms/cm

Figure 6.3

Thus,

$$\omega = 2\pi f \qquad (6.6)$$

This important equation states that the angular velocity in radians per second (rad/s) is equal to 2π times the frequency measured in hertz (Hz). For instance, the angular velocity of a 50 Hz sine wave is

$$\omega = 2\pi f = 2\pi \times 50 = 100\pi \text{ rad/s}$$

A summary of important points discussed in this section follows:

Definitions:

PERIOD—On a graph of a periodic function, the distance required to complete exactly one cycle. The period can be measured in degrees or radians as in Figure 6.1, or in seconds as in Figures 6.2 or 6.3.

TIME PERIOD (τ)—The period of a wave form measured in seconds.

FREQUENCY (f)—The number of cycles generated in one second. Measured in hertz (Hz).

Expressions:

$$\omega = \frac{\theta}{t} \qquad \theta = \omega t \qquad t = \frac{\theta}{\omega}$$

$$f = \frac{1}{\tau} \qquad \tau = \frac{1}{f} \qquad \omega = 2\pi f$$

Exercises 6.1

1. The shaft of an automobile alternator rotates with an angular velocity of 850 r/min. Determine the angular velocity in:
   (a) rad/s   (b) degrees/s

2. How many revolutions does a generator shaft complete in one second if its angular velocity is 500 rad/s?

3. What is the angular velocity in rad/s of a radar antenna that completes 1 revolution in 3 seconds?

4. Through what angle will a radar antenna rotate in a time of 0.5 second if it has an angular velocity of:
   (a) π rad/s   (b) 80°/s   (c) 0.3 revolution per second?

5. What is the angular velocity in rad/s of an antenna that rotates through an angle of 50° in 2 seconds?

6. Through what angle in degrees will a shaft rotate in 15 ms if its angular velocity is 100 rad/s?

7. What time is required for a generator shaft to make one revolution if it has an angular velocity of 10 rad/s? How much time is required to make 1/4 revolution?

8. What time is required for a second radar to scan over a 90° segment (out of 360°) if its angular velocity is known to be 100 r/min?

9. How much time is required for the shaft in Problem 1 to complete 1 revolution? 5 revolutions?

10. Calculate the time period τ of a sinusoidal wave form if its frequency is:
    (a)   5 Hz   (e)   87 kHz
    (b)   60 Hz   (f)   228 kHz
    (c)   500 Hz   (g)   1.8 MHz
    (d)   3 kHz   (h)   30 MHz

11. Calculate the frequency of a sinusoidal wave form having a time period τ equal to:
    (a) 1 s   (b) $\dfrac{1}{8}$ s   (c) 16.7 ms   (d) 2 ms   (e) 58 μs   (f) 0.22 μs

12. Calculate the time period τ and the frequency of the voltage wave form in Figure 6.3 if the horizontal scale is:
    (a) 1 ms/cm   (b) 0.2 ms/cm   (c) 5 μs/cm   (d) 1 μs/cm
    (e) 0.2 s/cm

13. Calculate the angular velocity ω in rad/s for each of the wave forms in Problem 10.

14. Calculate the angular velocity ω in rad/s for each of the wave forms in Problem 12.

6.2   SINUSOIDAL CURRENTS AND VOLTAGES

Virtually all of the electrical energy supplied to industrial and residential sites is provided in the form of *alternating current* (ac). Such ac currents and the accompanying ac voltages can be represented by sinusoidal wave forms.* Likewise, many of the electrical signals encountered in electronic circuits and systems are sinusoidal in nature. Furthermore, the analysis of an electric circuit or system often can be accomplished in terms of sinusoidal signals, even though the wave forms in the circuit or system are not themselves sinusoidal.[†] It should be clear then that a thorough understanding of the equations used to represent sinusoidal currents and voltages is essential.

Alternating Current

It is generally agreed that an alternating current is very nearly sinusoidal in nature, and thus an alternating current can be represented either in terms of a sine wave or a cosine wave. The units of measurement are amperes (A). It is customary to use a lower case letter i to represent a sinusoidal current, and subscripts are frequently used for clarification (e.g., i_1, i_a, etc.). For instance, $i = I_p \sin \theta$ is an equation that represents a sine wave of current having a peak value of I_p, as shown in Figure 6.4(a). Similarly, an ac current can be represented as the cosine wave $i = I_p \cos \theta$, as shown in Figure 6.4(b).

When solving electric circuit problems, it is usually more convenient to represent a sinusoidal variation of current as a function of time rather than as a function of θ, as in Figure 6.4. In the previous section, Eq. (6.2) specifies the angle θ in terms of angular velocity and time as

$$\theta = \omega t$$

Therefore, by substituting ωt for θ, the two current equations in Figure 6.4 can be written as

$$i = I_p \sin \omega t \qquad (6.7)$$
$$i = I_p \cos \omega t \qquad (6.8)$$

EXAMPLE 6.6. Write an equation for a sine wave of current having a peak value of 2 A and a frequency of 10 Hz.	SOLUTION. The desired expression is in the form of Eq. (6.7) $i = I_p \sin \omega t$ The peak value of the wave form is given as $$I_p = 2 \text{ A}$$ From Eq. (6.6), $\omega = 2\pi f = 2\pi \times 10 = 20\pi$ Then, by substitution, $i = 2 \sin 20\pi t$

---

*Notice that the two letters *ac* are the abbreviation for "alternating current." Thus to say "ac current" really means "alternating current current," which is not quite proper. Nevertheless it is common practice to speak of "ac current" and "ac voltage," and there is no loss of meaning.

[†]Fourier analysis provides the technique for representing nonsinusoidal wave forms in terms of sine and/or cosine waves.

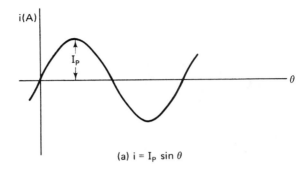

(a) i = I_p sin θ

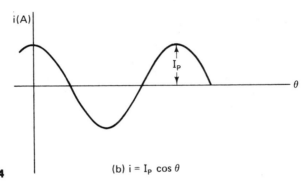

Figure 6.4

(b) i = I_p cos θ

Equations (6.7) and (6.8) can be graphed with the quantity ωt as the independent variable, where ω is a constant. Notice that t is in units of time, ω is in rad/s, and thus ωt has units of radians, as illustrated in Figure 6.5(a).

Alternatively, Eqs. (6.7) and (6.8) can be graphed with t as the independent variable, as in Figure 6.5(b). Again, ω is a constant. Notice that the scale on the horizontal axis is now time in seconds.

EXAMPLE 6.7. Sketch the sinusoidal current in Example 6.6 first as a function of ωt and then as a function of t.	SOLUTION. The given equation is $$i = I_p \sin \omega t = 2 \sin 20\pi t$$ First, since $\theta = \omega t$, the graph is exactly as shown in Figure 6.4(a), with θ replaced by ωt. The peak value is of course 2 A. Second, a plot with t as the independent variable appears in Figure 6.6. Since the frequency is $f = 10$ Hz, the time period is $$\tau = \frac{1}{f} = \frac{1}{10} = 0.1 \text{ s}$$ or 100 ms. Thus, one-quarter of the period is 25 ms; one-half period is 50 ms; and so on.

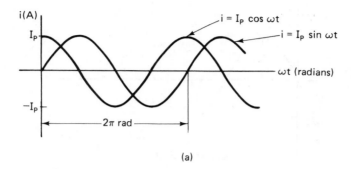

(a)

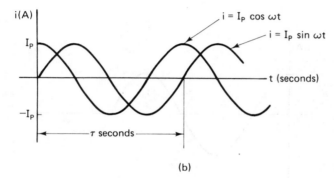

(b)

Figure 6.5

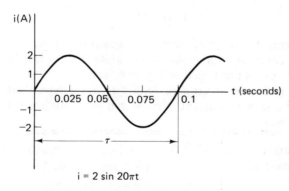

$i = 2 \sin 20\pi t$

Figure 6.6

Ac Voltages

An ac voltage can likewise be represented as either a sine wave or a cosine wave. The lower-case letter v is customarily used to represent a sinusoidal voltage, and the unit of measurement is the volt (V). For example, a sine wave of voltage having a peak value of V_p (volts) and a frequency f (Hz) can be written as

$$v = V_p \sin \omega t \tag{6.9}$$

where $\omega = 2\pi f$. Similarly, a cosine wave of voltage can be written as

$$v = V_p \cos \omega t \qquad (6.10)$$

For instance, a sine wave of voltage having a peak value of 15 V and a frequency of 50 Hz is written as

$$v = 15 \sin 100\pi t$$

where $\omega = 2\pi f = 2\pi \times 50 = 100\pi$.

EXAMPLE 6.8. Calculate the values of $v = 15 \sin 100\pi t$ for $t = 0$ to 20 ms in 5 ms intervals. Graph the wave form.

SOLUTION. For $t = 0$ (remember ωt is in radians):

$$v = 15 \sin(100\pi \times 0) = 15 \sin 0 = 0$$

For $t = 5$ ms:
$$v = 15 \sin(100\pi \times 5 \times 10^{-3})$$
$$= 15 \sin(1.571) = 15.0$$

For $t = 10$ ms:
$$v = 15 \sin(100\pi \times 10 \times 10^{-3})$$
$$= 15 \sin(\pi) = 0$$

For $t = 15$ ms:
$$v = 15 \sin(100\pi \times 15 \times 10^{-3})$$
$$= 15 \sin(4.712) = -15.0$$

For $t = 20$ ms:
$$v = 15 \sin(100\pi \times 20 \times 10^{-3})$$
$$= 15 \sin(2\pi) = 0$$

These points provide the wave form shown in Figure 6.7.

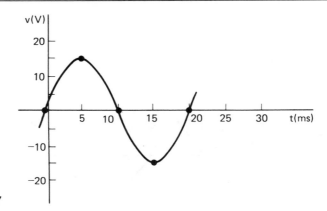

Figure 6.7

Phase Angle

In Figure 6.8(a), a sine wave of current and a cosine wave of voltage are drawn on the same scale. The rotating vector I_p is associated with the current, and the rotating vector V_p is associated with the voltage. Both of these vectors rotate CCW at the same constant angular velocity ω. Notice that there is a fixed angle ψ between these two rotating vectors. In this particular case, this angle ψ is exactly equal to 90° or $\pi/2$ radians. That ψ is indeed 90° can be seen by "sliding" the cosine wave to the right along the ωt-axis until it lies on top of the sine wave. A distance of $\pi/2$ rad or 90° along the ωt-axis will do it. This results in rotation of the V_p vector through an angle of exactly 90°, since V_p must now lie over I_p. Thus ψ must be equal to 90°.

In Figure 6.8(b), a sine wave of current and a sinusoidal wave form of voltage are drawn on the same scale. The angle ψ between the two vectors I_p and V_p is some arbitrary value between 0° and 360°. Such an angle is defined as a *phase angle*.

Notice that the voltage wave form is a sine wave that has been "shifted to the left" along the ωt-axis by a distance equal to the phase angle ψ. This is the same as "rotating" the vector V_p CCW such that it is *ahead* of I_p by an angle ψ. Thus V_p *leads* I_p by a phase angle ψ. The proper expression for v is then,

$$v = V_p \sin(\omega t + \psi)$$

This equation is seen to be valid, since letting $t=0$ yields

$$v = V_p \sin(\psi)$$

and this is the proper value for v when $t=0$.

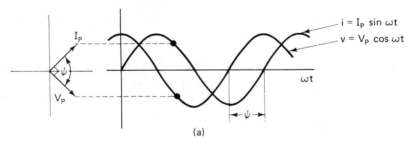

(a)

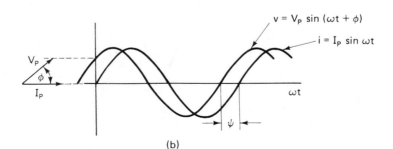

(b)

Figure 6.8

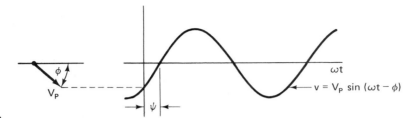

Figure 6.9

The voltage wave form in Figure 6.9 is a sine wave that has been "shifted to the right" along the ωt-axis by a distance equal to a phase angle ψ. This requires "rotating" the associated vector V_p CW through an angle ψ such that it is *behind* a reference vector for a sine wave having no phase angle. The appropriate expression for a sine wave of voltage that *lags* a reference sine wave is

$$v = V_p \sin(\omega t - \psi)$$

EXAMPLE 6.9. Calculate values of $$v = 5\sin\left(100\pi t + \frac{\pi}{4}\right)$$ for $t=0$ to 20 ms in 5 ms intervals. Graph the wave form.	SOLUTION. Notice that the phase angle is given in radians ($\psi = \pi/4$ radians), and remember that the angle $(100\pi t + \pi/4)$ is in radians. For $t=0$: $$v = 15\sin\left(100\pi \times 0 + \frac{\pi}{4}\right)$$ $$= 15\sin\left(\frac{\pi}{4}\right) = 15(0.707) = 10.6$$ For $t=5$ ms: $$v = 15\sin\left(100\pi \times 5 \times 10^{-3} + \frac{\pi}{4}\right)$$ $$= 15\sin\left(1.571 + \frac{\pi}{4}\right) = 10.6$$ For $t=10$ ms: $$v = 15\sin\left(100\pi \times 10 \times 10^{-3} + \frac{\pi}{4}\right)$$ $$= 15\sin\left(\pi + \frac{\pi}{4}\right) = -10.6$$ For $t=15$ ms: $$v = 15\sin\left(100\pi \times 15 \times 10^{-3} + \frac{\pi}{4}\right)$$ $$= 15\sin\left(4.712 + \frac{\pi}{4}\right) = -10.6$$ For $t=20$ ms: $$v = 15\sin\left(100\pi \times 20 \times 10^{-3} + \frac{\pi}{4}\right)$$ $$= 15\sin\left(2\pi + \frac{\pi}{4}\right) = 10.6$$

The first value of t where $v=0$ can be found by noting that at this point $v=15\sin\pi$, since $\sin\pi=0$. Thus,

$$\left(\omega t+\frac{\pi}{4}\right)=\left(100\pi t+\frac{\pi}{4}\right)=\pi$$

Solving for t yields

$$100\pi t+\frac{\pi}{4}=\pi$$

$$100\pi t=3\frac{\pi}{4}$$

$$t=\frac{3\dfrac{\pi}{4}}{100\pi}=\frac{3}{400}=7.5 \text{ ms}$$

The peak value of v is $V_p=15$ V, and the wave form thus appears as in Figure 6.10.

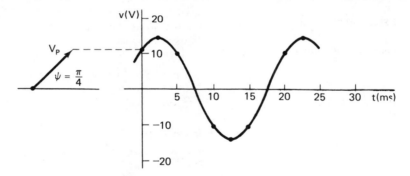

Figure 6.10

To summarize, ac currents and voltages are sinusoidal in nature. As such, they can be represented as either sine waves or as cosine waves. Each wave form has a peak value (I_p or V_p), a frequency ($\omega=2\pi f$), and possibly a phase angle ψ. Thus,

$$\qquad\qquad\text{(Currents)}\qquad\qquad\qquad\text{(Voltages)}$$
$$i=I_p\sin(\omega t+\psi)\ \leftarrow\text{lead}\rightarrow\ v=V_p\sin(\omega t+\psi)$$
$$i=I_p\sin(\omega t-\psi)\ \leftarrow\text{lag}\rightarrow\ v=V_p\sin(\omega t-\psi)$$

or

$$i=I_p\cos(\omega t+\psi)\ \leftarrow\text{lead}\rightarrow\ v=V_p\cos(\omega t+\psi)$$
$$i=I_p\cos(\omega t-\psi)\ \leftarrow\text{lag}\rightarrow\ v=V_p\cos(\omega t-\psi)$$

Lead implies a shifting of the wave form to the *left* on the ωt-axis. *Lag* implies a shifting of the wave form to the *right* on the ωt-axis.

Exercises 6.2

1. Write an expression for a sine wave of either current or voltage, given:
   (a) $I_p = 6$ A, $f = 35$ Hz
   (b) $V_p = 12$ V, $f = 50$ Hz
   (c) $V_p = 167$ V, $f = 60$ Hz
   (d) $I_p = 30$ mA, $f = 1$ kHz
   (e) $I_p = 7$ mA, $f = 3$ kHz
   (f) $V_p = 38$ mV, $f = 2$ MHz

2. Write an expression for a cosine wave of either current or voltage, given:
   (a) $I_p = 5$ A, $f = 40$ Hz
   (b) $V_p = 15$ V, $f = 60$ Hz
   (c) $V_p = 167$ V, $f = 60$ Hz
   (d) $I_p = 8$ mA, $f = 3$ kHz
   (e) $I_p = 21$ mA, $f = 50$ Hz
   (f) $V_p = 66$ mV, $f = 1$ MHz

3. Calculate enough points to accurately graph one period of the following wave forms:
   (a) $i = 16 \sin 30t$
   (b) $v = 6 \sin 200t$
   (c) $i = 1.5 \sin 1000t$
   (d) $v = 167 \sin 377t$
   (e) $i = 0.002 \sin 8000t$
   (f) $v = 0.14 \sin 10^6 t$
   (g) $i = 2.1 \cos 100t$
   (h) $v = 38 \cos 500t$
   (i) $i = 8.5 \cos 10^5 t$
   (j) $v = 38 \cos 300t$
   (k) $v = 167 \cos 377t$
   (l) $i = 0.007 \cos 10^3 t$

4. Write an expression for a sine wave of either current or voltage, given:
   (a) $I_p = 6$ A, $f = 35$ Hz, $\psi = +\pi/6$
   (b) $V_p = 12$ V, $f = 50$ Hz, $\psi = +\pi/4$
   (c) $I_p = 6$ A, $f = 35$ Hz, $\psi = -\pi/6$
   (d) $V_p = 167$ V, $f = 60$ Hz, $\psi = 35°$
   (e) $V_p = 167$ V, $f = 60$ Hz, $\psi = -35°$
   (f) $I_p = 35$ mA, $f = 1$ kHz, $\psi = 60°$

5. Write the expression for a cosine wave of either current or voltage, given:
   (a) $I_p = 5$ A, $f = 40$ Hz, $\psi = \pi/5$
   (b) $V_p = 15$ V, $f = 60$ Hz, $\psi = 53.7°$
   (c) $V_p = 167$ V, $f = 60$ Hz, $\psi = 45°$
   (d) $V_p = 167$ V, $f = 60$ Hz, $\psi = -45°$
   (e) $I_p = 35$ mA, $f = 1$ kHz, $\psi = -90°$
   (f) $I_p = 35$ mA, $f = 1$ kHz, $\psi = 90°$

6. Calculate enough points to accurately graph one period of the following wave forms:
   (a) $v = 167 \sin(377t + 30°)$
   (b) $v = 167 \sin(377t + 90°)$
   (c) $v = 167 \sin(377t - 30°)$
   (d) $i = 3.8 \sin(10^3 t - \pi/4)$
   (e) $i = 0.75 \cos(10^2 t + 40°)$
   (f) $i = 0.75 \cos(10^2 t - 40°)$
   (g) $v = 2.7 \cos(10^4 t - 90°)$
   (h) $v = 0.81 \cos(10^6 t + \pi/7)$

7. What is the relationship between $v = V_p \cos \omega t$ and $v = V_p \sin(\omega t + 90°)$? Make a sketch showing the wave forms.

8. What is the relationship between $i = I_p \sin \omega t$ and $i = I_p \sin(\omega t \pm 180°)$? Make a sketch.

9. The voltage supplied to a residence can be expressed as a sine wave with zero phase angle. Write an expression for this voltage if:

(a) $V_p = 167$ V, $f = 60$ Hz
(b) $V_p = 334$ V, $f = 60$ Hz
(c) $V_p = 167$ V, $f = 50$ Hz

6.3 SOME APPLICATIONS

Ohm's Law

The technique for expressing ac currents and voltages as sinusoidal wave forms has been discussed in the previous section. This technique can be used to find the ac currents and voltages in a "resistive circuit" according to Ohm's Law. Any circuit consisting of only sources and resistances (no inductance or capacitance) is termed a *resistive circuit*.

For a resistive circuit containing a single sinusoidal source, Ohm's Law can be expressed as

$$i = \frac{v}{R} \tag{6.11}$$

In such a circuit, the currents and voltages are always in phase with one another (the phase angle is zero) and they are all of the same frequency. Thus,

$$\text{If } v = V_p \sin \omega t, \qquad \text{then } i = I_p \sin \omega t$$

or

$$\text{If } v = V_p \cos \omega t, \qquad \text{then } i = I_p \cos \omega t$$

where $\omega = 2\pi f$.

EXAMPLE 6.10.	SOLUTION.
EXAMPLE 6.10. Figure 6.11(a) consists of an ac voltage source labeled v, and a resistance $R = 2$ kΩ. Use Ohm's Law to calculate i if $V_p = 30$ V and $f = 1$ kHz. Use a sine wave of voltage.	SOLUTION. The correct expression for v is $$v = V_p \sin \omega t$$ $$V_p = 30 \text{ V}, \omega = 2\pi f = 2\pi \times 10^3$$ Thus, $v = 30\sin(2\pi \times 10^3)t$ V. Using Ohm's Law, $$i = \frac{v}{R} = \frac{30\sin(2\pi \times 10^3)t}{2 \text{ k}\Omega}$$ $$= 15\sin(2\pi \times 10^3)t \text{ mA}$$ The period is found to be $\tau = 1/f = 1/10^3 = 1$ ms, and the two wave forms are sketched in Figure 6.11(b).

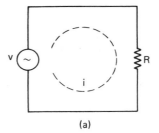

(a)

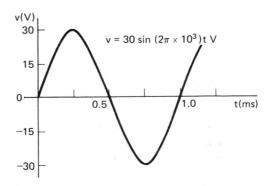

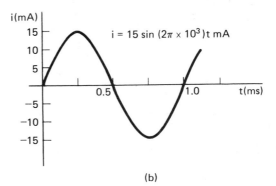

(b)

Figure 6.11

Average Value

The current wave form sketched in Figure 6.11(b) is a graphical representation of the current i in the circuit in that figure. During the time the wave form is positive (above the t-axis), the current in the circuit flows in one direction—say CW; and when the wave form is negative (below the t-axis), the current flows in the opposite direction—CCW, in the circuit. The important point is that the circuit current alternatively flows first in one direction and then in the other—thus the term *alternating current*.

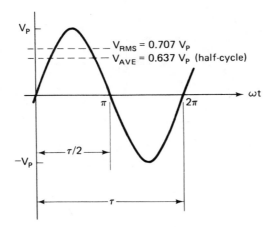

Figure 6.12

A careful examination of the current wave form illustrates that the current is positive for exactly the same amount of time it is negative over a single time period τ. Therefore, the *average value* of the wave form over a single time period is zero.

There are applications where it is useful to know the average value of a sinusoidal wave form over one-half of a cycle, for instance, an average reading voltmeter measures the average value of a sinusoidal voltage. In Figure 6.12 the voltage wave form is always positive for the entire half-cycle between $\omega t = 0$ and $\omega t = \pi$. The *average values* of the wave must be some place between 0 and V_p. In fact, it can be shown that the half-cycle average value of a sinusoidal wave form is

$$V_{\text{avg}} = \frac{2V_p}{\pi} = 0.637 V_p \qquad (6.12)$$

and

$$I_{\text{avg}} = \frac{2I_p}{\pi} = 0.637 I_p \qquad (6.13)$$

EXAMPLE 6.11. If the equation for the voltage wave form in Figure 6.12 is $v = 120 \sin 377t$, calculate the half-cycle average value for the wave.	SOLUTION. From the given equation, $$V_p = 120 \text{ V}$$ Using Eq. (6.12), $$V_{\text{avg}} = 0.637 V_p = 0.637 \times 120 = 76.4 \text{ V}$$
EXAMPLE 6.12. What is the peak value of a current wave form if its half-cycle average is 6.37 mA?	SOLUTION. Using Eq. (6.13), $$I_p = \frac{I_{\text{avg}}}{0.637} = \frac{6.37 \text{ mA}}{0.637} = 10.0 \text{ mA}$$

Rms Values

A very useful quantity associated with a sinusoidal wave form is its rms value. The rms is the abbreviation for Root-Mean-Square and refers to the technique used to derive the rms value of a wave form. The term *effective value* is also frequently used in place of rms. The effective or rms values of sinusoidal currents and voltages are used in ac circuit power calculations.

It is useful to be able to state that an alternating current has an effective value in amperes such that the power available from this current is exactly equal to the power available from an equal number of amperes of direct current. Clearly, the same statement can be made for an ac voltage, and it can be shown that the effective value or rms value of a sinusoidal wave form is

$$V_{rms} = V_{eff} = \frac{V_p}{\sqrt{2}} = 0.707 V_p \qquad (6.14)$$

and

$$I_{rms} = I_{eff} = \frac{I_p}{\sqrt{2}} = 0.707 I_p \qquad (6.15)$$

As shown in Figure 6.12, V_{rms} is just slightly greater than the half-cycle average V_{ave} for a sinusoidal wave form.

EXAMPLE 6.13. Calculate the rms value of $v = 167 \sin 377t$ V.

SOLUTION. From the given equation,
$$V_p = 167 \text{ V}$$
Using Eq. (6.14),
$$V_{rms} = 0.707 V_p = 0.707 \times 167 = 118 \text{ V}$$
The given equation represents the ac voltage commonly supplied to a residence. Notice that $\omega = 377$, and thus,
$$f = \frac{\omega}{2\pi} = \frac{377}{2\pi} = 60 \text{ Hz}$$

Power

In a resistive circuit having a single ac source, the currents and voltages in the circuit will all be in phase with each other, and they will all be of the same basic frequency. In this case, the rms values of circuit current and voltages are used to calculate the power in each circuit element. Thus,

$$P = V_{eff} I_{eff} \qquad (6.16)$$
$$P = I^2_{eff} R \qquad (6.17)$$
$$P = \frac{V^2_{eff}}{R} \qquad (6.18)$$

where

$$P = \text{power in watts}$$
$$V_{eff} = \text{rms voltage in volts}$$
$$I_{eff} = \text{rms current in amperes}$$
$$R = \text{element resistance in ohms}$$

EXAMPLE 6.14. Calculate the power dissipated in R in Figure 6.11.

SOLUTION. From the two wave forms, $V_p = 30$ V and $I_p = 15$ mA:

$$V_{\text{eff}} = 0.707 V_p = 0.707 \times 30 = 21.2 \text{ V}$$

$$I_{\text{eff}} = 0.707 I_p = 0.707 \times 15 \times 10^{-3} = 1.06 \times 10^{-3} \text{ A}$$

Using Eq. (6.16),

$$P = V_{\text{eff}} I_{\text{eff}} = 21.2 \times 1.06 \times 10^{-3} = 0.0225 \text{ W}$$

EXAMPLE 6.15. An electric baseboard heater is used to provide heat to a small work room. The heater is rated at 230 V rms, and 1.5 kW. What is the rms current drawn by the heater?

SOLUTION. From Eq. (6.16), $P = V_{\text{eff}} I_{\text{eff}}$. Thus,

$$I_{\text{eff}} = \frac{P}{V_{\text{eff}}} = \frac{1500 \text{ W}}{230 \text{ V}} = 6.52 \text{ A}$$

Exercises 6.3

1. What is the current in a lamp having a hot resistance of 52.5 Ω if the lamp terminal voltage is $v = 8.9 \sin \omega t$ V?

2. A 4700 Ω resistor is connected across the terminals of an ac voltage source. What is the resistor current if:
   (a) $v = 10 \sin \omega t$ V?
   (b) $V_p = 8 \sin V$, $f = 1$ kHz?
   (c) $v = 4.13 \sin \omega t$?
   (d) The peak-to-peak voltage is 33.0 V and $f = 167$ kHz?

3. Calculate the resistor current if each of the following resistors is connected across the terminals of an ac voltage source of $v = 35 \sin 10^4 t$ V:
   (a) 200 Ω   (b) 680 Ω   (c) 910 Ω   (d) 1.5 kΩ   (e) 3.3 kΩ
   (f) 56 kΩ   (g) 2.2 MΩ

4. Calculate the voltage across the terminals of a 6.8 kΩ resistor if the resistor current is $i = 3 \cos 10^5 t$ mA.

5. A 680 Ω resistor is connected across the terminals of an ac voltage source. Determine the proper source voltage to provide a current of:
   (a) $i = 0.25 \sin \omega t$ A   (c) $i = 8.71 \sin 10^3 t$ mA
   (b) $i = 62 \cos \omega t$ mA   (d) $I_p = 1.66$ mA, $f = 1$ kHz

6. What resistance must be connected across the terminals of a voltage source $v = 167\cos 377t$ V to provide a current of:

   (a) $i = 1.0\cos 377t$ mA?　　(c) $i = 1.0\cos 377t$ A?
   (b) $i = 56\cos 377t$ mA?　　(d) $I_p = 85$ mA?

7. Calculate the half-cycle average of a sinusoidal wave if:

   (a) $V_p = 167$ V　　(b) $I_p = 38.2$ mA　　(c) $V_p = 73.2$ mV

8. Calculate the rms or effective value of each of the following wave forms:

   (a) $v = 67\sin\omega t$ V　　　　(f) $i = 4.37\sin\omega t$ μA
   (b) $v = 1.33\cos\omega t$ V　　　(g) $v = V_p\sin 10^3 t$ V
   (c) $v = 382\sin\omega t$ mV　　　(h) $i = I_p\cos\omega t$ mA
   (d) $i = 0.711\cos\omega t$ A　　　(i) $v = 167\sin 377t$ V
   (e) $i = 8.31\sin 10^3 t$ mA

9. What current is required for a 1500 W electric heater if it is connected to a voltage source $V_{rms} = 117$ V?

10. What current is required for a 1500 W electric heater if it is connected to a voltage source $V_{rms} = 230$ V?

11. Calculate the power dissipated in a 6.3 V rms lamp if the lamp current is 0.12 A rms.

12. The nameplate data on an electric heater is 117 V rms @ 12.8 A rms. What is the power rating?

13. What is the power rating of an electric heater designed to operate with 230 V rms @ 6.52 A rms?

14. What is the power rating of a 117 V soldering iron if it operates with a current at 0.406 A rms?

15. Is it safe to connect a 100 Ω, 1/2 W resistor across the terminals of an ac source adjusted to $v = 70.7\sin\omega t$ V? Explain.

16. Calculate the power in each resistor in Problem 3.

17. Calculate the resistor power for each case in Problem 5.

18. Calculate the power dissipated in each case in Problem 6.

19. What must be the element resistance in an electric heater rated at 230 V rms and 750 W?

20. What must be the resistance of the heating element in a 47-1/2 W soldering iron designed to operate on 117 V rms?

UNIT 7
COMPLEX ALGEBRA

Sinusoidal voltages and currents have thus far been represented using equations of the form $v = V_p \sin(\omega t + \psi)$ and $i = I_p \cos(\omega t + \psi)$. These trigonometric equations are correct and quite precise, but they are somewhat unwieldly when dealing with ac circuits. Fortunately there is a much simpler mathematical representation for sinusoidal wave forms. An ac wave form can be described in terms of a *phasor*, and ac circuit problems can then be solved by using right triangles when the circuit voltages and currents are expressed as phasors.

7.1 PHASORS

The vector V_p in Figure 7.1 rotates CCW like the spoke of a wheel with a constant angular velocity ω. If t is the elapsed time since the vector was in position OA, then ωt represents the angle of rotation at time t. The projection of this vector on the vertical axis is a sinusoid expressed as $v = V_p \sin \omega t$. The projection on the horizontal axis is also a sinusoid that is expressed as $v = V_p \cos \omega t$. One complete revolution of this rotating vector corresponds exactly to one cycle of the generated sinusoidal wave form. The peak value of the sinusoid is equal to the length V_p of the vector, and the frequency f of the sinusoid is given in terms of the angular velocity, since $\omega = 2\pi f$. Thus the rotating vector in Figure 7.1 can be used to represent an ac voltage. Similarly, a vector having a length I_p can be used to represent an alternating current.

The voltages and currents in a given ac circuit are all assumed to have the same frequency, and they are all in phase with each other in a resistive circuit. However, in a circuit containing capacitance and/or inductance, there may be phase angles between the circuit currents and voltages. For example, in Figure 7.2(a) there is a phase angle ψ between the current vector and the voltage vector. Since rotation of the vectors is CCW, the current is said to *lead* the voltage by an angle ψ. They both rotate at the same angular velocity ω (they have the same frequency), but they are *spaced* or *separated* by the phase angle ψ. If these wave forms are expressed in terms of sine waves, then $v = V_p \sin \omega t$ and $i = I_p \sin(\omega t + \psi)$.

The vector V_p in Figure 7.2(a) thus represents a sine wave of voltage, and the vector I_p represents a sine wave of current that leads the voltage by an angle ψ. Since they both have the same frequency, all of the essential information is given in terms of the lengths of the two vectors and the angle ψ between them. The rest of the diagram can be omitted without any loss of information. Figure 7.2(b) shows all of the essential information regarding these two wave forms. Notice that not even the orientation of the two vectors is important, so long as the angle ψ between them is specified.

The main idea is to think of these two vectors as rotating at synchronous speed about their common origin, and any picture of them is simply a snapshot taken at some instant. Any snapshot would be equally valid, so it is usual to choose

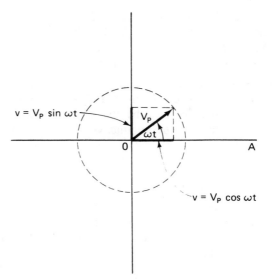

$v = V_P \sin \omega t$

V_P

ωt

0

A

$v = V_P \cos \omega t$

Figure 7.1

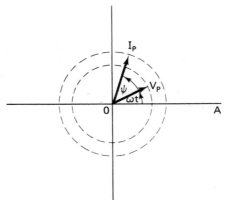

I_P

ψ V_P

ωt

0

A

(a) Rotating vector representation of a
sinusoidal voltage and current

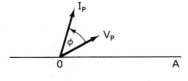

I_P

ϕ V_P

0   A

(b) Snapshot of the
rotating lines in (a)

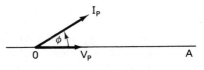

I_P

ϕ

0   V_P A

(c) Phasor diagram

Figure 7.2

a time when one of the vectors is on line *OA*, as seen in Figure 7.2(c).This picture is called a *phasor diagram*, and each of the vectors is defined as a *phasor*. The phasor $\mathbf{V}_p$ on line *OA* is called the *reference phasor*, and all other phasors are drawn with phase angles relative to this reference.

EXAMPLE 7.1. Write equations for sine waves of voltage represented by the phasor diagram in Figure 7.3. The angular velocity is given as $\omega = 2\pi f$.	**SOLUTION.** $\mathbf{V}_2$ is the reference phasor. It has a zero phase angle and a peak value of V_2. Thus, $$v_2 = V_2 \sin \omega t$$ The phasor $\mathbf{V}_1$ has a peak value of V_1 and a positive phase angle equal to θ. Thus, $$v_1 = V_1 \sin(\omega t + \theta)$$
EXAMPLE 7.2. Draw the phasor diagram showing the following currents as phasors: $$i_1 = 10 \sin \omega t$$ $$i_2 = 6 \sin(\omega t + 45°)$$ $$i_3 = 7 \sin(\omega t - 30°)$$	**SOLUTION.** $\mathbf{I}_1$ will be chosen as the reference phasor since it has a zero phase angle. Its peak value is 10, and it is drawn on line *OA* in Figure 7.4. $\mathbf{I}_2$ has a peak value of 6 and a positive phase angle of 45°. Thus it *leads* $\mathbf{I}_1$ as shown in Figure 7.4. $\mathbf{I}_3$ has a peak value of 7 and a negative phase angle of 30°. Thus it *lags* $\mathbf{I}_1$ by 30°.

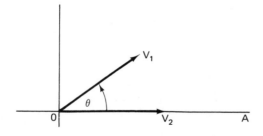

Figure 7.3

A phasor can be used to represent any sinusoidal voltage or current of the form $v = V_p \sin(\omega t + \psi)$. The phasor is simply a vector drawn on a convenient coordinate system, and it is completely specified by a magnitude and an angle. The magnitude is the peak value of the wave, and the angle is the phase angle. But the magnitude and angle of a radius vector are simply the *polar coordinates* of that vector. Thus the phasor representation of a sinusoidal wave form can be written as a set of polar coordinates. For instance, the phasor for the voltage $v = V_p \sin(\omega t + \psi)$ is written as $V_p \underline{/\psi}$.

It is customary to represent vectors with **boldface** type, such as **V** or **I**. Thus $\mathbf{V}_1 = V_1 \underline{/\psi_1}$ means the phasor $\mathbf{V}_1$ has a magnitude equal to V_1 and a phase angle of

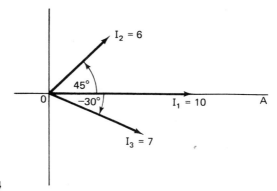

Figure 7.4

$+\psi_1$. Table 7.1 shows the proper phasors for different sinusoidal voltages and currents.

Table 7.1

SINUSOIDAL WAVE FORM	PHASOR REPRESENTATION
$v_1 = V_1\sin(\omega t + \psi_1)$	$\mathbf{V}_1 = V_1\underline{/\psi_1}$
$v_2 = V_2\cos(\omega t + \psi_2)$	$\mathbf{V}_2 = V_2\underline{/\psi_2}$
$i_3 = I_3\sin(\omega t + \psi_3)$	$\mathbf{I}_3 = I_3\underline{/\psi_3}$
$i_4 = I_4\cos(\omega t + \psi_4)$	$\mathbf{I}_4 = I_4\underline{/\psi_4}$

EXAMPLE 7.3. Write the phasor notation for: (a) $v_1 = 15\sin(\omega t + 37°)$ (b) $i_2 = 0.17\sin(377t - 40°)$	**SOLUTION.** (a) The magnitude is the peak value, or 15, and the phase angle is $+37°$. Thus, $$\mathbf{V}_1 = 15\underline{/37°}$$ (b) The magnitude is equal to 0.17, and the phase angle is $-40°$. Thus, $$\mathbf{I}_2 = 0.17\underline{/-40°}$$ Notice that $\omega = 2\pi f = 377$, but this does not appear in the phasor notation.
EXAMPLE 7.4. Write the sign wave expressions represented by the phasor: (a) $\mathbf{I}_1 = 2\underline{/21°}$ (b) $\mathbf{V}_2 = 12\underline{/-48°}$	**SOLUTION.** The frequency is specified by $\omega = 2\pi f$. (a) The peak value is 2, and the phase angle is $+21°$. Thus, $$i_1 = I_p\sin(\omega t + \psi)$$ $$= 2\sin(\omega t + 21°)$$

(b) The peak value is 12, and the phase angle is $-48°$. Thus,

$$v_2 = V_p \sin(\omega t + \psi)$$

$$= 12 \sin(\omega t - 48°)$$

Exercises 7.1

1. Write equations for sine waves of voltage for the phasors in Figure 7.3 for the following given values:

(a) $V_1 = 6$ V, $V_2 = 3$ V, $\theta = 30°$ (c) $V_1 = 17$ V, $V_2 = 9$ V, $\theta = 50°$
(b) $V_1 = 8$ mV, $V_2 = 22$ mV, $\theta = 17°$ (d) $V_1 = 71$ V, $V_2 = 95$ V, $\theta = 60°$

2. Draw phasor diagrams for the following wave forms:

(a) $v = 16 \sin \omega t$ (f) $v = 167 \sin(377t + 80°)$
(b) $v = 58 \sin(\omega t + 25°)$ (g) $i = 3 \sin \omega t$
(c) $v = 22 \sin(377t + 45°)$ (h) $i = 1.8 \sin(10t - 30°)$
(d) $v = 1.88 \sin(10^3 t - 30°)$ (i) $i = 6 \sin(\omega t + 45°)$
(e) $v = 9.5 \sin(\omega t = 45°)$ (j) $i = 0.11 \sin(10^4 t - 60°)$

3. Write the phasor representation for $v = V \sin(\omega t + \psi)$ for:

(a) $V = 7$ V, $\omega = 10^3$, $\psi = 16°$ (c) $V = 175$ V, $\psi = 40°$
(b) $V = 4.8$ mV, $\omega = 377$, $\psi = -80°$ (d) $V = V_p$, $\psi = -35°$

4. Write the phasor representation for $i = I \sin(\omega t + \psi)$ for:

(a) $I = 1.19$, $\omega = 10^4$, $\psi = 11°$ (c) $I = I_p$, $\psi = -73°$
(b) $I = 6$ mA, $\omega = 56$, $\psi = -40°$ (d) $I = 0.17$, $\psi = 63°$

5. Draw a phasor diagram showing $\mathbf{V}_1 = 6 \underline{/30°}$, $\mathbf{V}_2 = 11 \underline{/-45°}$, and $\mathbf{V}_3 = 4.8 \underline{/60°}$.

6. Draw a phasor diagram showing $v_1 = 17 \sin \omega t$, $v_2 = 9.6 \sin(\omega t - 30°)$, and $v_3 = 12.5 \sin(\omega t + 60°)$.

7.2   COMPLEX NUMBERS

The representation of alternating currents and voltages by means of phasors will greatly simplify the analysis of ac circuits. The phasor diagrams can be drawn on any convenient coordinate system, but a system that defines the *complex plane* will prove to be the most useful. In order to utilize the complex plane, it is first of all necessary to introduce the concept of a *complex number*.

Rectangular Form

Recall that the graphical representation of real numbers is given in terms of the real number line shown in Figure 7.5(a). In a similar fashion, imaginary numbers can be displayed in graphical form by means of the imaginary number

(a) The real number line

(b) The imaginary number line

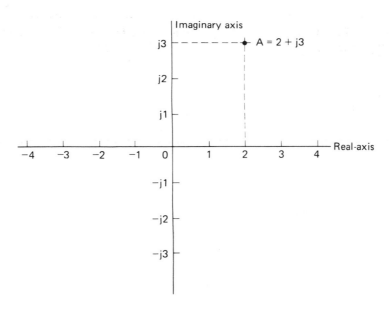

(c) The complex plane

Figure 7.5

line shown in Figure 7.5(b). Remember that an imaginary number was defined as ja, where $j \equiv \sqrt{-1}$.

The sum of any two real numbers yields another real number. Similarly, the sum of any two imaginary numbers yields another imaginary number. But the sum of a real number and an imaginary number yields a *complex number*. For instance, $2 + 3 = 5$ or $j2 + j3 = j5$; but the addition of the real number 2 and the imaginary number $j3$ yields the complex number $2 + j3$. In general, a complex number can be written as

$$\mathbf{Z} = a + jb$$

Notice that **boldface type** is used to express the complex number **Z**. The complex number **Z** is composed of two parts—the *real part a*, and the *imaginary part b*. For instance, the complex number $5-j6$ has a real part equal to 5 and an imaginary part equal to -6.

Complex Plane

Complex numbers can be conveniently plotted on a coordinate system formed by using the real number line for the horizontal axis, and the imaginary number line for the vertical axis, as shown in Figure 7.5(c). The plane formed by these two number lines is defined as the *complex plane*. A real number is plotted on the real axis. An imaginary number is plotted on the imaginary axis. A complex number is the unique point on the plane defined by combining the real part and the imaginary part. For instance, $A=2+j3$ is located as shown in Figure 7.5(c).

| EXAMPLE 7.5 Plot the following numbers on the complex plane:
(a) $A=3$
(b) $B=j2$
(c) $C=1-j2$
(d) $D=-1+j3$
(e) $E=-2-j2$ | SOLUTION. The numbers are plotted in Figure 7.6. |

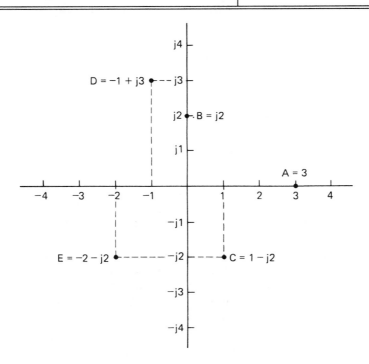

Figure 7.6

Polar Form

The complex number $\mathbf{Z} = a + jb$ can be plotted on the complex plane, and the rectangular coordinates of the point are simply the real part a and the imaginary part b. Any such unique point can also be uniquely specified in terms of polar coordinates ρ and θ. Using right-triangle relationships as shown in Figure 7.7,

$$\mathbf{Z} = a + jb \quad \text{(rectangular form)}$$
$$= \rho \angle \theta \quad \text{(polar form)}$$

where

$$\rho = \sqrt{a^2 + b^2}, \qquad \theta = \tan^{-1}\frac{b}{a} \tag{7.1}$$

and

$$a = \rho\cos\theta, \qquad b = \rho\sin\theta \tag{7.2}$$

Many electronic hand calculators will directly convert numbers given in rectangular form into polar form, and vice versa. But Eqs. (7.1) and (7.2) can be used if a calculator having polar/rectangular conversion is not available.

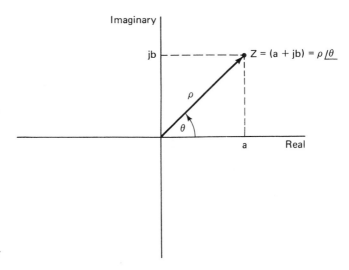

Figure 7.7

EXAMPLE 7.6.   Convert $\mathbf{Z} = 3 + j4$ into polar form.

SOLUTION.   If you have a calculator providing direct conversion, take this opportunity to learn how to do it in accordance with the instruction manual. Otherwise, use Eq. (7.1). From $\mathbf{Z} = 3 + j4$, $a = 3$ and $b = 4$. Thus,

$$\rho = \sqrt{a^2 + b^2} = \sqrt{3^2 + 4^2} = \sqrt{25} = 5$$

$$\theta = \tan^{-1}\frac{b}{a} = \tan^{-1}\frac{4}{3} = 53.1°$$

Thus,

$$\mathbf{Z}=3+j4=5\underline{/53.1^\circ}$$

This point is plotted in Figure 7.8.

EXAMPLE 7.7. Convert $\mathbf{Y}=10\underline{/45^\circ}$ into rectangular form.

SOLUTION. If you have a calculator providing direct conversion, take this opportunity to learn how to do it in accordance with the instruction manual. Otherwise, use Eq. (7.2). From $\mathbf{Y}=10\underline{/45^\circ}$, $\rho=10$ and $\theta=45^\circ$. Thus,

$$a=\rho\cos\theta=10\cos45^\circ=7.07$$
$$b=\rho\sin\theta=10\sin45^\circ=7.07$$

Thus,

$$\mathbf{Y}=10\underline{/45^\circ}=7.07+j7.07$$

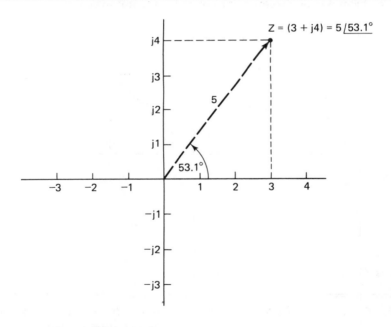

Figure 7.8

Euler's Equation

A precise mathematical expression for the vector $\mathbf{Z}$ drawn on the complex plane, as shown in Figure 7.7, is $\mathbf{Z}=\rho e^{j\theta}$. Take care to note that $e^{j\theta}$ is a complex number because of the j in the exponent. It is most definitely *not* a real number like e^3! In fact, $\rho e^{j\theta}$ means "a vector of length ρ drawn *on the complex plane* at an angle θ." The θ is measured in radians. Thus $\rho e^{j\theta}$ can be evaluated by noting that

$\mathbf{Z}=\rho e^{j\psi}=a+jb$. But using Eq. (7.2),

$$\mathbf{Z}=\rho e^{j\theta}=a+jb$$
$$=\rho\cos\theta+j\rho\sin\theta$$
$$=\rho(\cos\theta+j\sin\theta)$$

Thus,

$$e^{j\theta}=\cos\theta+j\sin\theta \qquad\qquad (7.3)$$

Equation (7.3) is a well-known mathematical expression referred to as *Euler's Equation*.

EXAMPLE 7.8.
(a) Write $\mathbf{Z}=7e^{j\pi/4}$ in rectangular form.
(b) Write $\mathbf{Z}=3+j4$ in the form $\rho e^{j\theta}$.

SOLUTION.
(a) $\rho=7$ and $\theta=\pi/4$ rad. Thus,

$$\mathbf{Z}=7e^{j\pi/4}=\rho(\cos\theta+j\sin\theta)$$

$$=7\left(\cos\frac{\pi}{4}+j\sin\frac{\pi}{4}\right)$$

$$=7(0.707+j0.707)=4.95+j4.95$$

(b) Using Eq. (7.1):

$$\rho=\sqrt{a^2+b^2}=\sqrt{3^2+4^2}=\sqrt{25}=5$$

$$\theta=\tan^{-1}\frac{b}{a}=\tan^{-1}\frac{4}{3}=0.927\text{ rad}$$

Thus,

$$\mathbf{Z}=3+j4=5e^{j0.927}$$

Phasors on the Complex Plane

Notice that Euler's Equation (7.3) has a real part that is a cosine function, and an imaginary part that is a sine function. Therefore a phasor of either current or voltage can be represented as a vector on the complex plane by using $e^{j\theta}$. For instance, the vector $\mathbf{V}=V_p e^{j\theta}$ shown on the complex plane in Figure 7.9 can be used to represent a voltage phasor. Using Euler's Equation,

$$\mathbf{V}=V_p e^{j\theta}=V_p\cos\theta+jV_p\sin\theta$$

The complex number $V_p e^{j\theta}$ contains both a sine function and a cosine function. The desired function is then selected by choosing either the *real part or* the *imarinary part*. Thus,

$$V_p\cos\theta=\text{"the real part of" }V_p e^{j\theta}=\text{Re}\left\{V_p e^{j\theta}\right\}$$
$$V_p\sin\theta=\text{"the imaginary part of" }V_p e^{j\theta}=\text{Im}\left\{V_p e^{j\theta}\right\}$$

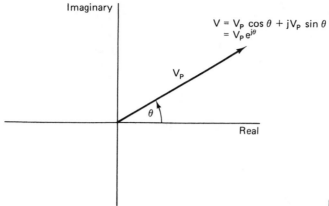

Figure 7.9

For instance, the ac voltage $v = V_p \sin(\omega t + \psi)$ can be written in phasor form as $\mathbf{V} = V_p\underline{/\psi}$. This phasor can be represented on the complex plane as a vector $\mathbf{V} = V_p e^{j\psi} = V_p \cos\psi + jV_p \sin\psi$. Notice that $\text{Im}\{V_p e^{j\psi}\} = V_p \sin\psi$ represents the desired phasor sine wave of voltage.

Remember, drawing phasors representing voltages and currents as vectors on the complex plane is simply a mathematical arrangement to aid in the analysis of ac circuits. To summarize:

TRIGONOMETRIC FORM	PHASOR FORM	COMPLEX FORM
$v = V_p \sin(\omega t + \psi)$	$\mathbf{V} = V_p\underline{/\psi}$	$\mathbf{V} = \text{Im}\{V_p e^{j\psi}\}$
$v = V_p \cos(\omega t + \psi)$	$\mathbf{V} = V_p\underline{/\psi}$	$\mathbf{V} = \text{Re}\{V_p e^{j\psi}\}$

Exercises 7.2

1. Plot the following numbers on a complex plane:

    (a) $1+j$ (b) $1-j$ (c) $-1+j$ (d) $-1-j$
    (e) $2-j3$ (f) $1.5+j2.2$ (g) $-3+j4$ (h) $j4$
    (i) $-j5$ (j) 5 (k) -3 (l) $-j+2$
    (m) $j-1$ (n) $2j-3$ (o) $j3-5$ (p) $4j$

2. Convert into polar form:

    (a) $1+j$ (b) $1-j$ (c) $-1+j$ (d) $-1-j$
    (e) $2-j3$ (f) $1.5+j2.2$ (g) $-3+j4$ (h) $j4$
    (i) $-j5$ (j) 5 (k) -3 (l) $-j+2$
    (m) $j-1$ (n) $2j-3$ (o) $j3-5$ (p) $4j$

3. Convert into rectangular form (it helps to plot the point first):

   (a) $3\,\underline{/30°}$ (b) $5\,\underline{/45°}$ (c) $10\,\underline{/-20°}$ (d) $6\,\underline{/130°}$
   (e) $6\,\underline{/145°}$ (f) $8\,\underline{/220°}$ (g) $2\,\underline{/-50°}$ (h) $3\,\underline{/340°}$
   (i) $1.7\,\underline{/300°}$ (j) $7.1\,\underline{/33°}$ (k) $3.8\,\underline{/-100°}$ (l) $5.6\,\underline{/225°}$

4. The impedance $\mathbf{Z}$ of an ac circuit can be defined as the ratio of ac voltage to ac current for that circuit, and $\mathbf{Z} = R \pm jX$, where R is the circuit resistance and X is the circuit reactance. Find R and X in each of the following:

   (a) $\mathbf{Z} = 15\,\underline{/30°}\ \Omega$ (d) $\mathbf{Z} = 220\,\underline{/-30°}\ \Omega$
   (b) $\mathbf{Z} = 3{,}800\,\underline{/40°}\ \Omega$ (e) $\mathbf{Z} = 470\,\underline{/-65°}\ \Omega$
   (c) $\mathbf{Z} = 680\,\underline{/58°}\ \Omega$ (f) $\mathbf{Z} = 51\,\underline{/-45°}\ \Omega$

5. Find $\mathbf{Z} = R \pm jX = \mathbf{Z}\underline{/\theta}$ for each of the following:

   (a) $R = 10\ \Omega,\ X = 20\ \Omega$ (d) $R = 4.7\ k\Omega,\ X = 2\ k\Omega$
   (b) $R = 1\ k\Omega,\ X = -1\ k\Omega$ (e) $R = 33\ k\Omega,\ X = -15\ k\Omega$
   (c) $R = 55\ \Omega,\ X = -750\ \Omega$ (f) $R = 3.9\ k\Omega,\ X = 39\ k\Omega$

6. Express in rectangular form:

   (a) $\mathbf{Z} = 3\,e^{j\pi/6}$ (b) $\mathbf{Z} = 15\,e^{j0.7}$ (c) $\mathbf{Z} = 3.8\,e^{j1.1}$
   (d) $\mathbf{Y} = 10^{-3}\,e^{j2\pi/5}$ (e) $\mathbf{Z} = 10^3\,e^{j3\pi/10}$ (f) $\mathbf{Z} = 80\,e^{j0}$

7. Express in the form $\rho e^{j\theta}$:

   (a) $2 + j4$ (b) $25 - j17$ (c) $10^3 + j2 \times 10^3$
   (d) $390 - j470$ (e) $93 + j22$ (f) $j6$
   (g) $-j88$ (h) $21.2 + j0$ (i) $5 + j5$

8. Write the complex form for:

   (a) $v = V_p \sin(\omega t + \pi/6)$ (d) $v = 167\cos(377t - 45°)$
   (b) $i = I_p \cos(\omega t - \pi/4)$ (e) $\mathbf{V} = 28\,\underline{/30°}$
   (c) $i = 0.17\sin(\omega t + \theta)$ (f) $\mathbf{I} = 0.071\,\underline{/-\pi/10}$

9. Write the trigonometric form. Assume $\omega = 2\pi f$.

   (a) $\mathbf{V} = \text{Im}\{21\,e^{j\pi/6}\}$ (c) $\mathbf{V} = \text{Re}\{167\,e^{-j1/5}\}$
   (b) $\mathbf{I} = \text{Re}\{2.1\,e^{j1.2}\}$ (d) $\mathbf{I} = \text{Im}\{0.77\,e^{-j0.1}\}$

7.3 ADDITION AND SUBTRACTION OF COMPLEX NUMBERS

Addition

A complex number is formed by combining a real number and an imaginary number. For instance, adding the real number 2 and the imaginary number $j3$ yields the complex number $2 + j3$. That the real part and the imaginary part must be kept separate is illustrated graphically by the complex plane.

The addition of two or more complex numbers is generally accomplished by first writing the number in rectangular form $(a + jb)$. The sum is then found by adding the real parts to find the real part of the sum, and adding the imaginary parts to find the imaginary part of the sum. For instance, if $\mathbf{A}=2+j3$ and $\mathbf{B}=4+j5$, the sum of $\mathbf{A}$ and $\mathbf{B}$ is found as

$$\begin{array}{r} 2+j3 \\ (+)\ 4+j5 \\ \hline 6+j8 \end{array}$$

In general, if $\mathbf{Z}=a+jb$ and $\mathbf{X}=c+jd$, then,

$$\mathbf{Z}+\mathbf{X}=(a+jb)+(c+jd)=(a+c)+j(b+d)$$

EXAMPLE 7.9. Find the sum of the following numbers:
(a) $(1+j)$ and $(2+j3)$
(b) 3 and $(4+j)$
(c) $j2$ and $(1-j)$
(d) 1 and $j3$

SOLUTION. In each case it is necessary to add the real parts and the imaginary parts.
(a) $(1+j)+(2+j3)=(1+2)+j(1+3)$
$$=3+j4$$
(b) Notice that 3 can be written as $3=3+j0$ (the imaginary part is zero). Then,
$$(3+j0)+(4+j)=(3+4)+j(0+1)$$
$$=7+j$$
(c) Notice that $j2$ is equivalent to $0+j2$ (the real part is zero). Thus,
$$(0+j2)+(1-j)=(0+1)+j(2-1)$$
$$=1+j$$
(d) Write $1=1+j0$, and $j3=0+j3$. Then,
$$(1+j0)+(0+j3)=1+j3$$

Subtraction

Clearly the subtraction of one complex number from another complex number is accomplished by operating on the real and imaginary parts separately just as in addition. In general, if $\mathbf{Z}=a+jb$ and $\mathbf{X}=c+jd$, then,

$$\mathbf{Z}-\mathbf{X}=(a+jb)-(c+jd)=(a-c)+j(b-d)$$

Take care with the signs!

EXAMPLE 7.10. Find the difference $\mathbf{Z}-\mathbf{X}$ if $\mathbf{Z}=3+j4$ and $\mathbf{X}=1-j2$.

SOLUTION. Careful with the signs!
$$\mathbf{Z}-\mathbf{X}=(3+j4)-(1-j2)$$
$$=(3-1)+j(4+2)$$
$$=2+j6$$

Equality

A well-known definition in mathematics states that "two complex numbers are equal if and only if their real parts are equal and their imaginary parts are equal." For instance, if $\mathbf{Z}=a+jb$ and $\mathbf{X}=c+jd$, then $\mathbf{Z}=\mathbf{X}$ if and only if $a=c$ and $b=d$. This definition is immediately apparent if you recall that any complex number $\mathbf{Z}=a+jb$ occupies a unique position on the complex plane (there is one and only one position defined by $a+jb$).

EXAMPLE 7.11. Is $3+j4$ equal to $5\underline{/40°}$?	SOLUTION. Convert $5\underline{/40°}$ to rectangular form:

$$5\underline{/40°}=a+jb$$

Using Eq. (7.2) and Figure 7.7, $\rho=5$ and $\theta=40°$. Thus,

$$a=\rho\cos\theta=5\cos 40°=3.83$$

$$b=\rho\sin\theta=5\sin 40°=3.21$$

Clearly $3+j4\neq 3.83+j3.21$

Series Impedance

The impedance $\mathbf{Z}$ of an ac circuit has been defined in the previous section as the ratio of ac voltage to ac current in a circuit, and $\mathbf{Z}=R\pm jX$, where R is the circuit resistance and X is the circuit reactance. For a number of impedances connected in series, the total equivalent impedance can be found as the sum of the individual impedances as illustrated in Figure 7.10. Thus, for n impedances connected in series,

$$\mathbf{Z}_T=\mathbf{Z}_1+\mathbf{Z}_2+\cdots+\mathbf{Z}_n \tag{7.4}$$

For instance, if $\mathbf{Z}_1=R_1+jX_1$ and $\mathbf{Z}_2=R_2+jX_2$, then the total equivalent impedance is given by

$$\mathbf{Z}_T=\mathbf{Z}_1+\mathbf{Z}_2=(R_1+jX_1)+(R_2+jX_2)$$
$$=(R_1+R_2)+j(X_1+X_2) \tag{7.5}$$

Notice that the first term is the sum of all the resistance in the circuit (the real part), and the second term is the sum of all the reactance in the circuit (the imaginary part).

EXAMPLE 7.12. Find the equivalent impedance if $\mathbf{Z}_1=(5+j7)$ Ω is connected in series with $\mathbf{Z}_2=(6-j4)$ Ω.	SOLUTION. Use Eq. (7.5) to obtain

$$\mathbf{Z}_T=(5+j7)+(6-j4)$$
$$=(5+6)+j(7-4)$$
$$=(11+j3)\ \Omega$$

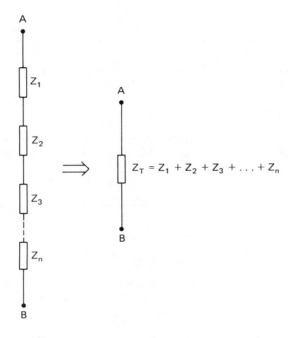

Figure 7.10

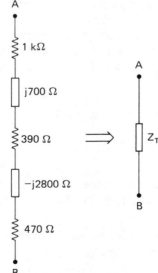

Figure 7.11

EXAMPLE 7.13.   Find the total equivalent impedance of the circuit in Figure 7.11.

SOLUTION.   Take each element to be an impedance **Z**, and add the results. Thus,

$$\mathbf{Z}_1 = 1000 + j0 \ \Omega$$
$$\mathbf{Z}_2 = \quad 0 \ + j700 \ \Omega$$
$$\mathbf{Z}_3 = \ 390 + j0 \ \Omega$$
$$\mathbf{Z}_4 = \quad 0 \ - j2800 \ \Omega$$
$$(+)\mathbf{Z}_5 = \ 470 + j0 \ \Omega$$

Then,

$$\mathbf{Z}_T = (1860 - j2100) \ \Omega$$

Exercises 7.3

1. Find the sums of the following numbers:

    (a) $1 + j, \ 2 + j3$
    (b) $1 - j, \ 3 - j$
    (c) $2 - j4, \ 4 - j2$
    (d) $j3 + 1, \ 2 - j$
    (e) $2, \ j7$
    (f) $-j3, \ 2 + j6$
    (g) $15 - j7.1, \ 17j + 6$
    (h) $1.8j + 2, \ 6 - j1.5$
    (i) $5.7, \ 3.1 - j6$
    (j) $1 + j2, \ 3\underline{/45°}$
    (k) $7\underline{/30°}, \ 5\underline{/60°}$
    (l) $j3, \ 6\underline{/-90°}$
    (m) $10\underline{/90°}, \ 3\underline{/-90°}$
    (n) $3, \ 7\underline{/-180°}$
    (o) $6, \ 12\underline{/180°}$
    (p) $1.5, \ 3\underline{/90°}$
    (q) $6\underline{/180°}, \ 3\underline{/-180°}$
    (r) $3 - j2, \ 4\underline{/45°}$
    (s) $3, j7, \ 2 - j4$
    (t) $3 - j, \ 2 + j3, \ 1 - j6$
    (u) $2\underline{/90°}, \ 6, \ 3\underline{/-90°}, \ -5$

2. Find the differences in each of the following:

    (a) $(1 - j) - (2 + j3)$
    (b) $(5 - j6) - (2 - j2)$
    (c) $3 - (2 + j3)$
    (d) $(6 + j7) - j5$
    (e) $j11 - (2 + j4)$
    (f) $(6 - j5) - 4\underline{/45°}$
    (g) $j4 - 2\underline{/-90°}$
    (h) $5 - 6\underline{/180°}$

3. If $\mathbf{Z} = 3 + j4$, what must be the value of θ if $5\underline{/\theta} = \mathbf{Z}$?

4. If $\mathbf{Z} = 10\underline{/-30°}$, what must be the values of a and b if $\mathbf{Z} = a - jb$?

5. Calculate the total equivalent resistance for the following elements connected in series:

    (a) $\mathbf{Z}_1 = 6 - j7 \ \Omega, \ \mathbf{Z}_2 = 32 \ \Omega$
    (b) $\mathbf{Z}_1 = -j15 \ \Omega, \ \mathbf{Z}_2 = 43 \ \Omega, \ \mathbf{Z}_3 = 11 + j5 \ \Omega$
    (c) $\mathbf{Z}_1 = 6\underline{/30°} \ k\Omega, \ \mathbf{Z}_2 = 3500 + j7100 \ \Omega$
    (d) $R_1 = 500 \ \Omega, \ jX_1 = j300 \ \Omega, \ R_2 = 390 \ \Omega, \ jX_2 = -j500 \ \Omega$
    (e) $500 \ \Omega, \ 390 \ \Omega, \ 750 \ \Omega, \ j700 \ \Omega, \ -j5000 \ \Omega$

6. An ac circuit consists of four impedances connected in series. Determine $\mathbf{Z}_1$ if:

$$\mathbf{Z}_T = 1000 \underline{/60°}$$
$$\mathbf{Z}_2 = 100$$
$$\mathbf{Z}_3 = j700$$
$$\mathbf{Z}_4 = 500 \underline{/45°}$$

7. What impedance must be connected in series with a resistance of 750 Ω and a reactance of $-j400$ Ω in order to provide a total impedance of $2000 \underline{/45°}\Omega$?

7.4 MULTIPLICATION OF COMPLEX NUMBERS

Rectangular Form

The multiplication of two complex numbers can be accomplished with both numbers written either in polar form or in rectangular form. When multiplying in rectangular form, it is necessary to make use of the fact that $j^2 = j \cdot j = -1$. The multiplication proceeds just as in ordinary algebra. Thus, if $\mathbf{Z}_1 = a + jb$ and $\mathbf{Z}_2 = c + jd$, then,

$$\mathbf{Z}_1 \cdot \mathbf{Z}_2 = (a+jb)(c+jd) = ac + jbc + jad + j^2bd$$

But since $j^2 = -1$,

$$\mathbf{Z}_1 \cdot \mathbf{Z}_2 = ac + jbc + jad + (-1)bd$$

Factoring and collecting terms,

$$\mathbf{Z}_1 \cdot \mathbf{Z}_2 = (ac - bd) + j(bc + ad) \tag{7.6}$$

For instance, if $\mathbf{Z}_1 = 1 + j2$ and $\mathbf{Z}_2 = 3 + j4$, then,

$$\mathbf{Z}_1 \cdot \mathbf{Z}_2 = (1+j2)(3+j4)$$
$$= (1 \cdot 3 - 2 \cdot 4) + j(2 \cdot 3 + 1 \cdot 4)$$
$$= -5 + j10$$

Polar Form

The product of two complex numbers can also be found when both numbers are written in polar form, and this is sometimes easier. The product is also given in polar form, and the *magnitude* of the product is equal to the *product of the magnitudes of the two numbers*, while the *angle* of the product is the *sum of the two given angles*. For instance, if $\mathbf{Z}_1 = Z_1 e^{j\theta_1}$ and $\mathbf{Z}_2 = Z_2 e^{j\theta_2}$, then the product is found to be

$$\mathbf{Z}_1 \cdot \mathbf{Z}_2 = Z_1 e^{j\theta_1} \cdot Z_2 e^{j\theta_2}$$

Using the Law of Exponents,

$$\mathbf{Z}_1 \cdot \mathbf{Z}_2 = Z_1 \cdot Z_2 e^{j(\theta_1 + \theta_2)}$$

Clearly $\mathbf{Z}_1 = Z_1 \underline{/\theta_1}$ and $\mathbf{Z}_2 = Z_2 \underline{/\theta_2}$, and thus,

$$\mathbf{Z}_1 \cdot \mathbf{Z}_2 = Z_1 \cdot Z_2 \underline{/\theta_1 + \theta_2} \tag{7.7}$$

For instance, if $\mathbf{Z_1} = 5\underline{/30°}$ and $\mathbf{Z_2} = 3\underline{/45°}$, then,

$$\mathbf{Z_1} \cdot \mathbf{Z_2} = 5 \cdot 3 \underline{/30° + 45°} = 15\underline{/75°}$$

EXAMPLE 7.14. Find the products of the following:
(a) $3 + j4$, $1 - j$
(b) $5\underline{/53.1°}$, $1.414\underline{/-45°}$
(c) $-j3$, $14.14\underline{/45°}$

SOLUTION.
(a) Use Eq. (7.6); take care with signs.

$$(3 + j4)(1 - j)$$
$$= [3 \cdot 1 - (4)(-1)] + j(4 \cdot 1 - 3 \cdot 1) = 7 + j$$

(b) Use Eq. (7.7):
$$5\underline{/53.1°} \times 1.414\underline{/-45°}$$
$$= 5 \times 1.414\underline{/53.1° + (-45°)}$$
$$= 7.07\underline{/8.1°}$$

(c) The multiplication may be done in either rectangular or in polar form.
Rectangular: Change $14.14\underline{/45°}$ into rectangular form.

$$14.14\underline{/45°} = 14.14(\cos 45° + j\sin 45°)$$
$$= 14.14(0.707 + j0.707)$$
$$= 10 + j10$$

Then,

$$(-j3)(10 + j10) = -j30 - (j^2)30$$
$$= -j30 - (-1)30$$
$$= 30 - j30$$

Polar: Change $-j3$ into polar form and sketch on the complex plane:

$$-j3 = 3\underline{/-90°}$$

Thus, using Eq. (7.7),
$$3\underline{/-90°} \times 14.14\underline{/45°} = 42.5\underline{/-45°}$$

As a check,
$$42.5\underline{/-45°} = 42.5(\cos 45° - j\sin 45°)$$
$$= 42.5(0.707 - j0.707)$$
$$= 30 - j30$$

When multiplying complex numbers, it will be very helpful to make a small sketch of the numbers on the complex plane. This will provide a quick visual check on the "reasonableness" of the answer, and will greatly increase your confidence.

Ohm's Law

The application of Ohm's Law to ac circuits requires the multiplication of complex numbers. That is,

$$\mathbf{V} = \mathbf{IZ} \tag{7.8}$$

where $\mathbf{V}$ and $\mathbf{I}$ are current and voltage phasors on the complex plane, and $\mathbf{Z}$ is the circuit impedance—a complex number that also can be drawn on the complex plane. In Figure 7.12(a), if the ac current $\mathbf{I}$ through the impedance $\mathbf{Z}$ is taken as the reference, then $\mathbf{I} = I_p \underline{/0°}$, and it is drawn on the complex plane as shown in Figure 7.12(b). Equation (7.8) can then be used to calculate the voltage $\mathbf{V}$ across the terminals of the impedance. For instance, if $\mathbf{Z} = 5\underline{/30°}$ kΩ and $\mathbf{I} = 6\underline{/0°}$ mA, then

$$\mathbf{V} = \mathbf{IZ} = 6 \times 10^{-3}\underline{/0°} \times 5 \times 10^{3}\underline{/30°} = 30\underline{/30°} \text{ V}$$

The voltage phasor $\mathbf{V}$ can then be drawn relative to $\mathbf{I}$ as in Figure 7.12(b).

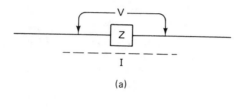

(a)

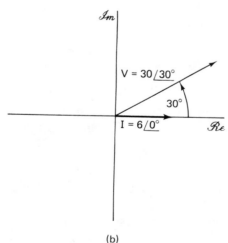

(b)

Figure 7.12

EXAMPLE 7.15. Find $\mathbf{V}$ if $\mathbf{I} = 0.2\underline{/0°}$ A and:
(a) $\mathbf{Z} = 30 \ \Omega$
(b) $\mathbf{Z} = j30 \ \Omega$
(c) $\mathbf{Z} = 30 + j30 \ \Omega$

SOLUTION. The solution may be found in either rectangular or polar form.
(a) *Rectangular*: $\mathbf{I} = 0.2\underline{/0°}$ A
$$= 0.2 + j0 = 0.2 \text{ A}$$

Thus,

$$V = IZ = 0.2(30) = 6 \text{ V}$$

Polar: $Z = 30 + j0 \; \Omega = 30\underline{/0°} \; \Omega$

Thus,

$$V = IZ = 0.2\underline{/0°} \times 30\underline{/0°}$$
$$= 6 \text{ V}$$

(b) *Rectangular*: $V = IZ = 0.2(j30) = j6 \text{ V}$

Polar: $Z = j30 = 30\underline{/90°}$

Thus,

$$V = IZ = 0.2\underline{/0°} \times 30\underline{/90°} = 6\underline{/90°} \; V$$

(c) *Rectangular*: $V = IZ = 0.2(30 + j30)$

$$= (6 + j6) \text{ V}$$

Polar: $Z = 30 + j30 = 42.4\underline{/45°} \; \Omega$

Thus,

$$V = IZ = 0.2\underline{/0°} \times 42.4\underline{/45°}$$
$$= 8.5\underline{/45°}$$

Exercises 7.4

1. Find the products of the following:

    (a) $1 + j2, \; 2 + j$
    (b) $3 - j, \; 2 + j2$
    (c) $2 - j2, \; 4 + j4$
    (d) $2 - j, \; j + 1$
    (e) $3j - 5, \; j4$
    (f) $3j, \; 2 - j3$
    (g) $2, \; j3$
    (h) $j5, \; j6$

    (i) $1.5j, \; -3j$
    (j) $\frac{1}{2}j + 1, \; \frac{2}{3} + j\frac{1}{4}$
    (k) $0.7 - j, \; 1.6 + 0.2j$
    (l) $10^3 j + 500, \; 2 \times 10^3 + j10^4$
    (m) $470j, \; 2200 - j1000$
    (n) $aj + b, \; 1 - j$
    (o) $x + jy, \; p - jq$

2. Find the products. Sketch each factor and the product on the complex plane.

    (a) $3\underline{/30°}, \; 5\underline{/20°}$
    (b) $1.7\underline{/20°}, \; 2.1\underline{/-10°}$
    (c) $16\underline{/40°}, \; 21\underline{/38°}$
    (d) $2\underline{/65°}, \; 1.7\underline{/38°}$
    (e) $3\underline{/90°}, \; 5, \; \underline{/-90°}$
    (f) $2\underline{/90°}, \; 1.8\underline{/90°}$

    (g) $11.3\underline{/0°}, \; 2.6\underline{/90°}$
    (h) $8.6\underline{/0°}, \; 3.9\underline{/0°}$
    (i) $71.2\underline{/180°}, \; 21.6\underline{/0°}$
    (j) $10^3\underline{/0°}, \; 2.7 \times 10^3\underline{/90°}$
    (k) $1.5 \times 10^3\underline{/20°}, \; 3 \times 10^3 \; \underline{45°}$
    (l) $I\underline{/\psi}, \; Z\underline{/\theta}, \; 0° < \theta < 90°, \; \psi = 2\theta$

3. Find the products first in rectangular form and then in polar form. Check solutions by changing the rectangular answer into polar form, or vice versa. Plot the polar factors and their product on the complex plane.

(a) $1+j2$, $3\underline{/30°}$
(b) $3-j$, $4\underline{/-30°}$
(c) $2-j2$, $4\underline{/45°}$
(d) $j3$, $2\underline{/45°}$
(e) 5, $6\underline{/90°}$

(f) $j7$, $2\underline{/-90°}$
(g) 1.8, $2.7\underline{/180°}$
(h) 9, $3\underline{/0°}$
(i) $1.2j$, $2.3\underline{/0°}$

4. Calculate the voltage **V** in Figure 7.12(a) and draw the current and voltage phasors on the complex plane of each of the following:

(a) $\mathbf{I}=2\underline{/0°}$ A, $\mathbf{Z}=10$ Ω
(b) $\mathbf{I}=10^{-3}\underline{/0°}$ A, $\mathbf{Z}=(2+j3)$ kΩ
(c) $\mathbf{I}=3.1\underline{/0°}$ mA, $\mathbf{Z}=2\underline{/10°}$ kΩ
(d) $\mathbf{I}=0.78\underline{/0°}$ A, $\mathbf{Z}=j7$ Ω

(e) $\mathbf{I}=7.1\underline{/0°}$ mA, $\mathbf{Z}=(1-j16)$ kΩ
(f) $\mathbf{I}=3.87\underline{/0°}$ μA, $\mathbf{Z}=1$ MΩ
(g) $\mathbf{I}=10^{-4}\underline{/0°}$ A, $\mathbf{Z}=(2-j)$ kΩ
(h) $\mathbf{I}=4\underline{/0°}$ mA, $\mathbf{Z}=31\underline{/40°}$ kΩ

7.5   DIVISION OF COMPLEX NUMBERS

Polar Form

Although the division of one complex number by another can be carried out in either rectangular form or in polar form, it is generally easier if the numbers are in polar form. The quotient will then appear in polar form also: the *magnitude* of the quotient is equal to the *quotient of the magnitude of the two factors*, and the *angle* of the quotient is equal to the *angle of the numerator factor minus the angle of the denominator factor*. For instance, if $\mathbf{Z}_1=Z_1e^{j\theta_1}$ and $\mathbf{Z}_2=Z_2e^{j\theta_2}$, the quotient is found to be

$$\frac{\mathbf{Z}_1}{\mathbf{Z}_2}=\frac{Z_1e^{j\theta_1}}{Z_2e^{j\theta_2}}$$

Using the Law of Exponents,

$$\frac{\mathbf{Z}_1}{\mathbf{Z}_2}=\frac{Z_1}{Z_2}e^{j(\theta_1-\theta_2)}$$

Thus, $\mathbf{Z}_1=Z_1\underline{/\theta_1}$ and $\mathbf{Z}_2=Z_2\underline{/\theta_2}$, and

$$\frac{\mathbf{Z}_1}{\mathbf{Z}_2}=\frac{Z_1\underline{/\theta_1}}{Z_2\underline{/\theta_2}}=\frac{Z_1}{Z_2}\underline{/\theta_1-\theta_2} \tag{7.9}$$

For instance, if $\mathbf{Z}_1=8\underline{/45°}$ and $\mathbf{Z}_2=4\underline{/30°}$, then

$$\frac{\mathbf{Z}_1}{\mathbf{Z}_2}=\frac{8}{4}\underline{/45°-30°}=2\underline{/15°}$$

Rectangular Form

In order to explain the division of one complex number by another in rectagular form, it is first of all necessary to define the *complex conjugate*. The *complex conjugate* of a complex number is the number obtained by changing the sign of the imaginary part. Thus, if $Z = a + jb$, then the *complex conjugate* of Z, written Z^* is $Z^* = a - jb$. For instance, the complex conjugate of $1 + j2$ is $1 - j2$; the conjugate of $2 - j3$ is $2 + j3$, and so on.

The useful thing here is that the product of a complex number and its conjugate is always a real number! That is, if $Z = a + jb$, $Z^* = a - jb$, and

$$Z \cdot Z^* = (a + jb)(a - jb) = (a^2 + b^2) + j(ab - ab) = (a^2 + b^2)$$

For instance, if $Z = 1 + j2$, then,

$$Z \cdot Z^* = (1 + j2)(1 - j2) = (1^2 + 2^2) = 5$$

Now, here is how the complex conjugate is used in division. As an illustration, consider the division of $(10 + j10)$ by $1 + j2$:

$$\frac{10 + j10}{1 + j2}$$

The first step is to multiply both the numerator and denominator by the complex conjugate of the denominator $(1 - j2)$. Thus,

$$\frac{10 + j10}{1 + j2} \times \frac{1 - j2}{1 - j2} = \frac{(10 + j10)(1 - j2)}{(1^2 + 2^2)} = \frac{30 - j10}{5}$$

The denominator will then always be a real number that can be divided into each part of the numerator. In this case,

$$\frac{30 - j10}{5} = 6 - j2$$

In a general form,

$$\frac{a + jb}{c + jd} = \frac{(a + jb)}{(c + jd)} \times \frac{(c - jd)}{(c - jd)} = \frac{(ac + bd) + j(bc - ad)}{c^2 + d^2} \qquad (7.10)$$

$$= \frac{ac + bd}{c^2 + d^2} + j\frac{bc - ad}{c^2 + d^2}$$

EXAMPLE 7.16. Find the quotient $7.07\underline{/45°}$ divided by $2 - j$ first in polar form and then in rectangular form.

SOLUTION. *Polar form*: First, change $2 - j$ into polar form as

$$2 - j = \sqrt{2^2 + 1^2} \ \underline{/\tan^{-1} -\tfrac{1}{2}}$$

$$= \sqrt{5} \ \underline{/-26.6°}$$

Then,

$$\frac{7.07\underline{/45^\circ}}{\sqrt{5}\ \underline{/-26.6^\circ}}$$

$$=\frac{7.07}{\sqrt{5}}\underline{/45^\circ+26.6^\circ}=3.16\underline{/71.6^\circ}$$

Rectangular form: First, change $7.07\underline{/45^\circ}$ into rectangular form as

$$7.07\underline{/45^\circ}=7.07(\cos 45^\circ+j\sin 45^\circ)$$

$$=7.07(0.707+j0.707)$$

$$=5+j5$$

Now,

$$\frac{5+j5}{2-j}=\frac{5+j5}{2-j}\times\frac{2+j}{2+j}=\frac{(5+j5)(2+j)}{2^2+1^2}$$

$$=\frac{(10-5)+j(10+5)}{5}=\frac{5+j15}{5}=1+j3$$

As a check,

$$1+j3=\sqrt{1^2+3^2}\ \underline{/\tan^{-1}\frac{3}{1}}=\sqrt{10}\ \underline{/71.6^\circ}$$

$$=3.16\underline{/71.6^\circ}$$

and the two results are seen to be equivalent.

Ohm's Law

The application of Ohm's Law to ac circuits can involve division of complex numbers. That is,

$$\mathbf{I}=\frac{\mathbf{V}}{\mathbf{Z}} \tag{7.11}$$

where $\mathbf{I}$ and $\mathbf{V}$ are phasors on the complex plane and $\mathbf{Z}$ is impedance. If the voltage phasor $\mathbf{V}$ is taken as the reference in Figure 7.13(a), then the current $\mathbf{I}$ is found by using Ohm's Law, and it is then drawn on the phasor diagram relative to $\mathbf{V}$. For instance, if $\mathbf{V}=10\underline{/0^\circ}$ V and $\mathbf{Z}=5\underline{/45^\circ}$ kΩ in Figure 7.13(a), then

$$\mathbf{I}=\frac{\mathbf{V}}{\mathbf{Z}}=\frac{10\underline{/0^\circ}}{5\times10^3\underline{/45^\circ}}=2\underline{/-45^\circ}\text{ mA}$$

The phasors are drawn on the complex plane as shown in Figure 7.13(b).

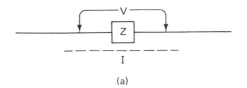

(a)

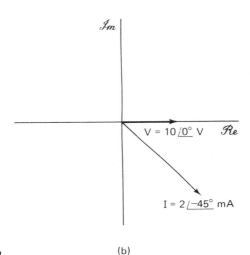

(b)

Figure 7.13

Parallel Impedances

Two impedances connected in parallel can be combined to form an equivalent impedance, as shown in Figure 7.14. The equivalent impedance is found by taking the product divided by the sum—just as with parallel resistors. Thus,

$$Z_{eq} = \frac{Z_1 \cdot Z_2}{Z_1 + Z_2} \qquad (7.12)$$

EXAMPLE 7.17.   Find Z_{eq} in Figure 7.14 if $Z_1 = 10\ \Omega$ and $Z_2 = 3 + j4\ \Omega$.

SOLUTION.   Use Eq. (7.12) to obtain:

$$Z_{eq} = \frac{10(3+j4)}{(10)+(3+j4)} = \frac{30+j40}{13+j4}$$

Convert to polar form:

$$Z_{eq} = \frac{\sqrt{30^2+40^2}\ \big/\tan^{-1}\frac{40}{30}}{\sqrt{13^2+4^2}\ \big/\tan^{-1}\frac{4}{13}} = \frac{50\ \big/53.1°}{13.6\ \big/17.1°}$$

$$= 3.68\ \big/36°\ \Omega$$

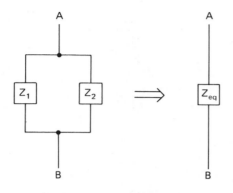

Figure 7.14

Exercises 7.5

1. Find the quotients. Sketch the factors and each quotient on the complex plane for the following:

   (a) $3\underline{/30°} \div 5\underline{/20°}$

   (b) $1.7\underline{/20°} \div 2.1\underline{/-10°}$

   (c) $16\underline{/40°} \div 21\underline{/38°}$

   (d) $2\underline{/65°} \div 1.7\underline{/38°}$

   (e) $3\underline{/90°} \div 5\underline{/-90°}$

   (f) $2\underline{/90°} \div 1.8\underline{/90°}$

   (g) $2.6\underline{/90°} \div 1.3\underline{/0°}$

   (h) $8.6\underline{/0°} \div 3.9\ \underline{0°}$

   (i) $71.2\underline{/180°} \div 21.6\underline{/0°}$

   (j) $10^3\underline{/0°} \div 2.7 \times 10^3\underline{/90°}$

   (k) $1.5 \times 10^3\underline{/20°} \div 3 \times 10^3\underline{/45°}$

   (l) $V\underline{/\psi} \div Z\underline{/\theta},\ 0° < \theta < 90°,\ \psi = 2\theta$

2. Write the complex conjugate:

   (a) $1+j$

   (b) $1-j$

   (c) $15+j7$

   (d) $j+3$

   (e) 2

   (f) $j4$

   (g) $-j7$

   (h) $j+6$

   (i) $10^3 - j10^3$

3. Find the quotients by using complex conjugates:

   (a) $(1+j2) \div (2+j)$

   (b) $(3-j) \div (2+j2)$

   (c) $(2-j2) \div (4+j4)$

   (d) $(2-j) \div (j+1)$

   (e) $(3j-5) \div (j4)$

   (f) $(3j) \div (2-j3)$

   (g) $(1.5j) \div (-3j)$

   (h) $(2) \div (j7)$

   (i) $(5) \div (2.5)$

4. Calculate the current **I** in Figure 7.13(a) and draw the current and voltage phasors on the complex plane:

   (a) $\mathbf{V} = 15\underline{/0°}$ V, $\mathbf{Z} = 3\underline{/0°}$ kΩ

   (b) $\mathbf{V} = 21\underline{/0°}$ V, $\mathbf{Z} = (3+j4)$ kΩ

   (c) $\mathbf{V} = 167\underline{/0°}$ V, $\mathbf{Z} = (21-j17)$ Ω

   (d) $\mathbf{V} = 12\underline{/0°}$ V, $\mathbf{Z} = (1+j2)$ kΩ

   (e) $\mathbf{V} = 2.18\underline{/0°}$ V, $\mathbf{Z} = (0.3-j0.1)$ MΩ

   (f) $\mathbf{V} = 21\underline{/0°}$ mV, $\mathbf{Z} = 7\underline{/45°}$ kΩ

5. Find Z_{eq} in Figure 7.14 for each of the following:

   (a) $Z_1 = 5\ \Omega,\ Z_2 = 10\ \Omega$ (e) $Z_1 = 4.7\ k\Omega,\ Z_2 = -j1.3\ k\Omega$

   (b) $Z_1 = 1 - j2,\ Z_2 = 2 - j$ (f) $Z_1 = 330\ \Omega,\ Z_2 = j680\ \Omega$

   (c) $Z_1 = 25\underline{/30°}\ \Omega,\ Z_2 = 15 - j20\ \Omega$ (g) $Z_1 = 65\underline{/-30°}\ \Omega,\ Z_2 = 28\underline{/44°}\ \Omega$

   (d) $Z_1 = j700\ \Omega,\ Z_2 = -j550\ \Omega$ (h) $Z_1 = 3\ M\Omega,\ Z_2 = 1.2 \times 10^6 \underline{/20°}\ \Omega$

UNIT 8
LOGARITHMS

Logarithms have traditionally been used as a mathematical aid to solve problems involving long and tedious computation. The electronic hand calculator has eliminated the need for logarithms as a computational aid, but logarithms have other important applications in science and engineering. They occur in a number of different equations and formulae, and they are useful in expressing calculated or measured data. It is therefore essential to know first of all exactly what a logarithm is, and then to know how to evaluate an equation or formula containing a logarithm.

8.1 DEFINITION—PROPERTIES

Definition

Simply stated, *a logarithm is an exponent*. The logarithm (log) of a number is revealed when that number is written in exponential form. For instance, in exponential form, $100,000 = 10^5$, $8 = 2^3$, $81 = 3^4$, and so on. In each case, the log of the number is equal to the *exponent*. Clearly the *base* of each exponential term must be known in order to uniquely determine the exponent and thus the log. That is,

---

LOG

The logarithm of a number N to a given base b is the exponent x, which must be placed on the base to provide the number. The base b can be any positive number other than 0 or 1.

The statement "The log to the base 10 of the number 100,000 is 5" can be written in mathematical shorthand as "$\log_{10} 100,000 = 5$." Similarly, $\log_2 8 = 3$ can be read as "The log of 8 to the base 2 is equal to 3," since $2^3 = 8$. And $\log_3 81 = 4$ can be read as "The log of 81 to the base 3 is equal to 4," since $3^4 = 81$. In general, if

$$b^x = N \tag{8.1}$$

then

$$\log_b N = x \tag{8.2}$$

Equation (8.1) is an *exponential equation* used to define a log, while Eq. (8.2) is a *logarithmic equation* that provides the same information. It should be clear that the base b must not be zero, since zero raised to any power is always zero ($0^x = 0$). Similarly, b must not be equal to 1, since 1 raised to any power is always 1 ($1^x = 1$).

EXAMPLE 8.1.   Determine the following logarithms:
(a) $10^2 = 100$
(b) $5^3 = 125$
(c) $25^{1/2} = 5$
(d) $e^x = y$

SOLUTION.   Using the definition of a log along with Eqs. (8.1) and (8.2),
(a) The exponent is 2 and the base is 10. Thus, "The log to the base 10 of 100 is equal to 2." Or,
$$\log_{10} 100 = 2$$

(b) "The log to the base 5 of 125 is equal to 3." Or,
$$\log_5 125 = 3$$

(c) "The log of 5 to the base 25 is equal to $\frac{1}{2}$." Or,
$$\log_{25} 5 = 0.5$$

(d) "The log of y to the base e is equal to x." That is,
$$\log_e y = x$$

Log of a Product

The Law of Exponents can be used to develop useful properties of logarithms. First of all, consider finding the log of a product of the two factors M and N. If $M = b^x$ and $N = b^y$, then, using the Law of Exponents,

$$(M \cdot N) = b^x \cdot b^y = b^{(x+y)}$$

Then from Eqs. (8.1) and (8.2), the log of this product is

$$\log_b(M \cdot N) = x + y$$

But $x = \log_b M$, and $y = \log_b N$. So by substitution,

$$\log_b(M \cdot N) = \log_b M + \log_b N \tag{8.3}$$

In words,

LOG OF A PRODUCT

The $\log_b$ of the product of two numbers is equal to the sum of the $\log_b$ of each individual number.

EXAMPLE 8.2.   Find the $\log_{10} 10^5$ by writing 10^5 as the product of 10^2 and 10^3.

SOLUTION.   Write $10^5 = 10^2 \cdot 10^3$. Let $M = 10^2$ and $N = 10^3$. Then $\log_{10} M = 2$ and $\log_{10} N = 3$. So, using Eq. (8.3),

$$\log_{10} M \cdot N = 2 + 3$$

or

$$\log_{10} 10^2 \cdot 10^3 = 5$$

Log of a Quotient

The Law of Exponents can also be used to develop an expression for the log of the quotient of two factors M and N. If $M = b^x$ and $N = b^y$, then,

$$\frac{M}{N} = \frac{b^x}{b^y} = b^{(x-y)}$$

Then using Eqs. (8.1) and (8.2), the log of this quotient is

$$\log_b\left(\frac{M}{N}\right) = x - y$$

Again, $x = \log_b M$ and $y = \log_b N$. So,

$$\log_b\left(\frac{M}{N}\right) = \log_b M - \log_b N \tag{8.4}$$

In words,

LOG OF A QUOTIENT

The $\log_b$ of the quotient of two numbers is equal to the $\log_b$ of the denominator subtracted from the $\log_b$ of the numerator.

EXAMPLE 8.3.

Find the $\log_{10}(10^3/10^2)$.

SOLUTION.   If $M = 10^3$ and $N = 10^2$, then $\log_{10} M = 3$ and $\log_{10} N = 2$. Using Eq. (8.4),

$$\log\left(\frac{10^3}{10^2}\right) = 3 - 2 = 1$$

Exercises 8.1

1. Express in exponential form:
   (a) $1{,}000{,}000 = 10^?$
   (b) $64 = 2^?$
   (c) $243 = 3^?$
   (d) $6.25 = 2.5^?$
   (e) $15{,}625 = 5^?$

2. Write as a logarithmic equation:

(a) $5^4 = 625$ (d) $64^{1/2} = 8$

(b) $3^0 = 1$ (e) $8^{1/3} = 2$

(c) $e^x = y$ (f) $16^3 = 4096$

3. Write as an exponential equation:

(a) $\log_2 8 = 3$ (d) $\log_{25} 5 = \dfrac{1}{2}$

(b) $\log_8 1 = 0$ (e) $\log_4 4 = 1$

(c) $\log_3 \dfrac{1}{3} = -1$ (f) $\log_2 \dfrac{1}{8} = -3$

4. Find the value of each of the following logarithms:

(a) $\log_{10} 10$ (i) $\log_3 1$

(b) $\log_{10} 10{,}000$ (j) $\log_9 3$

(c) $\log_2 4$ (k) $\log_{16} 4$

(d) $\log_2 16$ (l) $\log_{25} 5$

(e) $\log_2 64$ (m) $\log_{10} 0.01$

(f) $\log_2 1024$ (n) $\log_{10} 0.0001$

(g) $\log_2 1$ (o) $\log_2 \dfrac{1}{8}$

(h) $\log_{10} 1$

5. Find the unknown N, b, or x of the following:

(a) $\log_3 N = 4$ (f) $\log_6 N = -2$

(b) $\log_2 N = 6$ (g) $\log_b \dfrac{1}{9} = -\dfrac{1}{2}$

(c) $\log_b 16 = 4$ (h) $\log_b 14 = 1$

(d) $\log_{10} 0.1 = x$ (i) $\log_5 N = 0$

(e) $\log_b 81 = 4$

6. Find the logarithms:

(a) $\log_{10}(10^2 \cdot 10^4)$ (d) $\log_{10}(10^5 / 10^4)$

(b) $\log_2(8 \cdot 16)$ (e) $\log_2(16 \div 8)$

(c) $\log_3(27 \cdot 9)$ (f) $\log_3(81 / 3)$

8.2   COMMON LOGARITHMS—BASE 10

The *common* or *Briggs* system of logarithms employs the base 10. This system works nicely with decimal numbers, and it is so widely used that the subscript in Eq. (8.2) is frequently omitted. In general, when the base of a logarithm is not specified, it is *understood* that the base 10 is being used. Thus, to say "the log of N," implies "the log to the base 10 of N," and $\log N$ means $\log_{10} N$.

Log N

Any number can be written in *scientific notation* as the product of a number between 1 and 10 (the significant figures) and an appropriate power of 10 (to locate the decimal point). For example, the number 3162 can be written in scientific notation as

$$3162 = 3.162 \times 10^3$$

The significant figures are 3.162, and 10^3 is the appropriate power of 10. In this particular case, 3.162 happens to equal the square root of 10. That is, $3.162 = \sqrt{10} = 10^{1/2} = 10^{0.5}$. Thus, by substitution,

$$
\begin{aligned}
3162 &= 3.162 \times 10^3 \\
&= \sqrt{10} \times 10^3 \\
&= 10^{0.5} \times 10^3
\end{aligned}
$$

Using Eq. (8.3) for the log of a product of two numbers,

$$
\begin{aligned}
\log 3162 &= \log(10^{0.5} \times 10^3) \\
&= \log 10^{0.5} + \log 10^3 \\
&= 0.5 + 3 = 3.5
\end{aligned}
$$

This idea can of course be applied to any number A, and in general, if

$$A = 10^m \times 10^c = 10^{(c+m)} \tag{8.5}$$

then

$$\log A = (c + m) \tag{8.6}$$

where 10^m expresses the significant figures and 10^c locates the decimal point.

The log of any number then consists of the sum of two exponents, c and m. The c is defined as the *characteristic*, and it is always an integer; it can be either positive or negative. The m is defined as the *mantissa*, and it is always a positive number less than 1. From the above example,

$$3162 = 10^{0.5} \times 10^3$$

The characteristic is $c = 3$, the mantissa is $m = 0.5$, and thus the log of 3162 is

$$\log 3162 = (c + m) = (3 + 0.5) = 3.5$$

For a number smaller than 1.0, the characteristic will be negative. For example, $0.03162 = 3.162 \times 10^{-2} = 10^{0.5} \times 10^{-2}$. Thus,

$$0.03162 = 10^{0.5} \times 10^{-2}$$

and

$$\log 0.03162 = (c + m) = (-2 + 0.5) = -1.5$$

Most electronic hand calculators provide $\log N$ directly by entering the number N and then depressing "log." At this point, you should learn exactly how to determine $\log N$ using the instructions provided with your calculator.

In the event a hand calculator (or slide rule) is not available, the log of a number A can still be determined using Eqs. (8.5) and (8.6) and a table of logarithms as follows:

1. Write A in scientific notation.

2. The characteristic c is taken directly from the power of 10.

3. The mantissa m is determined using the Table of Common Logarithms found in the Appendix.

4. The logarithm is then the sum of c and m, as $\log A = c + m$.

EXAMPLE 8.4. Determine the log of 271 with a hand calculator and by using the Table of Common Logarithms in the Appendix.	SOLUTION. (1) Using a calculator: $\log 271 = 2.433$. (2) Using tables: First write 271 in scientific notation as $271 = 2.71 \times 10^2$. The characteristic is thus $c = 2$. In the common log tables, locate the significant figures 2.71 as follows: (2.7 in the left column and 0.01 across the top). The mantissa is found to be $m = 0.433$ (rounding to three places). Thus, $\log 271 = (c + m) = 2 + 0.433 = 2.433$
EXAMPLE 8.5. Determine the log of 0.0271 with a hand calculator and by using the Table of Common Logarithms in the Appendix.	SOLUTION. (1) Using a calculator: $\log 0.0271 = -1.567$. (2) In scientific notation: $0.0271 = 2.71 \times 10^{-2}$. The characteristic is $c = -2$. From the log tables, the mantissa is $m = 0.433$ (as in Example 8.4). Thus, $$\begin{aligned} \log 0.0271 &= (c + m) \\ &= (-2 + 0.433) \\ &= -1.567 \end{aligned}$$

Antilog A

There are occasions when the logarithm of a number is known, and it is desired to determine the number. The number A corresponding to a given log is called the *antilog*. For instance, in Example 8.4 it was determined that "the log of 271 is equal to 2.433." It can also be said that "the antilog of 2.433 is equal to the number 271." Or,

$$\text{antilog}\, 2.433 = 271$$

The antilog can be determined directly using an electronic hand calculator (usually by entering the log and then depressing the 10^x key). At this point, you should learn exactly how to determine the antilog using the instructions provided with your calculator.

In the event a hand calculator is not available, the antilog can be determined using Eqs. (8.5) and (8.6) along with a Table of Common Logarithms in the

Appendix as follows:

1. Identify the characteristic c and the mantissa m of the given log. The m must be a positive number.

2 Use c to write the appropriate power of ten (10^c).

3. Find the mantissa m in the body of the log tables and determine the significant figures.

4. The number A is then written in scientific notation as the product of the significant figures and the power of ten.

For instance, given that $\log A = 2.433,$

$$\underset{c}{2}.\underset{m}{433},$$

1. The characteristic is $c=2$, and the mantissa is $m=0.433$.

2. The power of ten is $10^c = 10^2$.

3. In the Appendix log tables, m is at the intersection of 2.7 and 0.01. The significant figures are thus 2.71.

4. The number A is then

$$A = 2.71 \times 10^2$$

In the case where the log is a negative quantity, it is first of all necessary to rewrite the log such that the mantissa m is a positive number having a magnitude less than 1. For instance, if $\log A = -1.5$, it is first of all necessary to rewrite -1.5 as $(\underset{c}{-2} + \underset{m}{0.5})$.

EXAMPLE 8.6. Find the antilog if $\log A = -1.567$.	SOLUTION. It is first of all necessary to express $\log A$ with a positive mantissa having a magnitude less than 1. So,

$$-1.567 = (\underset{c}{-2} + \underset{m}{0.433}).$$

The mantissa 0.433 provides the significant figures 2.71 in the Appendix log tables. Thus,

$$A = 2.71 \times 10^{-2}$$

Exercises 8.2

1. Use an electronic hand calculator to determine the log of each number. Check your results by using Eqs. (8.5) and (8.6), and the table of logs in the Appendix.

Use four digits in the mantissa.

(a) 1.788
(b) 291
(c) 4821
(d) 17.22
(e) 0.1388
(f) 0.00824

(g) 361.7
(h) 7.811
(i) 1/8
(j) 1/125
(k) 25×10^3
(l) 0.179×10^{-6}

2. Use an electronic calculator to determine the antilog of each of the following. Check results using Eqs. (8.5) and (8.6), and the table of logs in the Appendix. Use three significant figures.

(a) 1.7135
(b) 3.8116
(c) 2.1461
(d) -1.5003

(e) -2.1403
(f) -5.1394
(g) 6.7076
(h) -5.4802

3. Evaluate the following:

(a) $A_P = 10 \log\left(\dfrac{P_2}{P_1}\right)$:

  (i) $P_2 = 16$, $P_1 = 1.5$
  (ii) $P_2 = 185$, $P_1 = 1.00$
  (iii) $P_2 = 0.017$, $P_1 = 10^{-3}$
  (iv) $P_2 = 0.223$, $P_1 = 0.500$

(b) $A_V = 20 \log\left(\dfrac{V_2}{V_1}\right)$:

  (i) $V_2 = 1.88$, $V_1 = 0.070$
  (ii) $V_2 = 2 \times 10^{-3}$, $V_1 = 10^{-3}$
  (iii) $V_2 = 28.7$, $V_1 = 1.0$
  (iv) $V_2 = 0.028$, $V_1 = 1.00$

(c) $Z_0 = 276 \log\left(\dfrac{d}{r}\right)$:

  (i) $d = 2.0$ cm, $r = 2.1$ mm
  (ii) $d = 21.8$ cm, $r = 2.6$ mm
  (iii) $d = 125$ cm, $r = 2.59$ mm

8.3   NATURAL LOGARITHMS—BASE e

The *natural* or *Napierian* system of logarithms uses the irrational number 2.718281828459042... as a base. This number is encountered so frequently in mathematics that the script letter e is widely used as its symbol. Four siginficant figures are adequate for problems in this text, and thus

$$e \cong 2.718$$

Many naturally occurring phenomena behave in a fashion that is predictable by using the exponential equation (8.1) with e as the base. For example, the natural decay of the voltage across the terminals of a capacitor, or the natural decay of the current in an inductor. For this reason, logarithms found with e as the base are called *natural logs*.

$\ln N$

Using e as the base, the exponential equation (8.1) becomes

$$e^x = N \tag{8.7}$$

and the corresponding logarithmic equation becomes

$$\log_e N = \ln N = x \tag{8.8}$$

Notice that $\log_e$ has been given the special abbreviation ln. Thus, both $\log_e$ and ln have the same meaning; namely, "the logarithm to the base e," or simply "the natural logarithm." For instance, "ln 7" is read as "the natural log of seven."

The natural log of a number is a function frequently provided on an electronic hand calculator (usually by entering the number on the keyboard, and then depressing the "ln" key). You should determine exactly how to find $\ln N$ using your hand calculator. Here are some samples

$$\ln 3 = 1.0986$$

which is read as "The natural log of three is equal to 1.0986."

$$\ln 1.73 = 0.5481 \qquad \ln 4.77 = 1.5623$$
$$\ln 8.91 = 2.1872 \qquad \ln 21.3 = 3.0587$$
$$\ln 0.523 = -0.6482 \qquad \ln 0.216 = -1.5325$$

The natural log of a number can also be found by using the Table of Natural Logarithms given in the Appendix. This table includes numbers from 1.00 to 9.99. As an example, the natural log of the number $N = 1.73$ is found at the intersection of 1.7 in the left column and row 3 across the top ($N = 1.73$). Thus $\ln 1.73 = 0.5481$.

The ln of a number larger than 9.99 or smaller than 1.00 is found by first rewriting the number in scientific notation. Equation (8.3) is then used to find the log of the product. The ln of the significant figures are found in the table, and the ln of the power of ten are found written at the top of the table.

EXAMPLE 8.7. Find the natural log of:
(a) 370
(b) 0.0015

SOLUTION.
(a) In scientific notation:

$$370 = 3.70 \times 10^2$$

In the natural log table,

$$\ln 3.70 = 1.3083$$

At the top of table,

$$\ln 10^2 = 4.6052$$

Using Eq. (8.3),

$$\ln 3.70 \times 10^2 = \ln 3.370 + \ln 10^2$$
$$= 1.3083 + 4.605$$
$$= 5.9135$$

(b) In scientific notation:

$$0.0015 = 1.5 \times 10^{-3} = \frac{1.5}{10^3}$$

In the natural log table,

$$\ln 1.5 = 0.4055$$

At the top of table,

$$\ln 10^3 = 6.9078$$

Using Eq (8.4),

$$\ln \frac{1.5}{10^3} = \ln 1.5 - \ln 10^3$$

$$= 0.4055 - 6.9078$$

$$= -6.5023$$

Antiln N

Occasionally the natural log of a number N is known, and it is desired to find the number N. This number is called the *natural antilog*, written "antiln N." For instance, it is known that $\ln 370 = 5.9135$. Thus the natural antilog of 5.9135 is the number 370; or,

$$\text{antiln } 5.9135 = 370$$

The antiln can be found directly using an electronic hand calculator (usually by entering the ln and then depressing the e^x key). At this point, you should learn exactly how to determine antiln N using the instructions provided with your calculator.

In the event a hand calculator is not available, a table of natural logarithms can be used to determine the antiln N. The natural log table in the Appendix covers a range from 0.00 to 9.99 for N, and the corresponding range for $\ln N$ is from 0.0000 to 2.3016. The procedure is simply to locate $\ln N$ in the *body* of the table, and then determine the value of N at that position. For instance, the natural log

0.5481 is found in the body of the table to correspond to the number $N = 1.73$ (see the circled value in the natural log tables).

If the given natural logarithm is not in the table (it is not within the range $0.000 \leqslant \ln \leqslant 2.3016$), then the procedure is to use Eq. (8.3) and the natural log of a power of ten given at the top of the table. For instance, given $\ln(M - N) = 5.1533$: the number 5.1533 is not in the table, but using Eq. (8.3),

$$\ln(M - N) = \ln M + \ln N$$
$$(5.1533) = (4.6052) + (0.5481)$$

At the top of the natural log table, $\ln 10^2 = 4.6052$, and in the body of the table $\ln 1.73 = 0.5481$. Thus

$$M = 10^2, \quad N = 1.73$$

and

$$M \cdot N = 1.73 \times 10^2$$

that is,

$$\ln 1.73 \times 10^2 = 5.1533$$

EXAMPLE 8.8.   Find antiln 15.2167

SOLUTION.   15.2167 is not in the natural log table. The largest natural log of a power of ten smaller than 15.2167 is $\ln 10^6 = 13.8155$. Thus, subtracting,

$$\begin{array}{r} 15.2167 \\ (-)\underline{13.8155} \\ 1.4012 \end{array}$$

1.4012 is in the table and corresponds to $N = 4.06$. Thus,

$$\ln(M \cdot N) = \ln M + \ln N$$
$$15.2167 = 13.8155 + 1.4012.$$

Thus,

$$M = 10^6, \quad N = 4.06$$

and

$$M \cdot N = 4.06 \times 10^6$$

That is,

$$\ln(4.06 \times 10^6) = 15.2167$$

You should verify this result using an electronic hand calculator.

Exercises 8.3

1. Use an electronic hand calculator to find $\ln N$ to four decimal places. Check your results using the natural log tables in the Appendix.

    (a) 0.81
    (b) 1.31
    (c) 3.17
    (d) 6.02

    (e) 7.61
    (f) 8.00
    (g) 9.13
    (h) 0.00

2. Use an electronic hand calculator to determine antiln to three significant figures. Check your results using the natural log tables in the Appendix.

    (a) 0.0953
    (b) 0.5353
    (c) 1.1474
    (d) 1.6074
    (e) 1.9643
    (f) 2.2127

    (g) 2.2834
    (h) 2.3016
    (i) 6.9078
    (j) 9.7231
    (k) 18.2533
    (l) 20.7233

3. Evaluate the following:

    (a) $6.1 \ln 3.88$
    (b) $10^{-2} \ln 1.73$
    (c) $\dfrac{\ln 26.1}{3.88}$
    (d) $\dfrac{\ln 8.66}{\ln 1.33}$
    (e) $e^x = 17.3,\ x = ?$

    (f) $e^{-x} = 0.177,\ x = ?$
    (g) $17 = 25e^{-0.2t},\ t = ?$
    (h) $1.83 = 7e^{-3t},\ t = ?$
    (i) $16 = 29(1 - e^{-0.7t}),\ t = ?$

APPENDIX A
HEXADECIMAL, DECIMAL, OCTAL AND BINARY NUMBERS

HEXADECIMAL	DECIMAL	OCTAL	BINARY	HEXADECIMAL	DECIMAL	OCTAL	BINARY
0	0	00	000 000	20	32	40	100 000
1	1	01	000 001	21	33	41	100 001
2	2	02	000 010	22	34	42	100 010
3	3	03	000 011	23	35	43	100 011
4	4	04	000 100	24	36	44	100 100
5	5	05	000 101	25	37	45	100 101
6	6	06	000 110	26	38	46	100 110
7	7	07	000 111	27	39	47	100 111
8	8	10	001 000	28	40	50	101 000
9	9	11	001 001	29	41	51	101 001
A	10	12	001 010	2A	42	52	101 010
B	11	13	001 011	2B	43	53	101 011
C	12	14	001 100	2C	44	54	101 100
D	13	15	001 101	2D	45	55	101 101
E	14	16	001 110	2E	46	56	101 110
F	15	17	001 111	2F	47	57	101 111
10	16	20	010 000	30	48	60	110 000
11	17	21	010 001	31	49	61	110 001
12	18	22	010 010	32	50	62	110 010
13	19	23	010 011	33	51	63	110 011
14	20	24	010 100	34	52	64	110 100
15	21	25	010 101	35	53	65	110 101
16	22	26	010 110	36	54	66	110 110
17	23	27	010 111	37	55	67	110 111
18	24	30	011 000	38	56	70	111 000
19	25	31	011 001	39	57	71	111 001
1A	26	32	011 010	3A	58	72	111 010
1B	27	33	011 011	3B	59	73	111 011
1C	28	34	011 100	3C	60	74	111 100
1D	29	35	011 101	3D	61	75	111 101
1E	30	36	011 110	3E	62	76	111 110
1F	31	37	011 111	3F	63	77	111 111

APPENDIX B
COMMON LOGARITHMS OF NUMBERS 1–100

N	0	1	2	3	4	5	6	7	8	9
0	....	0000	3010	4771	6021	6990	7782	8451	9031	9542
1	0000	0414	0792	1139	1461	1761	2041	2304	2553	2788
2	3010	3222	3424	3617	3802	3979	4150	4314	4472	4624
3	4771	4914	5051	5185	5315	5441	5563	5682	5798	5911
4	6021	6128	6232	6335	6435	6532	6628	6721	6812	6902
5	6990	7076	7160	7243	7324	7404	7482	7559	7634	7709
6	7782	7853	7924	7993	8062	8129	8195	8261	8325	8388
7	8451	8513	8573	8633	8692	8751	8808	8865	8921	8976
8	9031	9085	9138	9191	9243	9294	9345	9395	9445	9494
9	9542	9590	9638	9685	9731	9777	9823	9868	9912	9956
10	0000	0043	0086	0128	0170	0212	0253	0294	0334	0374
11	0414	0453	0492	0531	0569	0607	0645	0682	0719	0755
12	0792	0828	0864	0899	0934	0969	1004	1038	1072	1106
13	1139	1173	1206	1239	1271	1303	1335	1367	1399	1430
14	1461	1492	1523	1553	1584	1614	1644	1673	1703	1732
15	1761	1790	1818	1847	1875	1903	1931	1959	1987	2014
16	2041	2068	2095	2122	2148	2175	2201	2227	2253	2279
17	2304	2330	2355	2380	2405	2430	2455	2480	2504	2529
18	2553	2577	2601	2625	2648	2672	2695	2718	2742	2765
19	2788	2810	2833	2856	2878	2900	2923	2945	2967	2989
20	3010	3032	3054	3075	3096	3118	3139	3160	3181	3201
21	3222	3243	3263	3284	3304	3324	3345	3365	3385	3404
22	3424	3444	3464	3483	3502	3522	3541	3560	3579	3598
23	3617	3636	3655	3674	3692	3711	3729	3747	3766	3784
24	3802	3820	3838	3856	3874	3892	3909	3927	3945	3962
25	3979	3997	4014	4031	4048	4065	4082	4099	4116	4133
26	4150	4166	4183	4200	4216	4232	4249	4265	4281	4298
27	4314	4330	4346	4362	4378	4393	4409	4425	4440	4456
28	4472	4487	4502	4518	4533	4548	4564	4579	4594	4609
29	4624	4639	4654	4669	4683	4698	4713	4728	4742	4757
30	4771	4786	4800	4814	4829	4843	4857	4871	4886	4900
31	4914	4928	4942	4955	4969	4983	4997	5011	5024	5038
32	5051	5065	5079	5092	5105	5119	5132	5145	5159	5172
33	5185	5198	5211	5224	5237	5250	5263	5276	5289	5302
34	5315	5328	5340	5353	5366	5378	5391	5403	5416	5428
35	5441	5453	5465	5478	5490	5502	5514	5527	5539	5551
36	5563	5575	5587	5599	5611	5623	5635	5647	5658	5670
37	5682	5694	5705	5717	5729	5740	5752	5763	5775	5786
38	5798	5809	5821	5832	5843	5855	5866	5877	5888	5899
39	5911	5922	5933	5944	5955	5966	5977	5988	5999	6010
40	6021	6031	6042	6053	6064	6075	6085	6096	6107	6117
41	6128	6138	6149	6160	6170	6180	6191	6201	6212	6222
42	6232	6243	6253	6263	6274	6284	6294	6304	6314	6325
43	6335	6345	6355	6365	6375	6385	6395	6405	6415	6425
44	6435	6444	6454	6464	6474	6484	6493	6503	6513	6522
45	6532	6542	6551	6561	6571	6580	6590	6599	6609	6618
46	6628	6637	6646	6656	6665	6675	6684	6693	6702	6712
47	6721	6730	6739	6749	6758	6767	6776	6785	6794	6803
48	6812	6821	6830	6839	6848	6857	6866	6875	6884	6893
49	6902	6911	6920	6928	6937	6946	6955	6964	6972	6981
50	6990	6998	7007	7016	7024	7033	7042	7050	7059	7067
N	0	1	2	3	4	5	6	7	8	9

N	0	1	2	3	4	5	6	7	8	9
50	6990	6998	7007	7016	7024	7033	7042	7050	7059	7067
51	7076	7084	7093	7101	7110	7118	7126	7135	7143	7152
52	7160	7168	7177	7185	7193	7202	7210	7218	7226	7235
53	7243	7251	7259	7267	7275	7284	7292	7300	7308	7316
54	7324	7332	7340	7348	7356	7364	7372	7380	7388	7396
55	7404	7412	7419	7427	7435	7443	7451	7459	7466	7474
56	7482	7490	7497	7505	7513	7520	7528	7536	7543	7551
57	7559	7566	7574	7582	7589	7597	7604	7612	7619	7627
58	7634	7642	7649	7657	7664	7672	7679	7686	7694	7701
59	7709	7716	7723	7731	7738	7745	7752	7760	7767	7774
60	7782	7789	7796	7803	7810	7818	7825	7832	7839	7846
61	7853	7860	7868	7875	7882	7889	7896	7903	7910	7917
62	7924	7931	7938	7945	7952	7959	7966	7973	7980	7987
63	7993	8000	8007	8014	8021	8028	8035	8041	8048	8055
64	8062	8069	8075	8082	8089	8096	8102	8109	8116	8122
65	8129	8136	8142	8149	8156	8162	8169	8176	8182	8189
66	8195	8202	8209	8215	8222	8228	8235	8241	8248	8254
67	8261	8267	8274	8280	8287	8293	8299	8306	8312	8319
68	8325	8331	8338	8344	8351	8357	8363	8370	8376	8382
69	8388	8395	8401	8407	8414	8420	8426	8432	8439	8445
70	8451	8457	8463	8470	8476	8482	8488	8494	8500	8506
71	8513	8519	8525	8531	8537	8543	8549	8555	8561	8567
72	8573	8579	8585	8591	8597	8603	8609	8615	8621	8627
73	8633	8639	8645	8651	8657	8663	8669	8675	8681	8686
74	8692	8698	8704	8710	8716	8722	8727	8733	8739	8745
75	8751	8756	8762	8768	8774	8779	8785	8791	8797	8802
76	8808	8814	8820	8825	8831	8837	8842	8848	8854	8859
77	8865	8871	8876	8882	8887	8893	8899	8904	8910	8915
78	8921	8927	8932	8938	8943	8949	8954	8960	8965	8971
79	8976	8982	8987	8993	8998	9004	9009	9015	9020	9025
80	9031	9036	9042	9047	9053	9058	9063	9069	9074	9079
81	9085	9090	9096	9101	9106	9112	9117	9122	9128	9133
82	9138	9143	9149	9154	9159	9165	9170	9175	9180	9186
83	9191	9196	9201	9206	9212	9217	9222	9227	9232	9238
84	9243	9248	9253	9258	9263	9269	9274	9279	9284	9289
85	9294	9299	9304	9309	9315	9320	9325	9330	9335	9340
86	9345	9350	9355	9360	9365	9370	9375	9380	9385	9390
87	9395	9400	9405	9410	9415	9420	9425	9430	9435	9440
88	9445	9450	9455	9460	9465	9469	9474	9479	9484	9489
89	9494	9499	9504	9509	9513	9518	9523	9528	9533	9538
90	9542	9547	9552	9557	9562	9566	9571	9576	9581	9586
91	9590	9595	9600	9605	9609	9614	9619	9624	9628	9633
92	9638	9643	9647	9652	9657	9661	9666	9671	9675	9680
93	9685	9689	9694	9699	9703	9708	9713	9717	9722	9727
94	9731	9736	9741	9745	9750	9754	9759	9763	9768	9773
95	9777	9782	9786	9791	9795	9800	9805	9809	9814	9818
96	9823	9827	9832	9836	9841	9845	9850	9854	9859	9863
97	9868	9872	9877	9881	9886	9890	9894	9899	9903	9908
98	9912	9917	9921	9926	9930	9934	9939	9943	9948	9952
99	9956	9961	9965	9969	9974	9978	9983	9987	9991	9996
100	0000	0004	0009	0013	0017	0022	0026	0030	0035	0039
N	0	1	2	3	4	5	6	7	8	9

APPENDIX C
NATURAL (NAPIERIAN) LOGARITHMS OF NUMBERS

N	0	1	2	3	4	5	6	7	8	9
1.0	0.0000	0.0100	0.0198	0.0296	0.0392	0.0488	0.0583	0.0677	0.0770	0.0862
1.1	0.0953	0.1044	0.1133	0.1222	0.1310	0.1398	0.1484	0.1570	0.1655	0.1740
1.2	0.1823	0.1906	0.1989	0.2070	0.2151	0.2231	0.2311	0.2390	0.2469	0.2546
1.3	0.2624	0.2700	0.2776	0.2852	0.2927	0.3001	0.3075	0.3148	0.3221	0.3293
1.4	0.3365	0.3436	0.3507	0.3577	0.3646	0.3716	0.3784	0.3853	0.3920	0.3988
1.5	0.4055	0.4121	0.4187	0.4253	0.4318	0.4383	0.4447	0.4511	0.4574	0.4637
1.6	0.4700	0.4762	0.4824	0.4886	0.4947	0.5008	0.5068	0.5128	0.5188	0.5247
1.7	0.5306	0.5365	0.5423	0.5481	0.5539	0.5596	0.5653	0.5710	0.5766	0.5822
1.8	0.5878	0.5933	0.5988	0.6043	0.6098	0.6152	0.6206	0.6259	0.6313	0.6366
1.9	0.6419	0.6471	0.6523	0.6575	0.6627	0.6678	0.6729	0.6780	0.6831	0.6881
2.0	0.6931	0.6981	0.7031	0.7080	0.7129	0.7178	0.7227	0.7275	0.7324	0.7372
2.1	0.7419	0.7467	0.7514	0.7561	0.7608	0.7655	0.7701	0.7747	0.7793	0.7839
2.2	0.7885	0.7930	0.7975	0.8020	0.8065	0.8109	0.8154	0.8198	0.8242	0.8286
2.3	0.8329	0.8372	0.8416	0.8459	0.8502	0.8544	0.8587	0.8629	0.8671	0.8713
2.4	0.8755	0.8796	0.8838	0.8879	0.8920	0.8961	0.9002	0.9042	0.9083	0.9123
2.5	0.9163	0.9203	0.9243	0.9282	0.9322	0.9361	0.9400	0.9439	0.9478	0.9517
2.6	0.9555	0.9594	0.9632	0.9670	0.9708	0.9746	0.9783	0.9821	0.9858	0.9895
2.7	0.9933	0.9969	1.0006	1.0043	1.0080	1.0116	1.0152	1.0188	1.0225	1.0260
2.8	1.0296	1.0332	1.0367	1.0403	1.0438	1.0473	1.0508	1.0543	1.0578	1.0613
2.9	1.0647	1.0682	1.0716	1.0750	1.0784	1.0818	1.0852	1.0886	1.0919	1.0953
3.0	1.0986	1.1019	1.1053	1.1086	1.1119	1.1151	1.1184	1.1217	1.1249	1.1282
3.1	1.1314	1.1346	1.1378	1.1410	1.1442	1.1474	1.1506	1.1537	1.1569	1.1600
3.2	1.1632	1.1663	1.1694	1.1725	1.1756	1.1787	1.1817	1.1848	1.1878	1.1909
3.3	1.1939	1.1969	1.2000	1.2030	1.2060	1.2090	1.2119	1.2149	1.2179	1.2208
3.4	1.2238	1.2267	1.2296	1.2326	1.2355	1.2384	1.2413	1.2442	1.2470	1.2499
3.5	1.2528	1.2556	1.2585	1.2613	1.2641	1.2669	1.2698	1.2726	1.2754	1.2782
3.6	1.2809	1.2837	1.2865	1.2892	1.2920	1.2947	1.2975	1.3002	1.3029	1.3056
3.7	1.3083	1.3110	1.3137	1.3164	1.3191	1.3218	1.3244	1.3271	1.3297	1.3324
3.8	1.3350	1.3376	1.3403	1.3429	1.3455	1.3481	1.3507	1.3533	1.3558	1.3584
3.9	1.3610	1.3635	1.3661	1.3686	1.3712	1.3737	1.3762	1.3788	1.3813	1.3838
4.0	1.3863	1.3888	1.3913	1.3938	1.3962	1.3987	1.4012	1.4036	1.4061	1.4085
4.1	1.4110	1.4134	1.4159	1.4183	1.4207	1.4231	1.4255	1.4279	1.4303	1.4327
4.2	1.4351	1.4375	1.4398	1.4422	1.4446	1.4469	1.4493	1.4516	1.4540	1.4563
4.3	1.4586	1.4609	1.4633	1.4656	1.4679	1.4702	1.4725	1.4748	1.4770	1.4793
4.4	1.4816	1.4839	1.4861	1.4884	1.4907	1.4929	1.4951	1.4974	1.4996	1.5019
4.5	1.5041	1.5063	1.5085	1.5107	1.5129	1.5151	1.5173	1.5195	1.5217	1.5239
4.6	1.5261	1.5282	1.5304	1.5326	1.5347	1.5369	1.5390	1.5412	1.5433	1.5454
4.7	1.5476	1.5497	1.5518	1.5539	1.5560	1.5581	1.5602	1.5623	1.5644	1.5665
4.8	1.5686	1.5707	1.5728	1.5748	1.5769	1.5790	1.5810	1.5831	1.5851	1.5872
4.9	1.5892	1.5913	1.5933	1.5953	1.5974	1.5994	1.6014	1.6034	1.6054	1.6074

N	0	1	2	3	4	5	6	7	8	9
5.0	**1.6094**	**1.6114**	**1.6134**	**1.6154**	**1.6174**	**1.6194**	**1.6214**	**1.6233**	**1.6253**	**1.6273**
5.1	1.6292	1.6312	1.6332	1.6351	1.6371	1.6390	1.6409	1.6429	1.6448	1.6467
5.2	1.6487	1.6506	1.6525	1.6544	1.6563	1.6582	1.6601	1.6620	1.6639	1.6658
5.3	1.6677	1.6696	1.6715	1.6734	1.6752	1.6771	1.6790	1.6808	1.6827	1.6845
5.4	1.6864	1.6882	1.6901	1.6919	1.6938	1.6956	1.6974	1.6993	1.7011	1.7029
5.5	1.7047	1.7066	1.7084	1.7102	1.7120	1.7138	1.7156	1.7174	1.7192	1.7210
5.6	1.7228	1.7246	1.7263	1.7281	1.7299	1.7317	1.7334	1.7352	1.7370	1.7387
5.7	1.7405	1.7422	1.7440	1.7457	1.7475	1.7492	1.7509	1.7527	1.7544	1.7561
5.8	1.7579	1.7596	1.7613	1.7630	1.7647	1.7664	1.7681	1.7699	1.7716	1.7733
5.9	1.7750	1.7766	1.7783	1.7800	1.7817	1.7834	1.7851	1.7867	1.7884	1.7901
6.0	**1.7918**	**1.7934**	**1.7951**	**1.7967**	**1.7984**	**1.8001**	**1.8017**	**1.8034**	**1.8050**	**1.8066**
6.1	1.8083	1.8099	1.8116	1.8132	1.8148	1.8165	1.8181	1.8197	1.8213	1.8229
6.2	1.8245	1.8262	1.8278	1.8294	1.8310	1.8326	1.8342	1.8358	1.8374	1.8390
6.3	1.8405	1.8421	1.8437	1.8453	1.8469	1.8485	1.8500	1.8516	1.8532	1.8547
6.4	1.8563	1.8579	1.8594	1.8610	1.8625	1.8641	1.8656	1.8672	1.8687	1.8703
6.5	1.8718	1.8733	1.8749	1.8764	1.8779	1.8795	1.8810	1.8825	1.8840	1.8856
6.6	1.8871	1.8886	1.8901	1.8916	1.8931	1.8946	1.8961	1.8976	1.8991	1.9006
6.7	1.9021	1.9036	1.9051	1.9066	1.9081	1.9095	1.9110	1.9125	1.9140	1.9155
6.8	1.9169	1.9184	1.9199	1.9213	1.9228	1.9242	1.9257	1.9272	1.9286	1.9301
6.9	1.9315	1.9330	1.9344	1.9359	1.9373	1.9387	1.9402	1.9416	1.9430	1.9445
7.0	**1.9459**	**1.9473**	**1.9488**	**1.9502**	**1.9516**	**1.9530**	**1.9544**	**1.9559**	**1.9573**	**1.9587**
7.1	1.9601	1.9615	1.9629	1.9643	1.9657	1.9671	1.9685	1.9699	1.9713	1.9727
7.2	1.9741	1.9755	1.9769	1.9782	1.9796	1.9810	1.9824	1.9838	1.9851	1.9865
7.3	1.9879	1.9892	1.9906	1.9920	1.9933	1.9947	1.9961	1.9974	1.9988	2.0001
7.4	2.0015	2.0028	2.0042	2.0055	2.0069	2.0082	2.0096	2.0109	2.0122	2.0136
7.5	2.0149	2.0162	2.0176	2.0189	2.0202	2.0215	2.0229	2.0242	2.0255	2.0268
7.6	2.0281	2.0295	2.0308	2.0321	2.0334	2.0347	2.0360	2.0373	2.0386	2.0399
7.7	2.0412	2.0425	2.0438	2.0451	2.0464	2.0477	2.0490	2.0503	2.0516	2.0528
7.8	2.0541	2.0554	2.0567	2.0580	2.0592	2.0605	2.0618	2.0631	2.0643	2.0656
7.9	2.0669	2.0681	2.0694	2.0707	2.0719	2.0732	2.0744	2.0757	2.0769	2.0782
8.0	**2.0794**	**2.0807**	**2.0819**	**2.0832**	**2.0844**	**2.0857**	**2.0869**	**2.0882**	**2.0894**	**2.0906**
8.1	2.0919	2.0931	2.0943	2.0956	2.0968	2.0980	2.0992	2.1005	2.1017	2.1029
8.2	2.1041	2.1054	2.1066	2.1078	2.1090	2.1102	2.1114	2.1126	2.1138	2.1150
8.3	2.1163	2.1175	2.1187	2.1199	2.1211	2.1223	2.1235	2.1247	2.1258	2.1270
8.4	2.1282	2.1294	2.1306	2.1318	2.1330	2.1342	2.1353	2.1365	2.1377	2.1389
8.5	2.1401	2.1412	2.1424	2.1436	2.1448	2.1459	2.1471	2.1483	2.1494	2.1506
8.6	2.1518	2.1529	2.1541	2.1552	2.1564	2.1576	2.1587	2.1599	2.1610	2.1622
8.7	2.1633	2.1645	2.1656	2.1668	2.1679	2.1691	2.1702	2.1713	2.1725	2.1736
8.8	2.1748	2.1759	2.1770	2.1782	2.1793	2.1804	2.1815	2.1827	2.1838	2.1849
8.9	2.1861	2.1872	2.1883	2.1894	2.1905	2.1917	2.1928	2.1939	2.1950	2.1961
9.0	**2.1972**	**2.1983**	**2.1994**	**2.2006**	**2.2017**	**2.2028**	**2.2039**	**2.2050**	**2.2061**	**2.2072**
9.1	2.2083	2.2094	2.2105	2.2116	2.2127	2.2138	2.2148	2.2159	2.2170	2.2181
9.2	2.2192	2.2203	2.2214	2.2225	2.2235	2.2246	2.2257	2.2268	2.2279	2.2289
9.3	2.2300	2.2311	2.2322	2.2332	2.2343	2.2354	2.2364	2.2375	2.2386	2.2395
9.4	2.2407	2.2418	2.2428	2.2439	2.2450	2.2460	2.2471	2.2481	2.2492	2.2502
9.5	2.2513	2.2523	2.2534	2.2544	2.2555	2.2565	2.2576	2.2586	2.2597	2.2607
9.6	2.2618	2.2628	2.2638	2.2649	2.2659	2.2670	2.2680	2.2690	2.2701	2.2711
9.7	2.2721	2.2732	2.2742	2.2752	2.2762	2.2773	2.2783	2.2793	2.2803	2.2814
9.8	2.2824	2.2834	2.2844	2.2854	2.2865	2.2875	2.2885	2.2895	2.2905	2.2915
9.9	2.2925	2.2935	2.2946	2.2956	2.2966	2.2976	2.2986	2.2996	2.3006	2.3016

The table gives the natural logarithms of numbers from 1.00 to 9.99 directly, and permits the finding of the logarithms of numbers outside of that range by the addition or subtraction of the natural logarithms of powers of 10.

EXAMPLES: $\log_e 679. = \log_c 6.79 + \log_c 10^2 = 1.9155 + 4.6052 = 6.5207.$
$\log_e .0679 = \log_e 6.79 - \log_e 10^2 = 1.9155 - 4.6052 = -2.6897.$

$\log_e 10 = 2.302\ 585$ $\log_e 10^4 = 9.210\ 340$ $\log_e 10^7 = 16.118\ 096$
$\log_e 10^2 = 4.605\ 170$ $\log_e 10^5 = 11.512\ 925$ $\log_e 10^8 = 18.420\ 681$
$\log_e 10^3 = 6.907\ 755$ $\log_e 10^6 = 13.815\ 511$ $\log_e 10^9 = 20.723\ 266$

APPENDIX D
VALUES OF e^x AND e^-x

x	Function	0.00	0.01	0.02	0.03	0.04	0.05	0.06	0.07	0.08	0.09
0.0	e^x	1.0000	1.0101	1.0202	1.0305	1.0408	1.0513	1.0618	1.0725	1.0833	1.0942
	e^{-x}	1.0000	0.9900	0.9802	0.9704	0.9608	0.9512	0.9418	0.9324	0.9231	0.9139
0.1	e^x	1.1052	1.1163	1.1275	1.1388	1.1503	1.1618	1.1735	1.1853	1.1972	1.2093
	e^{-x}	0.9048	0.8958	0.8869	0.8781	0.8694	0.8607	0.8521	0.8437	0.8353	0.8270
0.2	e^x	1.2214	1.2337	1.2461	1.2586	1.2712	1.2840	1.2969	1.3100	1.3231	1.3364
	e^{-x}	0.8187	0.8106	0.8025	0.7945	0.7866	0.7788	0.7711	0.7634	0.7558	0.7483
0.3	e^x	1.3499	1.3634	1.3771	1.3910	1.4049	1.4191	1.4333	1.4477	1.4623	1.4770
	e^{-x}	0.7408	0.7334	0.7261	0.7189	0.7118	0.7047	0.6977	0.6907	0.6839	0.6771
0.4	e^x	1.4918	1.5068	1.5220	1.5373	1.5527	1.5683	1.5841	1.6000	1.6161	1.6323
	e^{-x}	0.6703	0.6637	0.6570	0.6505	0.6440	0.6376	0.6313	0.6250	0.6188	0.6126
0.5	e^x	1.6487	1.6653	1.6820	1.6989	1.7160	1.7333	1.7507	1.7683	1.7860	1.8040
	e^{-x}	0.6065	0.6005	0.5945	0.5886	0.5827	0.5769	0.5712	0.5655	0.5599	0.5543
0.6	e^x	1.8221	1.8404	1.8589	1.8776	1.8965	1.9155	1.9348	1.9542	1.9739	1.9939
	e^{-x}	0.5488	0.5434	0.5379	0.5326	0.5273	0.5220	0.5169	0.5117	0.5066	0.5017
0.7	e^x	2.0138	2.0340	2.0544	2.0751	2.0959	2.1170	2.1383	2.1598	2.1815	2.2034
	e^{-x}	0.4966	0.4916	0.4868	0.4819	0.4771	0.4724	0.4677	0.4630	0.4584	0.4538
0.8	e^x	2.2255	2.2479	2.2705	2.2933	2.3164	2.3396	2.3632	2.3869	2.4109	2.4351
	e^{-x}	0.4493	0.4449	0.4404	0.4360	0.4317	0.4274	0.4232	0.4190	0.4148	0.4107
0.9	e^x	2.4596	2.4843	2.5093	2.5345	2.5600	2.5857	2.6117	2.6379	2.6645	2.6912
	e^{-x}	0.4066	0.4025	0.3985	0.3946	0.3906	0.3867	0.3829	0.3791	0.3753	0.3716
1.0	e^x	2.7183	2.7456	2.7732	2.8011	2.8292	2.8577	2.8864	2.9154	2.9447	2.9743
	e^{-x}	0.3679	0.3642	0.3606	0.3570	0.3535	0.3499	0.3465	0.3430	0.3396	0.3362
1.1	e^x	3.0042	3.0344	3.0649	3.0957	3.1268	3.1582	3.1899	3.2220	3.2544	3.2871
	e^{-x}	0.3329	0.3296	0.3263	0.3230	0.3198	0.3166	0.3135	0.3104	0.3073	0.3042
1.2	e^x	3.3201	3.3535	3.3872	3.4212	3.4556	3.4903	3.5254	3.5609	3.5966	3.6328
	e^{-x}	0.3012	0.2982	0.2952	0.2923	0.2894	0.2865	0.2837	0.2808	0.2780	0.2753
1.3	e^x	3.6693	3.7062	3.7434	3.7810	3.8190	3.8574	3.8962	3.9354	3.9749	4.0149
	e^{-x}	0.2725	0.2698	0.2671	0.2645	0.2618	0.2592	0.2567	0.2541	0.2516	0.2491
1.4	e^x	4.0552	4.0960	4.1371	4.1787	4.2207	4.2631	4.3060	4.3492	4.3929	4.4371
	e^{-x}	0.2466	0.2441	0.2417	0.2393	0.2369	0.2346	0.2322	0.2299	0.2276	0.2254
1.5	e^x	4.4817	4.5267	4.5722	4.6182	4.6646	4.7115	4.7588	4.8066	4.8550	4.9037
	e^{-x}	0.2231	0.2209	0.2187	0.2165	0.2144	0.2122	0.2101	0.2080	0.2060	0.2039
1.6	e^x	4.9530	5.0028	5.0531	5.1039	5.1552	5.2070	5.2593	5.3122	5.3656	5.4195
	e^{-x}	0.2019	0.1999	0.1979	0.1959	0.1940	0.1920	0.1901	0.1882	0.1864	0.1845
1.7	e^x	5.4739	5.5290	5.5845	5.6407	5.6973	5.7546	5.8124	5.8709	5.9299	5.9895
	e^{-x}	0.1827	0.1809	0.1791	0.1773	0.1755	0.1738	0.1720	0.1703	0.1686	0.1670
1.8	e^x	6.0496	6.1104	6.1719	6.2339	6.2965	6.3598	6.4237	6.4883	6.5535	6.6194
	e^{-x}	0.1653	0.1637	0.1620	0.1604	0.1588	0.1572	0.1557	0.1541	0.1526	0.1511
1.9	e^x	6.6859	6.7531	6.8210	6.8895	6.9588	7.0287	7.0993	7.1707	7.2427	7.3155
	e^{-x}	0.1496	0.1481	0.1466	0.1451	0.1437	0.1423	0.1409	0.1395	0.1381	0.1367

x	Function	0.00	0.01	0.02	0.03	0.04	0.05	0.06	0.07	0.08	0.09
2.0	ϵ^x	7.3891	7.4633	7.5383	7.6141	7.6906	7.7679	7.8460	7.9248	8.0045	8.0849
	ϵ^{-x}	0.1353	0.1340	0.1327	0.1313	0.1300	0.1287	0.1275	0.1262	0.1249	0.1237
2.1	ϵ^x	8.1662	8.2482	8.3311	8.4149	8.4994	8.5849	8.6711	8.7583	8.8463	8.9352
	ϵ^{-x}	0.1225	0.1212	0.1200	0.1188	0.1177	0.1165	0.1153	0.1142	0.1130	0.1119
2.2	ϵ^x	9.0250	9.1157	9.2073	9.2999	9.3933	9.4877	9.5831	9.6794	9.7767	9.8749
	ϵ^{-x}	0.1108	0.1097	0.1086	0.1075	0.1065	0.1054	0.1044	0.1033	0.1023	0.1013
2.3	ϵ^x	9.9742	10.074	10.176	10.278	10.381	10.486	10.591	10.697	10.805	10.913
	ϵ^{-x}	0.1003	0.0993	0.0983	0.0973	0.0963	0.0954	0.0944	0.0935	0.0926	0.0916
2.4	ϵ^x	11.023	11.134	11.246	11.359	11.473	11.588	11.705	11.822	11.941	12.061
	ϵ^{-x}	0.0907	0.0898	0.0889	0.0880	0.0872	0.0863	0.0854	0.0846	0.0837	0.0829
2.5	ϵ^x	12.182	12.305	12.429	12.554	12.680	12.807	12.936	13.066	13.197	13.330
	ϵ^{-x}	0.0821	0.0813	0.0805	0.0797	0.0789	0.0781	0.0773	0.0765	0.0758	0.0750
2.6	ϵ^x	13.464	13.599	13.736	13.874	14.013	14.154	14.296	14.440	14.585	14.732
	ϵ^{-x}	0.0743	0.0735	0.0728	0.0721	0.0714	0.0707	0.0699	0.0693	0.0686	0.0679
2.7	ϵ^x	14.880	15.029	15.180	15.333	15.487	15.643	15.800	15.959	16.119	16.281
	ϵ^{-x}	0.0672	0.0665	0.0659	0.0652	0.0646	0.0639	0.0633	0.0627	0.0620	0.0614
2.8	ϵ^x	16.445	16.610	16.777	16.945	17.116	17.288	17.462	17.637	17.814	17.993
	ϵ^{-x}	0.0608	0.0602	0.0596	0.0590	0.0584	0.0578	0.0573	0.0567	0.0561	0.0556
2.9	ϵ^x	18.174	18.357	18.541	18.728	18.916	19.106	19.298	19.492	19.688	19.886
	ϵ^{-x}	0.0550	0.0545	0.0539	0.0534	0.0529	0.0523	0.0518	0.0513	0.0508	0.0503
3.0	ϵ^x	20.086	20.287	20.491	20.697	20.905	21.115	21.328	21.542	21.758	21.977
	ϵ^{-x}	0.0498	0.0493	0.0488	0.0483	0.0478	0.0474	0.0469	0.0464	0.0460	0.0455
3.1	ϵ^x	22.198	22.421	22.646	22.874	23.104	23.336	23.571	23.807	24.047	24.288
	ϵ^{-x}	0.0450	0.0446	0.0442	0.0437	0.0433	0.0429	0.0424	0.0420	0.0416	0.0412
3.2	ϵ^x	24.533	24.779	25.028	25.280	25.534	25.790	26.050	26.311	26.576	26.843
	ϵ^{-x}	0.0408	0.0404	0.0400	0.0396	0.0392	0.0388	0.0384	0.0380	0.0376	0.0373
3.3	ϵ^x	27.113	27.385	27.660	27.938	28.219	28.503	28.789	29.079	29.371	29.666
	ϵ^{-x}	0.0369	0.0365	0.0362	0.0358	0.0354	0.0351	0.0347	0.0344	0.0340	0.0337
3.4	ϵ^x	29.964	30.265	30.569	30.877	31.187	31.500	31.817	32.137	32.460	32.786
	ϵ^{-x}	0.0334	0.0330	0.0327	0.0324	0.0321	0.0317	0.0314	0.0311	0.0308	0.0305
3.5	ϵ^x	33.115	33.448	33.784	34.124	34.467	34.813	35.163	35.517	35.874	36.234
	ϵ^{-x}	0.0302	0.0299	0.0296	0.0293	0.0290	0.0287	0.0284	0.0282	0.0279	0.0276
3.6	ϵ^x	36.598	36.966	37.338	37.713	38.092	38.475	38.861	39.252	39.646	40.045
	ϵ^{-x}	0.0273	0.0271	0.0268	0.0265	0.0263	0.0260	0.0257	0.0255	0.0252	0.0250
3.7	ϵ^x	40.447	40.854	41.264	41.679	42.098	42.521	42.948	43.380	43.816	44.256
	ϵ^{-x}	0.0247	0.0245	0.0242	0.0240	0.0238	0.0235	0.0233	0.0231	0.0228	0.0226
3.8	ϵ^x	44.701	45.150	45.604	46.063	46.525	46.993	47.465	47.942	48.424	48.911
	ϵ^{-x}	0.0224	0.0221	0.0219	0.0217	0.0215	0.0213	0.0211	0.0209	0.0207	0.0204
3.9	ϵ^x	49.402	49.899	50.400	50.907	51.419	51.935	52.457	52.985	53.517	54.055
	ϵ^{-x}	0.0202	0.0200	0.0198	0.0196	0.0195	0.0193	0.0191	0.0189	0.0187	0.0185

x	Function	0.00	0.01	0.02	0.03	0.04	0.05	0.06	0.07	0.08	0.09
4.0	ϵ^x	54.598	55.147	55.701	56.261	56.826	57.397	57.974	58.557	59.145	59.740
	ϵ^{-x}	0.0183	0.0181	0.0180	0.0178	0.0176	0.0174	0.0172	0.0171	0.0169	0.0167
4.1	ϵ^x	60.340	60.947	61.559	62.178	62.803	63.434	64.072	64.715	65.366	66.023
	ϵ^{-x}	0.0166	0.0164	0.0162	0.0161	0.0159	0.0158	0.0156	0.0155	0.0153	0.0151
4.2	ϵ^x	66.686	67.357	68.033	68.717	69.408	70.105	70.810	71.522	72.240	72.966
	ϵ^{-x}	0.0150	0.0148	0.0147	0.0146	0.0144	0.0143	0.0141	0.0140	0.0138	0.0137
4.3	ϵ^x	73.700	74.440	75.189	75.944	76.708	77.478	78.257	79.044	79.838	80.640
	ϵ^{-x}	0.0136	0.0134	0.0133	0.0132	0.0130	0.0129	0.0128	0.0127	0.0125	0.0124
4.4	ϵ^x	81.451	82.269	83.096	83.931	84.775	85.627	86.488	87.357	88.235	89.121
	ϵ^{-x}	0.0123	0.0122	0.0120	0.0119	0.0118	0.0117	0.0116	0.0114	0.0113	0.0112
4.5	ϵ^x	90.017	90.922	91.836	92.759	93.691	94.632	95.583	96.544	97.514	98.494
	ϵ^{-x}	0.0111	0.0110	0.0109	0.0108	0.0107	0.0106	0.0105	0.0104	0.0103	0.0102
4.6	ϵ^x	99.484	100.48	101.49	102.51	103.54	104.58	105.64	106.70	107.77	108.85
	ϵ^{-x}	0.0101	0.0100	0.0099	0.0098	0.0097	0.0096	0.0095	0.0094	0.0093	0.0092
4.7	ϵ^x	109.95	111.05	112.17	113.30	114.43	115.58	116.75	117.92	119.10	120.30
	ϵ^{-x}	0.0091	0.0090	0.0089	0.0088	0.0087	0.0087	0.0086	0.0085	0.0084	0.0083
4.8	ϵ^x	121.51	122.73	123.97	125.21	126.47	127.74	129.02	130.32	131.63	132.95
	ϵ^{-x}	0.0082	0.0081	0.0081	0.0080	0.0079	0.0078	0.0078	0.0077	0.0076	0.0075
4.9	ϵ^x	134.29	135.64	137.00	138.38	139.77	141.17	142.59	144.03	145.47	146.94
	ϵ^{-x}	0.0074	0.0074	0.0073	0.0072	0.0072	0.0071	0.0070	0.0069	0.0069	0.0068
5.0	ϵ^x	148.41	149.90	151.41	152.93	154.47	156.02	157.59	159.17	160.77	162.39
	ϵ^{-x}	0.0067	0.0067	0.0066	0.0065	0.0065	0.0064	0.0063	0.0063	0.0062	0.0062
5.1	ϵ	164.02	165.67	167.34	169.02	170.72	172.43	174.16	175.91	177.68	179.47
	ϵ^{-x}	0.0061	0.0060	0.0060	0.0059	0.0059	0.0058	0.0057	0.0057	0.0056	0.0056
5.2	ϵ^x	181.27	183.09	184.93	186.79	188.67	190.57	192.48	194.42	196.37	198.34
	ϵ^{-x}	0.0055	0.0055	0.0054	0.0054	0.0053	0.0052	0.0052	0.0051	0.0051	0.0050
5.3	ϵ^x	200.34	202.35	204.38	206.44	208.51	210.61	212.72	214.86	217.02	219.20
	ϵ^{-x}	0.0050	0.0049	0.0049	0.0048	0.0048	0.0047	0.0047	0.0047	0.0046	0.0046
5.4	ϵ^x	221.41	223.63	225.88	228.15	230.44	232.76	235.10	237.46	239.85	242.26
	ϵ^{-x}	0.0045	0.0045	0.0044	0.0044	0.0043	0.0043	0.0043	0.0042	0.0042	0.0041
5.5	ϵ^x	244.69	247.15	249.64	252.14	254.68	257.24	259.82	262.43	265.07	267.74
	ϵ^{-x}	0.0041	0.0040	0.0040	0.0040	0.0039	0.0039	0.0038	0.0038	0.0038	0.0037
5.6	ϵ^x	270.43	273.14	275.89	278.66	281.46	284.29	287.15	290.03	292.95	295.89
	ϵ^{-x}	0.0037	0.0037	0.0036	0.0036	0.0036	0.0035	0.0035	0.0034	0.0034	0.0034
5.7	ϵ^x	298.87	301.87	304.90	307.97	311.06	314.19	317.35	320.54	323.76	327.01
	ϵ^{-x}	0.0033	0.0033	0.0033	0.0032	0.0032	0.0032	0.0032	0.0031	0.0031	0.0031
5.8	ϵ^x	330.30	333.62	336.97	340.36	343.78	347.23	350.72	354.25	357.81	361.41
	ϵ^{-x}	0.0030	0.0030	0.0030	0.0029	0.0029	0.0029	0.0029	0.0028	0.0028	0.0028
5.9	ϵ^x	365.04	368.71	372.41	376.15	379.93	383.75	387.61	391.51	395.44	399.41
	ϵ^{-x}	0.0027	0.0027	0.0027	0.0027	0.0026	0.0026	0.0026	0.0026	0.0025	0.0025

APPENDIX E
TRIGONOMETRIC FUNCTIONS IN RADIAN MEASURE

Rad	Sin	Tan	Ctn	Cos	Rad	Sin	Tan	Ctn	Cos
.00	.0000	.0000	....	1.0000	.50	.4794	.5463	1.830	.8776
.01	.0100	.0100	99.997	1.0000	.51	.4882	.5594	1.788	.8727
.02	.0200	.0200	49.993	.9998	.52	.4969	.5726	1.747	.8678
.03	.0300	.0300	33.323	.9996	.53	.5055	.5859	1.707	.8628
.04	.0400	.0400	24.987	.9992	.54	.5141	.5994	1.668	.8577
.05	.0500	.0500	19.983	.9988	.55	.5227	.6131	1.631	.8525
.06	.0600	.0601	16.647	.9982	.56	.5312	.6269	1.595	.8473
.07	.0699	.0701	14.262	.9976	.57	.5396	.6410	1.560	.8419
.08	.0799	.0802	12.473	.9968	.58	.5480	.6552	1.526	.8365
.09	.0899	.0902	11.081	.9960	.59	.5564	.6696	1.494	.8309
.10	.0998	.1003	9.967	.9950	.60	.5646	.6841	1.462	.8253
.11	.1098	.1104	9.054	.9940	.61	.5729	.6989	1.431	.8196
.12	.1197	.1206	8.293	.9928	.62	.5810	.7139	1.401	.8139
.13	.1296	.1307	7.649	.9916	.63	.5891	.7291	1.372	.8080
.14	.1395	.1409	7.096	.9902	.64	.5972	.7445	1.343	.8021
.15	.1494	.1511	6.617	.9888	.65	.6052	.7602	1.315	.7961
.16	.1593	.1614	6.197	.9872	.66	.6131	.7761	1.288	.7900
.17	.1692	.1717	5.826	.9856	.67	.6210	.7923	1.262	.7838
.18	.1790	.1820	5.495	.9838	.68	.6288	.8087	1.237	.7776
.19	.1889	.1923	5.200	.9820	.69	.6365	.8253	1.212	.7712
.20	.1987	.2027	4.933	.9801	.70	.6442	.8423	1.187	.7648
.21	.2085	.2131	4.692	.9780	.71	.6518	.8595	1.163	.7584
.22	.2182	.2236	4.472	.9759	.72	.6594	.8771	1.140	.7518
.23	.2280	.2341	4.271	.9737	.73	.6669	.8949	1.117	.7452
.24	.2377	.2447	4.086	.9713	.74	.6743	.9131	1.095	.7385
.25	.2474	.2553	3.916	.9689	.75	.6816	.9316	1.073	.7317
.26	.2571	.2660	3.759	.9664	.76	.6889	.9505	1.052	.7248
.27	.2667	.2768	3.613	.9638	.77	.6961	.9697	1.031	.7179
.28	.2764	.2876	3.478	.9611	.78	.7033	.9893	1.011	.7109
.29	.2860	.2984	3.351	.9582	.79	.7104	1.009	.9908	.7038
.30	.2955	.3093	3.233	.9553	.80	.7174	1.030	.9712	.6967
.31	.3051	.3203	3.122	.9523	.81	.7243	1.050	.9520	.6895
.32	.3146	.3314	3.018	.9492	.82	.7311	1.072	.9331	.6822
.33	.3240	.3425	2.920	.9460	.83	.7379	1.093	.9146	.6749
.34	.3335	.3537	2.827	.9428	.84	.7446	1.116	.8964	.6675
.35	.3429	.3650	2.740	.9394	.85	.7513	1.138	.8785	.6600
.36	.3523	.3764	2.657	.9359	.86	.7578	1.162	.8609	.6524
.37	.3616	.3879	2.578	.9323	.87	.7643	1.185	.8437	.6448
.38	.3709	.3994	2.504	.9287	.88	.7707	1.210	.8267	.6372
.39	.3802	.4111	2.433	.9249	.89	.7771	1.235	.8100	.6294
.40	.3894	.4228	2.365	.9211	.90	.7833	1.260	.7936	.6216
.41	.3986	.4346	2.301	.9171	.91	.7895	1.286	.7774	.6137
.42	.4078	.4466	2.239	.9131	.92	.7956	1.313	.7615	.6058
.43	.4169	.4586	2.180	.9090	.93	.8016	1.341	.7458	.5978
.44	.4259	.4708	2.124	.9048	.94	.8076	1.369	.7303	.5898
.45	.4350	.4831	2.070	.9004	.95	.8134	1.398	.7151	.581
.46	.4439	.4954	2.018	.8961	.96	.8192	1.428	.7001	.573
.47	.4529	.5080	1.969	.8916	.97	.8249	1.459	.6853	.565
.48	.4618	.5206	1.921	.8870	.98	.8305	1.491	.6707	.55
.49	.4706	.5334	1.875	.8823	.99	.8360	1.524	.6563	.54
.50	.4794	.5463	1.830	.8776	1.00	.8415	1.557	.6421	

π radians = 180°. π = 3.14159 26536
1 radian = 57°17'44".80625 = 57°.29577 95131
1° = 0.01745 32925 19943 radian = 60' = 3600"

Rad	Sin	Tan	Ctn	Cos
1.00	.8415	1.557	.6421	.5403
1.01	.8468	1.592	.6281	.5319
1.02	.8521	1.628	.6142	.5234
1.03	.8573	1.665	.6005	.5148
1.04	.8624	1.704	.5870	.5062
1.05	.8674	1.743	.5736	.4976
1.06	.8724	1.784	.5604	.4889
1.07	.8772	1.827	.5473	.4801
1.08	.8820	1.871	.5344	.4713
1.09	.8866	1.917	.5216	.4625
1.10	.8912	1.965	.5090	.4536
1.11	.8957	2.014	.4964	.4447
1.12	.9001	2.066	.4840	.4357
1.13	.9044	2.120	.4718	.4267
1.14	.9086	2.176	.4596	.4176
1.15	.9128	2.234	.4475	.4085
1.16	.9168	2.296	.4356	.3993
1.17	.9208	2.360	.4237	.3902
1.18	.9246	2.427	.4120	.3809
1.19	.9284	2.498	.4003	.3717
1.20	.9320	2.572	.3888	.3624
1.21	.9356	2.650	.3773	.3530
1.22	.9391	2.733	.3659	.3436
1.23	.9425	2.820	.3546	.3342
1.24	.9458	2.912	.3434	.3248
1.25	.9490	3.010	.3323	.3153
1.26	.9521	3.113	.3212	.3058
1.27	.9551	3.224	.3102	.2963
1.28	.9580	3.341	.2993	.2867
1.29	.9608	3.467	.2884	.2771
1.30	.9636	3.602	.2776	.2675

Rad	Sin	Tan	Ctn	Cos
1.30	.9636	3.602	.2776	.2675
1.31	.9662	3.747	.2669	.2579
1.32	.9687	3.903	.2562	.2482
1.33	.9711	4.072	.2456	.2385
1.34	.9735	4.256	.2350	.2288
1.35	.9757	4.455	.2245	.2190
1.36	.9779	4.673	.2140	.2092
1.37	.9799	4.913	.2035	.1994
1.38	.9819	5.177	.1931	.1896
1.39	.9837	5.471	.1828	.1798
1.40	.9854	5.798	.1725	.1700
1.41	.9871	6.165	.1622	.1601
1.42	.9887	6.581	.1519	.1502
1.43	.9901	7.055	.1417	.1403
1.44	.9915	7.602	.1315	.1304
1.45	.9927	8.238	.1214	.1205
1.46	.9939	8.989	.1113	.1106
1.47	.9949	9.887	.1011	.1006
1.48	.9959	10.983	.0910	.0907
1.49	.9967	12.350	.0810	.0807
1.50	.9975	14.101	.0709	.0707
1.51	.9982	16.428	.0609	.0608
1.52	.9987	19.670	.0508	.0508
1.53	.9992	24.498	.0408	.0408
1.54	.9995	32.461	.0308	.0308
1.55	.9998	48.078	.0208	.0208
1.56	.9999	92.620	.0108	.0108
1.57	1.0000	1255.8	.0008	.0008
1.58	1.0000	− 108.65	−.0092	−.0092
1.59	.9998	− 52.067	−.0192	−.0192
1.60	.9996	− 34.233	−.0292	−.0292

Radians to Degrees, Minutes, and Seconds

Rad		Rad		Rad		Rad		Rad	
1	57°17'44".8	.1	5°43'46".5	.01	0°34'22".6	.001	0° 3'26".3	.0001	0°0'20".6
2	114°35'29".6	.2	11°27'33".0	.02	1° 8'45".3	.002	0° 6'52".5	.0002	0°0'41".3
3	171°53'14".4	.3	17°11'19".4	.03	1°43'07".9	.003	0°10'18".8	.0003	0°1'01".9
4	229°10'59".2	.4	22°55'05".9	.04	2°17'30".6	.004	0°13'45".1	.0004	0°1'22".3
5	286°28'44".0	.5	28°38'52".4	.05	2°51'53".2	.005	0°17'11".3	.0005	0°1'43".1
6	343°46'28".8	.6	34°22'38".9	.06	3°26'15".9	.006	0°20'37".6	.0006	0°2'03".6
7	401° 4'13".6	.7	40° 6'25".4	.07	4° 0'38".5	.007	0°24'03".9	.0007	0°2'24".4
8	458°21'58".4	.8	45°50'11".8	.08	4°35'01".2	.008	0°27'30".1	.0008	0°2'45".6
9	515°39'43".3	.9	51°33'58".3	.09	5° 9'23".8	.009	0°30'56".4	.0009	0°3'05".6

Radians to Degrees

Rad	Degrees	Rad	Degrees	Rad	Degrees
1	57.2958	4	229.1831	7	401.0705
2	114.5916	5	286.4789	8	458.3662
3	171.8873	6	343.7747	9	515.6620

NATURAL SINES, COSINES, AND TANGENTS

Degs.	Function	0.0°	0.1°	0.2°	0.3°	0.4°	0.5°	0.6°	0.7°	0.8°	0.9°
0	sin	0.0000	0.0017	0.0035	0.0052	0.0070	0.0087	0.0105	0.0122	0.0140	0.0157
	cos	1.0000	1.0000	1.0000	1.0000	1.0000	1.0000	0.9999	0.9999	0.9999	0.9999
	tan	0.0000	0.0017	0.0035	0.0052	0.0070	0.0087	0.0105	0.0122	0.0140	0.0157
1	sin	0.0175	0.0192	0.0209	0.0227	0.0244	0.0262	0.0279	0.0297	0.0314	0.0332
	cos	0.9998	0.9998	0.9998	0.9997	0.9997	0.9997	0.9996	0.9996	0.9995	0.9995
	tan	0.0175	0.0192	0.0209	0.0227	0.0244	0.0262	0.0279	0.0297	0.0314	0.0332
2	sin	0.0349	0.0366	0.0384	0.0401	0.0419	0.0436	0.0454	0.0471	0.0488	0.0506
	cos	0.9994	0.9993	0.9993	0.9992	0.9991	0.9990	0.9990	0.9989	0.9988	0.9987
	tan	0.0349	0.0367	0.0384	0.0402	0.0419	0.0437	0.0454	0.0472	0.0489	0.0507
3	sin	0.0523	0.0541	0.0558	0.0576	0.0593	0.0610	0.0628	0.0645	0.0663	0.0680
	cos	0.9986	0.9985	0.9984	0.9983	0.9982	0.9981	0.9980	0.9979	0.9978	0.9977
	tan	0.0524	0.0542	0.0559	0.0577	0.0594	0.0612	0.0629	0.0647	0.0664	0.0682
4	sin	0.0698	0.0715	0.0732	0.0750	0.0767	0.0785	0.0802	0.0819	0.0837	0.0854
	cos	0.9976	0.9974	0.9973	0.9972	0.9971	0.9969	0.9968	0.9966	0.9965	0.9963
	tan	0.0699	0.0717	0.0734	0.0752	0.0769	0.0787	0.0805	0.0822	0.0840	0.0857
5	sin	0.0872	0.0889	0.0906	0.0924	0.0941	0.0958	0.0976	0.0993	0.1011	0.1028
	cos	0.9962	0.9960	0.9959	0.9957	0.9956	0.9954	0.9952	0.9951	0.9949	0.9947
	tan	0.0875	0.0892	0.0910	0.0928	0.0945	0.0963	0.0981	0.0998	0.1016	0.1033
6	sin	0.1045	0.1063	0.1080	0.1097	0.1115	0.1132	0.1149	0.1167	0.1184	0.1201
	cos	0.9945	0.9943	0.9942	0.9940	0.9938	0.9936	0.9934	0.9932	0.9930	0.9928
	tan	0.1051	0.1069	0.1086	0.1104	0.1122	0.1139	0.1157	0.1175	0.1192	0.1210
7	sin	0.1219	0.1236	0.1253	0.1271	0.1288	0.1305	0.1323	0.1340	0.1357	0.1374
	cos	0.9925	0.9923	0.9921	0.9919	0.9917	0.9914	0.9912	0.9910	0.9907	0.9905
	tan	0.1228	0.1246	0.1263	0.1281	0.1299	0.1317	0.1334	0.1352	0.1370	0.1388
8	sin	0.1392	0.1409	0.1426	0.1444	0.1461	0.1478	0.1495	0.1513	0.1530	0.1547
	cos	0.9903	0.9900	0.9898	0.9895	0.9893	0.9890	0.9888	0.9885	0.9882	0.9880
	tan	0.1405	0.1423	0.1441	0.1459	0.1477	0.1495	0.1512	0.1530	0.1548	0.1566
9	sin	0.1564	0.1582	0.1599	0.1616	0.1633	0.1650	0.1668	0.1685	0.1702	0.1719
	cos	0.9877	0.9874	0.9871	0.9869	0.9866	0.9863	0.9860	0.9857	0.9854	0.9851
	tan	0.1584	0.1602	0.1620	0.1638	0.1655	0.1673	0.1691	0.1709	0.1727	0.1745
10	sin	0.1736	0.1754	0.1771	0.1788	0.1805	0.1822	0.1840	0.1857	0.1874	0.1891
	cos	0.9848	0.9845	0.9842	0.9839	0.9836	0.9833	0.9829	0.9826	0.9823	0.9820
	tan	0.1763	0.1781	0.1799	0.1817	0.1835	0.1853	0.1871	0.1890	0.1908	0.1926
11	sin	0.1908	0.1925	0.1942	0.1959	0.1977	0.1994	0.2011	0.2028	0.2045	0.2062
	cos	0.9816	0.9813	0.9810	0.9806	0.9803	0.9799	0.9796	0.9792	0.9789	0.9785
	tan	0.1944	0.1962	0.1980	0.1998	0.2016	0.2035	0.2053	0.2071	0.2089	0.2107
12	sin	0.2079	0.2096	0.2113	0.2130	0.2147	0.2164	0.2181	0.2198	0.2215	0.2232
	cos	0.9781	0.9778	0.9774	0.9770	0.9767	0.9763	0.9759	0.9755	0.9751	0.9748
	tan	0.2126	0.2144	0.2162	0.2180	0.2199	0.2217	0.2235	0.2254	0.2272	0.2290
13	sin	0.2250	0.2267	0.2284	0.2300	0.2318	0.2334	0.2351	0.2368	0.2385	0.2402
	cos	0.9744	0.9740	0.9736	0.9732	0.9728	0.9724	0.9720	0.9715	0.9711	0.9707
	tan	0.2309	0.2327	0.2345	0.2364	0.2382	0.2401	0.2419	0.2438	0.2456	0.2475
14	sin	0.2419	0.2436	0.2453	0.2470	0.2487	0.2504	0.2521	0.2538	0.2554	0.2571
	cos	0.9703	0.9699	0.9694	0.9690	0.9686	0.9681	0.9677	0.9673	0.9668	0.9664
	tan	0.2493	0.2512	0.2530	0.2549	0.2568	0.2586	0.2605	0.2623	0.2642	0.2661
Degs.	Function	0'	6'	12'	18'	24'	30'	36'	42'	48'	54'

Degs.	Function	0.0°	0.1°	0.2°	0.3°	0.4°	0.5°	0.6°	0.7°	0.8°	0.9°
15	sin	0.2588	0.2605	0.2622	0.2639	0.2656	0.2672	0 2689	0.2706	0.2723	0.2740
	cos	0 9659	0.9655	0.9650	0.9646	0.9641	0.9636	0 9632	0.9627	0 9622	0.9617
	tan	0.2679	0.2698	0.2717	0.2736	0.2754	0.2773	0.2792	0.2811	0.2830	0.2849
16	sin	0.2756	0.2773	0.2790	0.2807	0.2823	0.2840	0.2857	0.2874	0.2890	0.2907
	cos	0.9613	0.9608	0.9603	0 9598	0.9593	0.9588	0.9583	0.9578	0.9573	0.9568
	tan	0.2867	0.2886	0.2905	0.2924	0.2943	0.2962	0.2981	0.3000	0.3019	0.3038
17	sin	0.2924	0 2940	0.2957	0.2974	0.2990	0.3007	0.3024	0.3040	0.3057	0.3074
	cos	0.9563	0.9558	0.9553	0.9548	0.9542	0.9537	0.9532	0.9527	0.9521	0.9516
	tan	0.3057	0.3076	0.3096	0.3115	0.3134	0.3153	0.3172	0.3191	0.3211	0.3230
18	sin	0.3090	0.3107	0.3123	0.3140	0.3156	0.3173	0.3190	0.3206	0.3223	0.3239
	cos	0.9511	0.9505	0.9500	0.9494	0.9489	0.9483	0.9478	0.9472	0.9466	0.9461
	tan	0.3249	0.3269	0.3288	0.3307	0.3327	0.3346	0.3365	0.3385	0.3404	0.3424
19	sin	0.3256	0.3272	0.3289	0.3305	0.3322	0.3338	0.3355	0.3371	0.3387	0.3404
	cos	0.9455	0.9449	0.9444	0.9438	0.9432	0.9426	0.9421	0.9415	0.9409	0.9403
	tan	0.3443	0.3463	0.3482	0.3502	0.3522	0.3541	0.3561	0.3581	0.3600	0.3620
20	sin	0.3420	0.3437	0.3453	0.3469	0.3486	0.3502	0.3518	0.3535	0.3551	0.3567
	cos	0.9397	0.9391	0.9385	0.9379	0.9373	0.9367	0.9361	0.9354	0.9348	0.9342
	tan	0.3640	0.3659	0.3679	0.3699	0.3719	0.3739	0.3759	0.3779	0.3799	0.3819
21	sin	0.3584	0.3600	0.3616	0.3633	0.3649	0.3665	0.3681	0.3697	0.3714	0.3730
	cos	0.9336	0.9330	0 9323	0.9317	0.9311	0.9304	0.9298	0.9291	0.9285	0.9278
	tan	0.3839	0.3859	0.3879	0.3899	0.3919	0.3939	0.3959	0.3979	0.4000	0.4020
22	sin	0.3746	0.3762	0.3778	0.3795	0.3811	0.3827	0.3843	0.3859	0.3875	0.3891
	cos	0.9272	0.9265	0.9259	0.9252	0.9245	0.9239	0.9232	0.9225	0.9219	0.9212
	tan	0.4040	0.4061	0.4081	0.4101	0.4122	0.4142	0.4163	0.4183	0.4204	0.4224
23	sin	0.3907	0.3923	0.3939	0.3955	0.3971	0.3987	0.4003	0.4019	0.4035	0.4051
	cos	0.9205	0.9198	0.9191	0.9184	0.9178	0.9171	0.9164	0.9157	0.9150	0.9143
	tan	0.4245	0.4265	0.4286	0.4307	0.4327	0.4348	0.4369	0.4390	0.4411	0.4431
24	sin	0.4067	0.4083	0.4099	0.4115	0.4131	0.4147	0.4163	0.4179	0.4195	0.4210
	cos	0.9135	0.9128	0.9121	0.9114	0.9107	0.9100	0.9092	0.9085	0.9078	0.9070
	tan	0.4452	0.4473	0.4494	0.4515	0.4536	0.4557	0.4578	0.4599	0.4621	0.4642
25	sin	0.4226	0.4242	0.4258	0.4274	0.4289	0.4305	0.4321	0.4337	0.4352	0.4368
	cos	0.9063	0.9056	0.9048	0.9041	0.9033	0.9026	0.9018	0.9011	0.9003	0.8996
	tan	0.4663	0.4684	0.4706	0.4727	0.4748	0.4770	0.4791	0.4813	0.4834	0.4856
26	sin	0.4384	0.4399	0.4415	0.4431	0.4446	0.4462	0.4478	0.4493	0.4509	0.4524
	cos	0.8988	0.8980	0.8973	0.8965	0.8957	0.8949	0.8942	0.8934	0.8926	0.8918
	tan	0.4877	0.4899	0.4921	0.4942	0.4964	0.4986	0.5008	0.5029	0.5051	0.5073
27	sin	0.4540	0.4555	0.4571	0.4586	0.4602	0.4617	0.4633	0.4648	0.4664	0.4679
	cos	0.8910	0.8902	0.8894	0.8886	0.8878	0.8870	0.8862	0.8854	0.8846	0.8838
	tan	0.5095	0.5117	0.5139	0.5161	0.5184	0.5206	0.5228	0.5250	0.5272	0.5295
28	sin	0.4695	0.4710	0.4726	0.4741	0.4756	0.4772	0.4787	0.4802	0.4818	0.4833
	cos	0.8829	0.8821	0.8813	0.8805	0.8796	0.8788	0.8780	0.8771	0.8763	0.8755
	tan	0.5317	0.5340	0.5362	0.5384	0.5407	0.5430	0.5452	0.5475	0.5498	0.5520
29	sin	0.4848	0.4863	0.4879	0.4894	0.4909	0.4924	0.4939	0.4955	0.4970	0.4985
	cos	0.8746	0.8738	0.8729	0.8721	0.8712	0.8704	0.8695	0.8686	0.8678	0.8669
	tan	0.5543	0.5566	0.5589	0.5612	0.5635	0.5658	0.5681	0.5704	0.5727	0.5750
Degs.	Function	0′	6′	12′	18′	24′	30′	36′	42′	48′	54′

Degs.	Function	0.0°	0.1°	0.2°	0.3°	0.4°	0.5°	0.6°	0.7°	0.8°	0.9°
30	sin	0.5000	0.5015	0.5030	0.5045	0.5060	0.5075	0.5090	0.5105	0.5120	0.5135
	cos	0.8660	0.8652	0.8643	0.8634	0.8625	0.8616	0.8607	0.8599	0.8590	0.8581
	tan	0.5774	0.5797	0.5820	0.5844	0.5867	0.5890	0.5914	0.5938	0.5961	0.5985
31	sin	0.5150	0.5165	0.5180	0.5195	0.5210	0.5225	0.5240	0.5255	0.5270	0.5284
	cos	0.8572	0.8563	0.8554	0.8545	0.8536	0.8526	0.8517	0.8508	0.8499	0.8490
	tan	0.6009	0.6032	0.6056	0.6080	0.6104	0.6128	0.6152	0.6176	0.6200	0.6224
32	sin	0.5299	0.5314	0.5329	0.5344	0.5358	0.5373	0.5388	0.5402	0.5417	0.5432
	cos	0.8480	0.8471	0.8462	0.8453	0.8443	0.8434	0.8425	0.8415	0.8406	0.8396
	tan	0.6249	0.6273	0.6297	0.6322	0.6346	0.6371	0.6395	0.6420	0.6445	0.6469
33	sin	0.5446	0.5461	0.5476	0.5490	0.5505	0.5519	0.5534	0.5548	0.5563	0.5577
	cos	0.8387	0.8377	0.8368	0.8358	0.8348	0.8339	0.8329	0.8320	0.8310	0.8300
	tan	0.6494	0.6519	0.6544	0.6569	0.6594	0.6619	0.6644	0.6669	0.6694	0.6720
34	sin	0.5592	0.5606	0.5621	0.5635	0.5650	0.5664	0.5678	0.5693	0.5707	0.5721
	cos	0.8290	0.8281	0.8271	0.8261	0.8251	0.8241	0.8231	0.8221	0.8211	0.8202
	tan	0.6745	0.6771	0.6796	0.6822	0.6847	0.6873	0.6899	0.6924	0.6950	0.6976
35	sin	0.5736	0.5750	0.5764	0.5779	0.5793	0.5807	0.5821	0.5835	0.5850	0.5864
	cos	0.8192	0.8181	0.8171	0.8161	0.8151	0.8141	0.8131	0.8121	0.8111	0.8100
	tan	0.7002	0.7028	0.7054	0.7080	0.7107	0.7133	0.7159	0.7186	0.7212	0.7239
36	sin	0.5878	0.5892	0.5906	0.5920	0.5934	0.5948	0.5962	0.5976	0.5990	0.6004
	cos	0.8090	0.8080	0.8070	0.8059	0.8049	0.8039	0.8028	0.8018	0.8007	0.7997
	tan	0.7265	0.7292	0.7319	0.7346	0.7373	0.7400	0.7427	0.7454	0.7481	0.7508
37	sin	0.6018	0.6032	0.6046	0.6060	0.6074	0.6088	0.6101	0.6115	0.6129	0.6143
	cos	0.7986	0.7976	0.7965	0.7955	0.7944	0.7934	0.7923	0.7912	0.7902	0.7891
	tan	0.7536	0.7563	0.7590	0.7618	0.7646	0.7673	0.7701	0.7729	0.7757	0.7785
38	sin	0.6157	0.6170	0.6184	0.6198	0.6211	0.6225	0.6239	0.6252	0.6266	0.6280
	cos	0.7880	0.7869	0.7859	0.7848	0.7837	0.7826	0.7815	0.7804	0.7793	0.7782
	tan	0.7813	0.7841	0.7869	0.7898	0.7926	0.7954	0.7983	0.8012	0.8040	0.8069
39	sin	0.6293	0.6307	0.6320	0.6334	0.6347	0.6361	0.6374	0.6388	0.6401	0.6414
	cos	0.7771	0.7760	0.7749	0.7738	0.7727	0.7716	0.7705	0.7694	0.7683	0.7672
	tan	0.8098	0.8127	0.8156	0.8185	0.8214	0.8243	0.8273	0.8302	0.8332	0.8361
40	sin	0.6428	0.6441	0.6455	0.6468	0.6481	0.6494	0.6508	0.6521	0.6534	0.6547
	cos	0.7660	0.7649	0.7638	0.7627	0.7615	0.7604	0.7593	0.7581	0.7570	0.7559
	tan	0.8391	0.8421	0.8451	0.8481	0.8511	0.8541	0.8571	0.8601	0.8632	0.8662
41	sin	0.6561	0.6574	0.6587	0.6600	0.6613	0.6626	0.6639	0.6652	0.6665	0.6678
	cos	0.7547	0.7536	0.7524	0.7513	0.7501	0.7490	0.7478	0.7466	0.7455	0.7443
	tan	0.8693	0.8724	0.8754	0.8785	0.8816	0.8847	0.8878	0.8910	0.8941	0.8972
42	sin	0.6691	0.6704	0.6717	0.6730	0.6743	0.6756	0.6769	0.6782	0.6794	0.6807
	cos	0.7431	0.7420	0.7408	0.7396	0.7385	0.7373	0.7361	0.7349	0.7337	0.7325
	tan	0.9004	0.9036	0.9067	0.9099	0.9131	0.9163	0.9195	0.9228	0.9260	0.9293
43	sin	0.6820	0.6833	0.6845	0.6858	0.6871	0.6884	0.6896	0.6909	0.6921	0.6934
	cos	0.7314	0.7302	0.7290	0.7278	0.7266	0.7254	0.7242	0.7230	0.7218	0.7206
	tan	0.9325	0.9358	0.9391	0.9424	0.9457	0.9490	0.9523	0.9556	0.9590	0.9623
44	sin	0.6947	0.6959	0.6972	0.6984	0.6997	0.7009	0.7022	0.7034	0.7046	0.7059
	cos	0.7193	0.7181	0.7169	0.7157	0.7145	0.7133	0.7120	0.7108	0.7096	0.7083
	tan	0.9657	0.9691	0.9725	0.9759	0.9793	0.9827	0.9861	0.9896	0.9930	0.9965
Degs.	Function	0'	6'	12'	18'	24'	30'	36'	42'	48'	54'

Degs.	Function	0.0°	0.1°	0.2°	0.3°	0.4°	0.5°	0.6°	0.7°	0.8°	0.9°
45	sin	0.7071	0.7083	0.7096	0.7108	0.7120	0.7133	0.7145	0.7157	0.7169	0.7181
	cos	0.7071	0.7059	0.7046	0.7034	0.7022	0.7009	0.6997	0.6984	0.6972	0.6959
	tan	1.0000	1.0035	1.0070	1.0105	1.0141	1.0176	1.0212	1.0247	1.0283	1.0319
46	sin	0.7193	0.7206	0.7218	0.7230	0.7242	0.7254	0.7266	0.7278	0.7290	0.7302
	cos	0.6947	0.6934	0.6921	0.6909	0.6896	0.6884	0.6871	0.6858	0.6845	0.6833
	tan	1.0355	1.0392	1.0428	1.0464	1.0501	1.0538	1.0575	1.0612	1.0649	1.0686
47	sin	0.7314	0.7325	0.7337	0.7349	0.7361	0.7373	0.7385	0.7396	0.7408	0.7420
	cos	0.6820	0.6807	0.6794	0.6782	0.6769	0.6756	0.6743	0.6730	0.6717	0.6704
	tan	1.0724	1.0761	1.0799	1.0837	1.0875	1.0913	1.0951	1.0990	1.1028	1.1067
48	sin	0.7431	0.7443	0.7455	0.7466	0.7478	0.7490	0.7501	0.7513	0.7524	0.7536
	cos	0.6691	0.6678	0.6665	0.6652	0.6639	0.6626	0.6613	0.6600	0.6587	0.6574
	tan	1.1106	1.1145	1.1184	1.1224	1.1263	1.1303	1.1343	1.1383	1.1423	1.1463
49	sin	0.7547	0.7559	0.7570	0.7581	0.7593	0.7604	0.7615	0.7627	0.7638	0.7649
	cos	0.6561	0.6547	0.6534	0.6521	0.6508	0.6494	0.6481	0.6468	0.6455	0.6441
	tan	1.1504	1.1544	1.1585	1.1626	1.1667	1.1708	1.1750	1.1792	1.1833	1.1875
50	sin	0.7660	0.7672	0.7683	0.7694	0.7705	0.7716	0.7727	0.7738	0.7749	0.7760
	cos	0.6428	0.6414	0.6401	0.6388	0.6374	0.6361	0.6347	0.6334	0.6320	0.6307
	tan	1.1918	1.1960	1.2002	1.2045	1.2088	1.2131	1.2174	1.2218	1.2261	1.2305
51	sin	0.7771	0.7782	0.7793	0.7804	0.7815	0.7826	0.7837	0.7848	0.7859	0.7869
	cos	0.6293	0.6280	0.6266	0.6252	0.6239	0.6225	0.6211	0.6198	0.6184	0.6170
	tan	1.2349	1.2393	1.2437	1.2482	1.2527	1.2572	1.2617	1.2662	1.2708	1.2753
52	sin	0.7880	0.7891	0.7902	0.7912	0.7923	0.7934	0.7944	0.7955	0.7965	0.7976
	cos	0.6157	0.6143	0.6129	0.6115	0.6101	0.6088	0.6074	0.6060	0.6046	0.6032
	tan	1.2799	1.2846	1.2892	1.2938	1.2985	1.3032	1.3079	1.3127	1.3175	1.3222
53	sin	0.7986	0.7997	0.8007	0.8018	0.8028	0.8039	0.8049	0.8059	0.8070	0.8080
	cos	0.6018	0.6004	0.5990	0.5976	0.5962	0.5948	0.5934	0.5920	0.5906	0.5892
	tan	1.3270	1.3319	1.3367	1.3416	1.3465	1.3514	1.3564	1.3613	1.3663	1.3713
54	sin	0.8090	0.8100	0.8111	0.8121	0.8131	0.8141	0.8151	0.8161	0.8171	0.8181
	cos	0.5878	0.5864	0.5850	0.5835	0.5821	0.5807	0.5793	0.5779	0.5764	0.5750
	tan	1.3764	1.3814	1.3865	1.3916	1.3968	1.4019	1.4071	1.4124	1.4176	1.4229
55	sin	0.8192	0.8202	0.8211	0.8221	0.8231	0.8241	0.8251	0.8261	0.8271	0.8281
	cos	0.5736	0.5721	0.5707	0.5693	0.5678	0.5664	0.5650	0.5635	0.5621	0.5606
	tan	1.4281	1.4335	1.4388	1.4442	1.4496	1.4550	1.4605	1.4659	1.4715	1.4770
56	sin	0.8290	0.8300	0.8310	0.8320	0.8329	0.8339	0.8348	0.8358	0.8368	0.8377
	cos	0.5592	0.5577	0.5563	0.5548	0.5534	0.5519	0.5505	0.5490	0.5476	0.5461
	tan	1.4826	1.4882	1.4938	1.4994	1.5051	1.5108	1.5166	1.5224	1.5282	1.5340
57	sin	0.8387	0.8396	0.8406	0.8415	0.8425	0.8434	0.8443	0.8453	0.8462	0.8471
	cos	0.5446	0.5432	0.5417	0.5402	0.5388	0.5373	0.5358	0.5344	0.5329	0.5314
	tan	1.5399	1.5458	1.5517	1.5577	1.5637	1.5697	1.5757	1.5818	1.5880	1.5941
58	sin	0.8480	0.8490	0.8499	0.8508	0.8517	0.8526	0.8536	0.8545	0.8554	0.8563
	cos	0.5299	0.5284	0.5270	0.5255	0.5240	0.5225	0.5210	0.5195	0.5180	0.5165
	tan	1.6003	1.6066	1.6128	1.6191	1.6255	1.6319	1.6383	1.6447	1.6512	1.6577
59	sin	0.8572	0.8581	0.8590	0.8599	0.8607	0.8616	0.8625	0.8634	0.8643	0.8652
	cos	0.5150	0.5135	0.5120	0.5105	0.5090	0.5075	0.5060	0.5045	0.5030	0.5015
	tan	1.6643	1.6709	1.6775	1.6842	1.6909	1.6977	1.7045	1.7113	1.7182	1.7251
Degs.	Function	0'	6'	12'	18'	24'	30'	36'	42'	48'	54'

Degs.	Function	0.0°	0.1°	0.2°	0.3°	0.4°	0.5°	0.6°	0.7°	0.8°	0.9°
60	sin	0.8660	0.8669	0.8678	0.8686	0.8695	0.8704	0.8712	0.8721	0.8729	0.8738
	cos	0.5000	0.4985	0.4970	0.4955	0.4939	0.4924	0.4909	0.4894	0.4879	0.4863
	tan	1.7321	1.7391	1.7461	1.7532	1.7603	1.7675	1.7747	1.7820	1.7893	1.7966
61	sin	0.8746	0.8755	0.8763	0.8771	0.8780	0.8788	0.8796	0.8805	0.8813	0.8821
	cos	0.4848	0.4833	0.4818	0.4802	0.4787	0.4772	0.4756	0.4741	0.4726	0.4710
	tan	1.8040	1.8115	1.8190	1.8265	1.8341	1.8418	1.8495	1.8572	1.8650	1.8728
62	sin	0.8829	0.8838	0.8846	0.8854	0.8862	0.8870	0.8878	0.8886	0.8894	0.8902
	cos	0.4695	0.4679	0.4664	0.4648	0.4633	0.4617	0.4602	0.4586	0.4571	0.4555
	tan	1.8807	1.8887	1.8967	1.9047	1.9128	1.9210	1.9292	1.9375	1.9458	1.9542
63	sin	0.8910	0.8918	0.8926	0.8934	0.8942	0.8949	0.8957	0.8965	0.8973	0.8980
	cos	0.4540	0.4524	0.4509	0.4493	0.4478	0.4462	0.4446	0.4431	0.4415	0.4399
	tan	1.9626	1.9711	1.9797	1.9883	1.9970	2.0057	2.0145	2.0233	2.0323	2.0413
64	sin	0.8988	0.8996	0.9003	0.9011	0.9018	0.9026	0.9033	0.9041	0.9048	0.9056
	cos	0.4384	0.4368	0.4352	0.4337	0.4321	0.4305	0.4289	0.4274	0.4258	0.4242
	tan	2.0503	2.0594	2.0686	2.0778	2.0872	2.0965	2.1060	2.1155	2.1251	2.1348
65	sin	0.9063	0.9070	0.9078	0.9085	0.9092	0.9100	0.9107	0.9114	0.9121	0.9128
	cos	0.4226	0.4210	0.4195	0.4179	0.4163	0.4147	0.4131	0.4115	0.4099	0.4083
	tan	2.1445	2.1543	2.1642	2.1742	2.1842	2.1943	2.2045	2.2148	2.2251	2.2355
66	sin	0.9135	0.9143	0.9150	0.9157	0.9164	0.9171	0.9178	0.9184	0.9191	0.9198
	cos	0.4067	0.4051	0.4035	0.4019	0.4003	0.3987	0.3971	0.3955	0.3939	0.3923
	tan	2.2460	2.2566	2.2673	2.2781	2.2889	2.2998	2.3109	2.3220	2.3332	2.3445
67	sin	0.9205	0.9212	0.9219	0.9225	0.9232	0.9239	0.9245	0.9252	0.9259	0.9265
	cos	0.3907	0.3891	0.3875	0.3859	0.3843	0.3827	0.3811	0.3795	0.3778	0.3762
	tan	2.3559	2.3673	2.3789	2.3906	2.4023	2.4142	2.4262	2.4383	2.4504	2.4627
68	sin	0.9272	0.9278	0.9285	0.9291	0.9298	0.9304	0.9311	0.9317	0.9323	0.9330
	cos	0.3746	0.3730	0.3714	0.3697	0.3681	0.3665	0.3649	0.3633	0.3616	0.3600
	tan	2.4751	2.4876	2.5002	2.5129	2.5257	2.5386	2.5517	2.5649	2.5782	2.5916
69	sin	0.9336	0.9342	0.9348	0.9354	0.9361	0.9367	0.9373	0.9379	0.9385	0.9391
	cos	0.3584	0.3567	0.3551	0.3535	0.3518	0.3502	0.3486	0.3469	0.3453	0.3437
	tan	2.6051	2.6187	2.6325	2.6464	2.6605	2.6746	2.6889	2.7034	2.7179	2.7326
70	sin	0.9397	0.9403	0.9409	0.9415	0.9421	0.9426	0.9432	0.9438	0.9444	0.9449
	cos	0.3420	0.3404	0.3387	0.3371	0.3355	0.3338	0.3322	0.3305	0.3289	0.3272
	tan	2.7475	2.7625	2.7776	2.7929	2.8083	2.8239	2.8397	2.8556	2.8716	2.8878
71	sin	0.9455	0.9461	0.9466	0.9472	0.9478	0.9483	0.9489	0.9494	0.9500	0.9505
	cos	0.3256	0.3239	0.3223	0.3206	0.3190	0.3173	0.3156	0.3140	0.3123	0.3107
	tan	2.9042	2.9208	2.9375	2.9544	2.9714	2.9887	3.0061	3.0237	3.0415	3.0595
72	sin	0.9511	0.9516	0.9521	0.9527	0.9532	0.9537	0.9542	0.9548	0.9553	0.9558
	cos	0.3090	0.3074	0.3057	0.3040	0.3024	0.3007	0.2990	0.2974	0.2957	0.2940
	tan	3.0777	3.0961	3.1146	3.1334	3.1524	3.1716	3.1910	3.2106	3.2305	3.2506
73	sin	0.9563	0.9568	0.9573	0.9578	0.9583	0.9588	0.9593	0.9598	0.9603	0.9608
	cos	0.2924	0.2907	0.2890	0.2874	0.2857	0.2840	0.2823	0.2807	0.2790	0.2773
	tan	3.2709	3.2914	3.3122	3.3332	3.3544	3.3759	3.3977	3.4197	3.4420	3.4646
74	sin	0.9613	0.9617	0.9622	0.9627	0.9632	0.9636	0.9641	0.9646	0.9650	0.9655
	cos	0.2756	0.2740	0.2723	0.2706	0.2689	0.2672	0.2656	0.2639	0.2622	0.2605
	tan	3.4874	3.5105	3.5339	3.5576	3.5816	3.6059	3.6305	3.6554	3.6806	3.7062
Degs.	Function	0′	6′	12′	18′	24′	30′	36′	42′	48′	54′

Degs.	Function	0.0°	0.1°	0.2°	0.3°	0.4°	0.5°	0.6°	0.7°	0.8°	0.9°
75	sin	0.9659	0.9664	0.9668	0.9673	0.9677	0.9681	0.9686	0.9690	0.9694	0.9699
	cos	0.2588	0.2571	0.2554	0.2538	0.2521	0.2504	0.2487	0.2470	0.2453	0.2436
	tan	3.7321	3.7583	3.7848	3.8118	3.8391	3.8667	3.8947	3.9232	3.9520	3.9812
76	sin	0.9703	0.9707	0.9711	0.9715	0.9720	0.9724	0.9728	0.9732	0.9736	0.9740
	cos	0.2419	0.2402	0.2385	0.2368	0.2351	0.2334	0.2317	0.2300	0.2284	0.2267
	tan	4.0108	4.0408	4.0713	4.1022	4.1335	4.1653	4.1976	4.2303	4.2635	4.2972
77	sin	0.9744	0.9748	0.9751	0.9755	0.9759	0.9763	0.9767	0.9770	0.9774	0.9778
	cos	0.2250	0.2232	0.2215	0.2198	0.2181	0.2164	0.2147	0.2130	0.2113	0.2096
	tan	4.3315	4.3662	4.4015	4.4374	4.4737	4.5107	4.5483	4.5864	4.6252	4.6646
78	sin	0.9781	0.9785	0.9789	0.9792	0.9796	0.9799	0.9803	0.9806	0.9810	0.9813
	cos	0.2079	0.2062	0.2045	0.2028	0.2011	0.1994	0.1977	0.1959	0.1942	0.1925
	tan	4.7046	4.7453	4.7867	4.8288	4.8716	4.9152	4.9594	5.0045	5.0504	5.0970
79	sin	0.9816	0.9820	0.9823	0.9826	0.9829	0.9833	0.9836	0.9839	0.9842	0.9845
	cos	0.1908	0.1891	0.1874	0.1857	0.1840	0.1822	0.1805	0.1788	0.1771	0.1754
	tan	5.1446	5.1929	5.2422	5.2924	5.3435	5.3955	5.4486	5.5026	5.5578	5.6140
80	sin	0.9848	0.9851	0.9854	0.9857	0.9860	0.9863	0.9866	0.9869	0.9871	0.9874
	cos	0.1736	0.1719	0.1702	0.1685	0.1668	0.1650	0.1633	0.1616	0.1599	0.1582
	tan	5.6713	5.7297	5.7894	5.8502	5.9124	5.9758	6.0405	6.1066	6.1742	6.2432
81	sin	0.9877	0.9880	0.9882	0.9885	0.9888	0.9890	0.9893	0.9895	0.9898	0.9900
	cos	0.1564	0.1547	0.1530	0.1513	0.1495	0.1478	0.1461	0.1444	0.1426	0.1409
	tan	6.3138	6.3859	6.4596	6.5350	6.6122	6.6912	6.7720	6.8548	6.9395	7.0264
82	sin	0.9903	0.9905	0.9907	0.9910	0.9912	0.9914	0.9917	0.9919	0.9921	0.9923
	cos	0.1392	0.1374	0.1357	0.1340	0.1323	0.1305	0.1288	0.1271	0.1253	0.1236
	tan	7.1154	7.2066	7.3002	7.3962	7.4947	7.5958	7.6996	7.8062	7.9158	8.0285
83	sin	0.9925	0.9928	0.9930	0.9932	0.9934	0.9936	0.9938	0.9940	0.9942	0.9943
	cos	0.1219	0.1201	0.1184	0.1167	0.1149	0.1132	0.1115	0.1097	0.1080	0.1063
	tan	8.1443	8.2636	8.3863	8.5126	8.6427	8.7769	8.9152	9.0579	9.2052	9.3572
84	sin	0.9945	0.9947	0.9949	0.9951	0.9952	0.9954	0.9956	0.9957	0.9959	0.9960
	cos	0.1045	0.1028	0.1011	0.0993	0.0976	0.0958	0.0941	0.0924	0.0906	0.0889
	tan	9.5144	9.6768	9.8448	10.02	10.20	10.39	10.58	10.78	10.99	11.20
85	sin	0.9962	0.9963	0.9965	0.9966	0.9968	0.9969	0.9971	0.9972	0.9973	0.9974
	cos	0.0872	0.0854	0.0837	0.0819	0.0802	0.0785	0.0767	0.0750	0.0732	0.0715
	tan	11.43	11.66	11.91	12.16	12.43	12.71	13.00	13.30	13.62	13.95
86	sin	0.9976	0.9977	0.9978	0.9979	0.9980	0.9981	0.9982	0.9983	0.9984	0.9985
	cos	0.0698	0.0680	0.0663	0.0645	0.0628	0.0610	0.0593	0.0576	0.0558	0.0541
	tan	14.30	14.67	15.06	15.46	15.89	16.35	16.83	17.34	17.89	18.46
87	sin	0.9986	0.9987	0.9988	0.9989	0.9990	0.9990	0.9991	0.9992	0.9993	0.9993
	cos	0.0523	0.0506	0.0488	0.0471	0.0454	0.0436	0.0419	0.0401	0.0384	0.0366
	tan	19.08	19.74	20.45	21.20	22.02	22.90	23.86	24.90	26.03	27.27
88	sin	0.9994	0.9995	0.9995	0.9996	0.9996	0.9997	0.9997	0.9997	0.9998	0.9998
	cos	0.0349	0.0332	0.0314	0.0297	0.0279	0.0262	0.0244	0.0227	0.0209	0.0192
	tan	28.64	30.14	31.82	33.69	35.80	38.19	40.92	44.07	47.74	52.08
89	sin	0.9998	0.9999	0.9999	0.9999	0.9999	1.000	1.000	1.000	1.000	1.000
	cos	0.0175	0.0157	0.0140	0.0122	0.0105	0.0087	0.0070	0.0052	0.0035	0.0017
	tan	57.29	63.66	71.62	81.85	95.49	114.6	143.2	191.0	286.5	573.0
Degs.	Function	0′	6′	12′	18′	24′	30′	36′	42′	48′	54′

Fractions	Decimals	Fractions	Decimals	Fractions	Decimals	Fractions	Decimals
$\frac{1}{64}$	0.015625	$\frac{17}{64}$	0.265625	$\frac{33}{64}$	0.515625	$\frac{49}{64}$	0.765625
$\frac{1}{32}$	0.03125	$\frac{9}{32}$	0.28125	$\frac{17}{32}$	0.53125	$\frac{25}{32}$	0.78125
$\frac{3}{64}$	0.046875	$\frac{19}{64}$	0.296875	$\frac{35}{64}$	0.546875	$\frac{51}{64}$	0.796875
$\frac{1}{16}$	0.0625	$\frac{5}{16}$	0.3125	$\frac{9}{16}$	0.5625	$\frac{13}{16}$	0.8125
$\frac{5}{64}$	0.078125	$\frac{21}{64}$	0.328125	$\frac{37}{64}$	0.578125	$\frac{53}{64}$	0.828125
$\frac{3}{32}$	0.09375	$\frac{11}{32}$	0.34375	$\frac{19}{32}$	0.59375	$\frac{27}{32}$	0.84375
$\frac{7}{64}$	0.109375	$\frac{23}{64}$	0.359375	$\frac{39}{64}$	0.609375	$\frac{55}{64}$	0.859375
$\frac{1}{8}$	0.125	$\frac{3}{8}$	0.375	$\frac{5}{8}$	0.625	$\frac{7}{8}$	0.875
$\frac{9}{64}$	0.140625	$\frac{25}{64}$	0.390625	$\frac{41}{64}$	0.640625	$\frac{57}{64}$	0.890625
$\frac{5}{32}$	0.15625	$\frac{13}{32}$	0.40625	$\frac{21}{32}$	0.65625	$\frac{29}{32}$	0.90625
$\frac{11}{64}$	0.171875	$\frac{27}{64}$	0.421875	$\frac{43}{64}$	0.671875	$\frac{59}{64}$	0.921875
$\frac{3}{16}$	0.1875	$\frac{7}{16}$	0.4375	$\frac{11}{16}$	0.6875	$\frac{15}{16}$	0.9375
$\frac{13}{64}$	0.203125	$\frac{29}{64}$	0.453125	$\frac{45}{64}$	0.703125	$\frac{61}{64}$	0.953125
$\frac{7}{32}$	0.21875	$\frac{15}{32}$	0.46875	$\frac{23}{32}$	0.71875	$\frac{31}{32}$	0.96875
$\frac{15}{64}$	0.234375	$\frac{31}{64}$	0.484375	$\frac{47}{64}$	0.734375	$\frac{63}{64}$	0.984375
$\frac{1}{4}$	0.25	$\frac{1}{2}$	0.5	$\frac{3}{4}$	0.75	1	1

Greek Alphabet

A	α	Alpha	N	ν	Nu
B	β	Beta	Ξ	ξ	Xi
Γ	γ	Gamma	O	o	Omicron
Δ	δ	Delta	Π	π	Pi
E	ϵ	Epsilon	P	ρ	Rho
Z	ζ	Zeta	Σ	σ	Sigma
H	η	Eta	T	τ	Tau
Θ	θ	Theta	Y	υ	Upsilon
I	ι	Iota	Φ	ϕ	Phi
K	κ	Kappa	X	χ	Chi
Λ	λ	Lambda	Ψ	ψ	Psi
M	μ	Mu	Ω	ω	Omega

APPENDIX H
INTERNATIONAL ATOMIC WEIGHTS

	Symbol	Atomic No.	Atomic Weight		Symbol	Atomic No.	Atomic Weight
Actinium	Ac	89	(227)	Mercury	Hg	80	200.59
Aluminum	Al	13	26.9815	Molybdenum	Mo	42	95.94
Americium	Am	95	(243)	Neodymium	Nd	60	144.24
Antimony	Sb	51	121.75	Neon	Ne	10	20.183
Argon	Ar	18	39.948	Neptunium	Np	93	(237)
Arsenic	As	33	74.9216	Nickel	Ni	28	58.71
Astatine	At	85	(210)	Niobium	Nb	41	92.906
Barium	Ba	56	137.34	Nitrogen	N	7	14.0067
Berkelium	Bk	97	(249)	Nobelium	No	102	(253)
Beryllium	Be	4	9.0122	Osmium	Os	76	190.2
Bismuth	Bi	83	208.980	Oxygen	O	8	15.9994
Boron	B	5	10.811	Palladium	Pd	46	106.4
Bromine	Br	35	79.909	Phosphorus	P	15	30.9738
Cadmium	Cd	48	112.40	Platinum	Pt	78	195.09
Calcium	Ca	20	40.08	Plutonium	Pu	94	(244)
Californium	Cf	98	(249)	Polonium	Po	84	(210)
Carbon	C	6	12.01115	Potassium	K	19	39.102
Cerium	Ce	58	140.12	Praseodymium	Pr	59	140.907
Cesium	Cs	55	132.905	Promethium	Pm	61	(145)
Chlorine	Cl	17	35.453	Protactinium	Pa	91	(231)
Chromium	Cr	24	51.996	Radium	Ra	88	(226)
Cobalt	Co	27	58.9332	Radon	Rn	86	(222)
Copper	Cu	29	63.54	Rhenium	Re	75	186.2
Curium	Cm	96	(245)	Rhodium	Rh	45	102.905
Dysprosium	Dy	66	162.50	Rubidium	Rb	37	85.47
Einsteinium	Es	99	(254)	Ruthenium	Ru	44	101.07
Erbium	Er	68	167.26	Samarium	Sm	62	150.35
Europium	Eu	63	151.96	Scandium	Sc	21	44.956
Fermium	Fm	100	(252)	Selenium	Se	34	78.96
Fluorine	F	9	18.9984	Silicon	Si	14	28.086
Francium	Fr	87	(223)	Silver	Ag	47	107.870
Gadolinium	Gd	64	157.25	Sodium	Na	11	22.9898
Gallium	Ga	31	69.72	Strontium	Sr	38	87.62
Germanium	Ge	32	72.59	Sulfur	S	16	32.064
Gold	Au	79	196.967	Tantalum	Ta	73	180.948
Hafnium	Hf	72	178.49	Technetium	Tc	43	(99)
Helium	He	2	4.0026	Tellurium	Te	52	127.60
Holmium	Ho	67	164.930	Terbium	Tb	65	158.924
Hydrogen	H	1	1.00797	Thallium	Tl	81	204.37
Indium	In	49	114.82	Thorium	Th	90	232.038
Iodine	I	53	126.9044	Thulium	Tm	69	168.934
Iridium	Ir	77	192.2	Tin	Sn	50	118.69
Iron	Fe	26	55.847	Titanium	Ti	22	47.90
Krypton	Kr	36	83.80	Tungsten	W	74	183.85
Lanthanum	La	57	138.91	Uranium	U	92	238.03
Lead	Pb	82	207.19	Vanadium	V	23	50.942
Lithium	Li	3	6.939	Xenon	Xe	54	131.30
Lutetium	Lu	71	174.97	Ytterbium	Yb	70	173.04
Magnesium	Mg	12	24.312	Yttrium	Y	39	88.905
Manganese	Mn	25	54.9380	Zinc	Zn	30	65.37
Mendelevium	Md	101	(256)	Zirconium	Zr	40	91.22

Value in parenthesis denotes isotope of longest half-life.

APPENDIX I
STANDARD COLOR CODE OF RESISTORS AND CAPACITORS

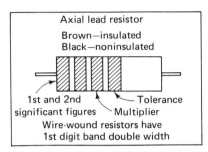

Axial lead resistor

Brown—insulated
Black—noninsulated

1st and 2nd significant figures — Tolerance — Multiplier

Wire-wound resistors have 1st digit band double width

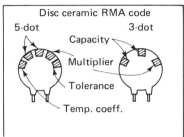

Disc ceramic RMA code

5-dot   3-dot

Capacity

Multiplier

Tolerance

Temp. coeff.

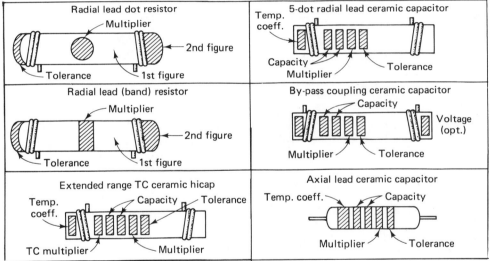

Radial lead dot resistor

Multiplier — 2nd figure — Tolerance — 1st figure

5-dot radial lead ceramic capacitor

Temp. coeff. — Capacity — Multiplier — Tolerance

Radial lead (band) resistor

Multiplier — 2nd figure — Tolerance — 1st figure

By-pass coupling ceramic capacitor

Capacity — Voltage (opt.) — Multiplier — Tolerance

Extended range TC ceramic hicap

Temp. coeff. — Capacity — Tolerance — TC multiplier — Multiplier

Axial lead ceramic capacitor

Temp. coeff. — Capacity — Multiplier — Tolerance

Insulated uninsulated color	First ring body color first figure	Second ring end color second figure	Third ring dot color multiplier
Black	0	0	None
Brown	1	1	0
Red	2	2	00
Orange	3	3	,000
Yellow	4	4	0,000
Green	5	5	00,000
Blue	6	6	000,000
Violet	7	7	0,000,000
Gray	8	8	00,000,000
White	9	9	000,000,000

Resistor tolerance is indicated as follows: gold = 5 percent, silver = 10 percent and absence of a fourth band = 20 percent. If the resistor is wire-wound the first band is double width.

APPENDIX J
MOULDED MICA TYPE CAPACITORS

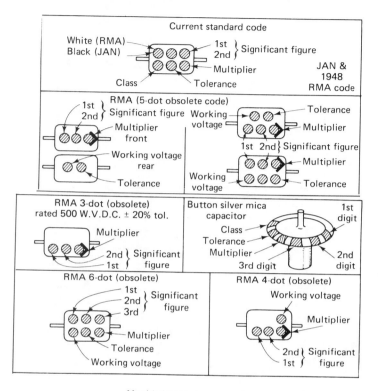

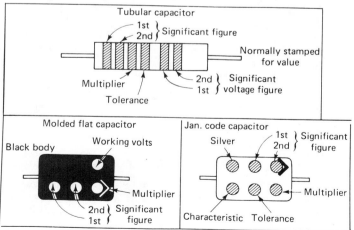

Tolerance ratings of capacitors are read from the color code directly. For example: red = 2 percent, yellow = 4 percent. The voltage rating is found by multiplying the color value by 100.

APPENDIX K
ELECTRONIC SYMBOLS

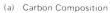

(a) Carbon Composition   (b) Wire Wound   (c) Tapped

RESISTORS

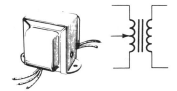

(a) Air Core   (b) Iron Core

INDUCTORS   **TRANSFORMER**

(a) Non-polarized   (b) Polarized   (c) Variable

CAPACITORS

1.5 V   9 V V_{dc} v

CHEMICAL CELL   **BATTERY**   **dc VOLTAGE SOURCE**   **ac VOLTAGE SOURCE**

CIRCUIT BREAKER   **INDICATOR LAMP**   **FUSE**

417

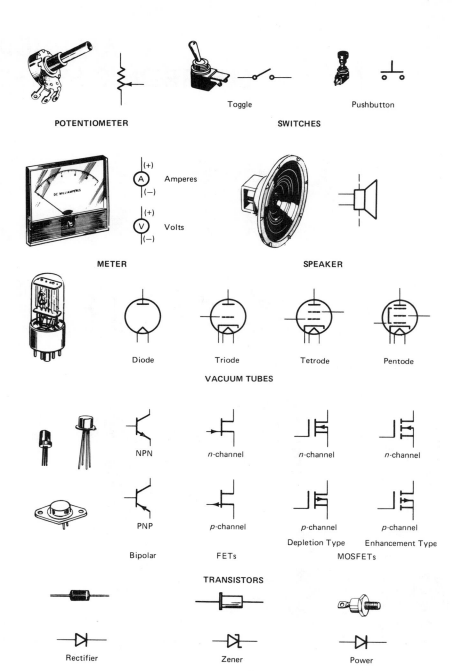

POTENTIOMETER

Toggle Pushbutton

SWITCHES

A Amperes

(+)

(−)

V Volts

(+)

(−)

METER **SPEAKER**

Diode Triode Tetrode Pentode

VACUUM TUBES

NPN n-channel n-channel n-channel

PNP p-channel p-channel p-channel

Depletion Type Enhancement Type

Bipolar FETs MOSFETs

TRANSISTORS

Rectifier Zener Power

SEMICONDUCTOR DIODES

1 Horsepower 117 Volts ac

INTEGRATED CIRCUITS **ELECTRIC MOTOR**

ANSWERS TO ODD-NUMBERED PROBLEMS

PART I   FUNDAMENTAL OPERATIONS

1.2   ARITHMETIC

1. Even   10, 12, 18, 22, 56, 68, 82
   Odd   15, 27, 33, 39, 47
3. Even   50, 60, 90, 110, 280, 340, 410
   Odd   75, 135, 165, 195, 235
5. (a) $3+11=14$
   (b) $28.3-17.1=11.2$
   (c) $7 \cdot 18=126$
   (d) $88 \div 4=22$
   (e) $11+1/8 < 15$
   (f) $9 \cdot 11 > 90$
   (g) Vol. < 33 liters
   (h) $4700 \leqslant R \leqslant 4900$
7. (a) 3   (d) 6
   (b) 39   (e) 77
   (c) 46   (f) 50
11. (a) The magnitude of 3.7.
    (b) The magnitude of negative 16 is equal to 16.
    (c) The magnitude of negative 11 is greater than 5.
    (d) The magnitude of 6 divided by 5 is less than 2.
    (e) The magnitude of the voltage divided by the resistance is equal to the current.

1.3   ALGEBRA

1. (a) $V < 220$ volts
   (b) $P \geqslant 0$
   (c) $I < 1.5$ amperes
   (d) $450 \leqslant R \leqslant 550$
3. $V = hwd$
5. (a) 5   (f) 6
   (b) 45   (g) 44 5/7
   (c) 0   (h) 10.2
   (d) 157.5   (i) 16,500
   (e) 1.9
7. (a) 3600 watts
   (b) 10 watts
9. (a) 40
   (b) 25
   (c) 13
11. (a) 49   (e) 100
    (b) 8   (f) 1,000
    (c) 16   (g) 10,000
    (d) 27

2.1   SIGNIFICANT FIGURES

1. (a) 4.25 cm
   (b) 8.50 cm
   (c) 2.10 cm
   (d) 6.40 cm
   (e) 0.40 cm

2.2   ARITHMETIC OPERATIONS

3. (a) 10   (b) 17
   (c) 9   (d) 10, 14
   (e) 49   (f) 46
   (g) 135   (h) 171
   (i) 162   (j) 645
   (k) 431   (l) 1021
   (m) 1039   (n) 745
   (o) 1049   (p) 613
   (q) 54   (r) 181
   (s) 1780   (t) 86.91
   (u) 198.7   (v) 121.4
   (w) 0.723   (x) 1.269
   (y) 0.00750
5. (a) 5   (b) 15
   (c) 9   (d) 0.120
   (e) 7.021   (f) 51
   (g) 4.612   (h) 43.80
7. (a) 1.833   (b) 0.3333
   (c) 0.3500   (d) 0.3721
   (e) 0.7854   (f) 5.093
   (g) 0.4474   (h) 3.962
   (i) 0.5236   (j) 64.48
   (k) 18.71   (l) 0.6093
   (m) 0.4451   (n) 1.545
   (o) 0.7500   (p) 0.2345
   (q) 0.2572   (r) 0.0004606
   (s) 9038   (t) 21.01
   (u) 0.3526   (v) 16.29
   (w) 0.01574   (x) 0.5235
   (y) 0.04506   (z) 0.1127
   (aa) 2.148   (bb) 7.583

2.3   SOME APPLICATIONS

1. 0.1200 A
3. 0.003636 A
5. 6.516 A
7. 0.001636 A

9. (a) 0.1037 A
   (b) 0.04118 A
   (c) 0.005000 A
   (d) 0.0008485 A
   (e) 0.00004118 A
11. 108.3 V
13. 131.3 V
15. (a) 12.21 V   (b) 36.63 V
    (c) 299.7 V   (d) 621.6 V
    (e) 1110 V
17. 554.5 Ω
19. (a) 12,000 Ω (b) 363.6 Ω
    (c) 28.92 Ω (d) 8.889 Ω
    (e) 1.535 Ω
21. 0.7560 W
23. 1498 W
25. 47.50 W
27. $P = 25$ W
29. (a) 0.1960 W   (b) 1.064 W
    (c) 2.660 W   (d) 13.22 W
    (e) 108.6 W
31. $P = 0.5319$ W Not safe
33. 288.2 Ω
35. 70.53 Ω
37. (a) 0 W
    (b) 0.05556 W
    (c) 1.125 W
    (d) 4.014 W
    (e) 34.72 W
39. (a) 1.355 W   (b) 4.066 W
    (c) 33.27 W   (d) 69.00 W
    (e) 123.2 W

3.1   ADDITION AND SUBTRACTION

1. (a) 28   (b) -22
   (c) -9.6 (d) 8
   (e) -8 (f) -63.2
   (g) -5.60 (h) -0.04
   (i) 0.0085   (j) -2455
   (k) -28.5 (l) -1.54
   (m) -10 (n) -0.30
   (o) 0.60   (p) -0.0091
3. (a) 21   (b) -9.70
   (c) 0.58   (d) -19.0
   (e) 752   (f) -33.9
5. (a) 24.5°C   (b) -16.5°C
   (c) -22.5°C   (d) -32°C

3.2 SIGNED ALGEBRAIC TERMS

1. (a) x (b) $-i^2R$
   (c) $20iv$ (d) $-\pi r^2$
   (e) $3a^2b$ (f) $-4p-2vi$
3. (a) $-IR+6V-9X$
   (b) $20ab-1.5\pi r^2+y$
   (c) $3.6x^2+y$
   (d) $0.2229W-2.1V^2/R$
   (e) $-3abc-4ac-ab$
   (f) $1.4a^2-10b^2$
   (g) $v^2/R-7P$
   (h) $-2XY-3$
   (i) $1-3x^2y-7a$
   (j) $-4V/R+6$
   (k) $-x^2+5x+6$
   (l) 0
   (m) $21a-2ab+3ac-10bc$
   (n) $0.3y^3-0.6y^2$

3.3 MULTIPLICATION

1. (a) 28 (b) -45
   (c) 18 (d) -165
   (e) -0.07 (f) 0.335
   (g) -24.0 (h) -0.0252
   (i) 35.49 (j) -9.546
   (k) $-448,600$ (l) -0.01
   (m) a^6 (n) $-x^5$
   (o) 10^{11} (p) 2^{12}
   (q) $-x^4$ (r) $-I^2R$
   (s) vi (t) $-V^4$
   (u) $-0.66R^2$ (v) $-9z$
   (w) $21xy$ (x) $-14a^2b$
   (y) I^2R (z) $z^{(a-b)}$
   (aa) $-Q^3$ (bb) $a^{(3+m)}$
3. (a) $40\ W$ (b) $-11.70\ W$
   (c) $-2.100\ W$ (d) $13.86\ W$
5. (a) $15\,v$ (b) $15x^2$
   (c) $-35m^3n^3$ (d) $12a^2b^3$
   (e) $24x^{(a+b)}y$ (f) $-65ac^3d^2$
   (g) $-27p^3$ (h) $-9x^5yz$
   (i) $12i^3Rv$ (j) e^3x^{-1}
   (k) $-45xy^2t^3$
   (l) $12x^{(a+1)}y^{(b+c)}$
   (m) $-1.4zw^{-1}$
   (n) $-21a^4$
   (o) $10zy^{-1}$
   (p) $a^{(x+1)}b^{(n+m)}c^{(p+1)}$
   (q) $8\times10^{-10}xy$
   (r) $-57.60gt^2v$
   (s) $70x^3$
   (t) -343×10^{-3}

3.4 DIVISION

1. (a) 4.667 (b) -17.40
   (c) -5.938 (d) -0.5500

   (e) 1.797 (f) 0.2386
   (g) -2.069 (h) -0.7802
   (i) -0.1200 (j) -1.401
3. (a) 0.1250 (b) -0.01471
   (c) 0.01407
5. (a) -2.441 (b) 3.989
   (c) 14.92 (d) -24.38
   (e) 0.004235 (f) 0.001067
   (g) -0.02588 (h) -0.006519
   (i) 0.01220 (j) 0.003072
   (k) 0.4555 (l) 0.1147
7. (a) $5x$ (b) $0.3333ab^{-1}$
   (c) $1.5v$ (d) $-0.3000x^{-1}y^{-1}$
   (e) $2.1a^{-1}z^{-2}$ (f) $-0.25i$
   (g) $4x^2$ (h) 2
   (i) $0.2x^{-2}y^{-1}$
9. 4 times
11. (a) multiplied by 4
    (b) divided by 4

4.1 POWERS OF TEN

1. (a) 10^2 (b) 10^0
   (c) 10^5 (d) 10^7
   (e) 10^9 (f) 10^{-1}
   (g) 10^{-3} (h) 10^{-7}
   (i) 10^{-9} (j) 10^{-12}
3. (a) deka, da   (b) kilo, k
   (c) milli, m   (d) mega, M
   (e) micro, μ (f) nano, n
   (g) pico, p   (h) giga, G
5. (a) millivolt 10^{-3} V
   (b) milliamp 10^{-3} A
   (c) milliwatt 10^{-3} W
   (d) kilohm 10^3 Ω
   (e) megohm 10^6 Ω
   (f) kilovolt 10^3 V
   (g) megavolt 10^6 V
   (h) kilowatt 10^3 W

4.2 SCIENTIFIC NOTATION

1. (a) 461 (b) 856
   (c) 21070 (d) 107
   (e) 1070
3. (a) 4.38×10^2 (b) 9.8×10^3
   (c) 1.006×10^{-1} (d) 2.710×10
   (e) 8.2×10^{-5} (f) 7.61×10^6
   (g) 2×10^{-4} (h) 6.72
5. (a) 7.7×10^5
   (b) 3.86×10^{-4}
   (c) 3.81×10^4
   (d) 1.17×10^{-4}

4.3 SOME APPLICATIONS

3. (a) 14.10 V   (b) 3.300 V
   (c) 1.156×10^2 V
   (d) 8.019×10 V

---

5. (a) 16.50 V   (b) 2.387 V
   (c) 349.2 V   (d) 50.52 V
   (e) 2.553 mA   (f) 17.65 mA
   (g) 1.064 mA   (h) 7.353 mA
   (i) 3.419 kΩ (j) 161.5 Ω
   (k) 1.425 kΩ (l) 67.30 Ω
   (m) 57.90 mW   (n) 8.378 mW
   (o) 25.95 W   (p) 3.754 W
   (q) 30.62 mW   (r) 211.8 mW
   (s) 5.319 mW   (t) 36.77 mW
   (u) 42.12 mW   (v) 17.55 mW
   (w) 0.8916 W   (x) 0.3715 W

5.1 SQUARE ROOTS

1. (a) 7 (b) 3.5
   (c) 15 (d) 5.5
   (e) 13 (f) a
   (g) $a^{1/2}$ (h) 66

5.2 CUBE ROOTS

1. (a) 3 (b) -6
   (c) 7 (d) 11
   (e) 10 (f) 10^2
   (g) 10^{-3} (h) $-R^2$

5.3 EVALUATION

1. (a) 5.657 (b) 8.660
   (c) $2.450x$ (d) 0.75
   (e) 0.4714 (f) $0.8485xy^{-2}$
   (g) $2.080y^{-1}$ (h) $-2.571z^{-2}$
   (i) $4.796p^3$ (j) $2.125v$
   (k) 0.8889 (l) 0.4427
   (m) 0.8707 (n) 0.7910
   (o) 2.450×10^2
   (p) -2.080×10^2
   (q) 1.732×10^{-3}
   (r) 2.828×10^{-6}
   (s) 0.9054
   (t) $-ab^2c^3 10^3$
3. 114.9 V
5. (a) 10 A   (b) 1.414 A
   (c) 5.477 mA   (d) 23.06 mA
7. 6.25 A
9. (a) 5.642 in.   (b) 1.382 ft
   (c) 3 in.   (d) 7.071 cm
11. (a) 4.472 (b) 4
    (c) 3 (d) 12.49

6.1 PERCENT

1. (a) 38% (b) 71.6%
   (c) 11.29% (d) 99.99%
   (e) 31.25% (f) 0%
   (g) 36.36% (h) 13%
   (i) 7.5% (j) 2.209%
   (k) 100% (l) 56.25%
   (m) 87.50% (n) 78.13%
   (o) 10% (p) 6.667%

6.2   EFFICIENCY

1. (a) 80%   (b)   75%
   (c) 92.86%   (d)   77.78%
3. 91.67%
5. 100
7. (b)

6.3   TOLERANCE

1. $0.24 \leqslant l \leqslant 0.26$
3. $1.73 \leqslant d \leqslant 1.77$
5. $475 \leqslant R \leqslant 525$
7. $\pm 5\%$
9. (a) $37{,}600 \leqslant R \leqslant 56{,}400$
   (b) $42{,}300 \leqslant R \leqslant 51{,}700$
   (c) $44{,}650 \leqslant R \leqslant 49{,}350$
   (d) $46{,}530 \leqslant R \leqslant 47{,}470$
11. 5% or 1%
13. $\pm 20\%$

6.4   PERCENT ERROR

1. (a) 3%   (b) 1.724%
   (c) 6.667%   (d) 10.71%
   (e) 5.970%   (f) 23.08%
   (g) 12.50%   (h) 1.515%
   (i) 5.067%

7.1   FORMATION

1. (a) $16 + P$ (b) $x + 9$
   (c) $Q + q + r$ (d) $z + Y - 4$
   (e) $R_1 - R_2$ (f) $17x - 11y$
   (g) $3x - 5y$ (h) $F \cdot d$
   (i) $4(x - 7) - 6P$
   (j) $80 \cdot 4$
   (k) $\dfrac{R_a R_b}{R_a + R_b}$ (l) $\dfrac{15 \text{ mi}}{1 \text{ hr}}$

7.2   MANIPULATION OF EQUATIONS

1. (a) $V = IR$ (b) $R = V/I$
3. (a) $I = P/V$ (b) $V = P/I$
5. (a) $Q_1 = \dfrac{Fd^2}{kQ_2}$ (b) $k = \dfrac{Fd^2}{Q_1 Q_2}$
7. (a) $W = VQ$ (b) $Q = W/V$
9. (a) $P = W/t$ (b) $t = W/P$
11. (a) $P = VQ/t$ (b) $V = Pt/Q$
13. (a) $P = I^2 R$ (b) $R = P/I^2$
15. (a) $P_{in} = P_{out}/\eta$
17. (a) $\rho = RA/L$ (b) $L = RA/\rho$
    (c) $A = \rho L/R$
19. (a) $N = \dfrac{1}{q(\mu_n + \mu_p)\rho}$
    (b) $\mu_n = \dfrac{1}{Nq\rho} - \mu_p$

21. $R_1 = \dfrac{RR_2}{R_2 - R}$

23. (a) $I_T = \dfrac{I_x(R_x + R_{opp})}{R_{opp}}$
    (b) $R_{opp} = \dfrac{I_x R_x}{(I_{opp} - I_x)}$
    (c) $R_x = \dfrac{I_t R_{opp} - I_x R_{opp}}{I_x}$

25. (a) $f = X_L/2\pi L$ (b) $L = \dfrac{X_L}{2\pi f}$
27. (a) $\omega = QR/L$
    (b) $R = \omega L/Q$
29. $d = \text{Vol}/hl$
31. (a) $Z = \sqrt{R^2 + X^2}$
    (b) $R = \sqrt{Z^2 - X^2}$
    (c) $X = \sqrt{Z^2 - R^2}$
33. (a) $I = \dfrac{T}{BlNd}$ (b) $N = \dfrac{T}{BlId}$
35. (a) $k = \dfrac{L_m}{\sqrt{L_1 L_2}}$
    (b) $L_1 = \dfrac{L_m^2}{k^2 L_2}$
37. $t = \dfrac{r^2 N^2 - 8Lr}{11L}$
39. $V_0 = \sqrt{2W/C}$
41. (a) $L = \dfrac{1}{4\pi^2 f^2 C}$ (b) $c = \dfrac{1}{4\pi^2 f^2 L}$
43. (a) $x = 3$ (b) $V = 10$
    ck: $3 + 2 = 5$ ck: $10 - 3 = 7$
    (c) $P = 5.5$ (d) $P = 5$
    ck: $5.5 + 1.5 = 7$ ck: $7 = 5 + 2$
    (e) $x = 3$ (f) $V = 9$
    ck: $3 \cdot 3 = 9$ ck: $2 \cdot 9 - 3 = 15$
    (g) $P = -14$
    ck: $-\dfrac{14}{2} = -14 + 7$
    (h) $i = 24$
    ck: $\dfrac{24}{4} = 6$
    (i) $x = \dfrac{3}{10}$
    ck: $2\frac{1}{2} \cdot \frac{3}{10} = \frac{3}{4}$
    (j) $v = 8$
    ck: $8 + 7 = 3 \cdot 8 - 9$
    (k) $v = 40$
    ck: $\dfrac{40 + 2}{6} = 7$
    (l) $z = -12$
    ck: $-12 - 5 = \dfrac{(3)(-12) + 2}{2}$
    (m) $x = 2.58$ (n) $y = -26$
    (o) $x = 4/27$

8.1   CIRCUIT SYMBOLS

1.

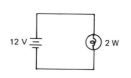

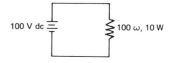

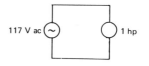

8.2   OHM'S LAW

1. 17.58 mA
3. 63Ω
5. 6.383 mA
7. $10.8 \text{M}\Omega$
9. 9.00 V
11. 29.33 kΩ

8.3   POWER

1. 1.17 kW
3. 158.7 mA
5. 22.31 mW
7. 105.8 V
9. $\dfrac{117^2}{50} = 273.8\text{W}$ not safe.
11. (a) 3.243 A   (b) 2 hp
    (c) 1.942 hp   (d) 91.2%
13. (a) 15kWh   (b) 24¢
    (c) $22.96
15. 1.389×10^{-6}¢

PART II ALGEBRA

3.1 REDUCTION

1. (a) $\dfrac{1}{2}$ (b) $\dfrac{3}{4}$

   (c) $\dfrac{1}{4}$ (d) $\dfrac{11}{15}$

   (e) $\dfrac{3}{x^2}$ (f) $\dfrac{I^2}{2}$

   (g) $\dfrac{4x}{7yz}$ (h) $\dfrac{5a}{9bc^2}$

   (i) $\dfrac{5P^2R^2}{7T^2}$ (j) $-\dfrac{2IR}{5V}$

   (k) $-\dfrac{11xy}{13z}$ (l) $\dfrac{1}{21x}$

3.2 MULTIPLICATION AND DIVISION

1. (a) $\dfrac{1}{3}$ (b) $\dfrac{3ax}{8}$

   (c) $\dfrac{-6y}{x^2}$ (d) $\dfrac{TIR^2}{21a}$

   (e) $7bv$ (f) $-\dfrac{3xb}{8ay^2}$

   (g) $\dfrac{14nP^2}{3m}$ (h) $\dfrac{3P}{5x}$

   (i) $\dfrac{3}{20}$ (j) $\dfrac{7a^2}{3x}$

   (k) $\dfrac{ab}{4x}$ (l) $18x$

4.1 LOWEST COMMON DENOMINATOR

1. $\dfrac{14}{21}, \dfrac{15}{21}$

3. $\dfrac{42}{48}, \dfrac{9}{48}, \dfrac{10}{48}$

5. $\dfrac{9v}{3pv}, \dfrac{4ap}{3pv}$

7. $\dfrac{mi}{2piv}, \dfrac{2pn}{2piv}$

9. $\dfrac{R_2R_3}{R_1R_2R_3}, \dfrac{R_1R_3}{R_1R_2R_3}, \dfrac{R_1R_2}{R_1R_2R_3}$

11. $\dfrac{i_1G_2}{G_1G_2}, \dfrac{G_1i_2}{G_1G_2}$

13. $\dfrac{G_2}{G_1G_2}, \dfrac{G_1}{G_1G_2}$

15. $\dfrac{V_1}{R_1}, \dfrac{-3R_1}{R_1}$

4.2 ADDITION AND SUBTRACTION

1. $\dfrac{29}{21}$

3. $\dfrac{7}{12}$

5. $\dfrac{9v+4ap}{3pv}$

7. $\dfrac{mi+2np}{2piv}$

9. $\dfrac{R_2R_3+R_1R_3+R_1R_2}{R_1R_2R_3}$

11. $\dfrac{i_1G_2+i_2G_1}{G_1G_2}$

13. $\dfrac{6I_1+V_1}{I_1}$

15. $\dfrac{5p+3ab}{5p}$

17. $\dfrac{v^2RR_T-7v_T}{R_T}$

19. $\dfrac{V_1^2R_2R_3+V_2^2R_1R_3+V_3^2R_1R_2}{R_1R_2R_3}$

21. $\dfrac{V_1R_2+V_2R_1}{R_1R_2}$

23. (a) $\dfrac{2}{3}$ kΩ, (b) 250Ω

   (c) $\dfrac{75}{8}\,\Omega$

25. (a) 2.5 kΩ (b) 50Ω,
   (c) 400 kΩ

31. 2.4Ω, 240Ω

5.1 POLYNOMIALS— ADDITION AND SUBTRACTION

1. (a) $4R_1-2R_2$
   (b) $3V_1+2V_2$
   (c) I_a+I_b
   (d) i^2R
   (e) $\dfrac{V_a}{18}+\dfrac{V_b}{6}$ or $\dfrac{V_a+3V_b}{18}$
   (f) $\dfrac{16V^2}{21}$
   (g) $2i_1+i_2-2i_3$
   (h) $2x^2y+4x-2$
   (i) $3xy^2-x^2y+12$
   (j) $12-7t$

3. (a) $5i_2^2R+2$
   (b) $.61ax-1.19by^2$
   (c) $\dfrac{-23t}{4}+\dfrac{10u}{3}+\dfrac{2s}{15}$ or
     $\dfrac{-345t+200u+8s}{60}$
   (d) $-V_a-3V_b-3V_c$

5.2 POLYNOMIALS —MULTIPLICA-TION

1. IR_1+IR_2
3. $G_AV_A+G_BV_A+G_CV_A$
5. $\dfrac{V_A}{R_1}+\dfrac{V_A}{R_2}+\dfrac{V_A}{R_3}$
7. $3I+3j$
9. $10j-15j^2$
11. $\dfrac{aj}{2}-\dfrac{3jb}{4}$
13. $1-j^2$
15. $v_1i_1+v_1i_2-v_2i_1-v_2i_2$
17. $\dfrac{V_A}{R_1}+\dfrac{V_A}{R_2}+\dfrac{V_B}{R_1}+\dfrac{V_B}{R_2}$
19. $3I_1R_1-I_2R_1+2I_3R_1+3I_1R_2$
     $-I_2R_2+2I_3R_2+3I_1R_3-I_2R_3$
     $+2I_3R_3$
21. $I_1I_3R_1+I_1I_3R_2+I_2I_3R_1$
     $+I_2I_3R_2-I_1I_4R_1-I_1I_4R_2$
     $-I_2I_4R_1-I_2I_4R_2$
23. $\dfrac{4j-7j^2+2j^3+4}{6}$
25. $3a^4+2a^3b+4a^2c$
     $-3a^2b^2-2ab^3-4b^2c$
     $+3a^2c^2+2abc^2+4c^3$

5.3 POLYNOMIAL— DIVISION

1. $3-5y$
3. $\dfrac{R_1}{R_T}+\dfrac{R_2}{R_T}$
5. $\dfrac{6iR_1}{R}-\dfrac{7iR_2}{R}+1$
7. $\dfrac{3}{R_1v}+\dfrac{3}{R_2v}+\dfrac{3}{R_3v}$
9. $6x+14-22y$
11. $x+2$
13. $3i-6$
15. $3j+1$
17. x^2+3x+3
19. $x+y$

6.1 KVL

1. (a) 23 V (b) 16.5 V
   (c) 81.8 V (d) 3.33 mA
   (e) .64 mA (f) 1 kΩ
   (g) 6.66 kΩ (h) 153.33 V
   (i) 50.14 V (j) 43.75 V
3. 1.32 V
5. (a) $\doteq$ 9 V (b) $\doteq$ 12 V
   (c) 0 V

7. 114.06 V

6.2 KCL

1. (a) 2A (b) 8A
   (c) 5.48 mA (d) −.85 mA
3. 93.33Ω

7.1 COMMON FACTORS

1. $5(R_1 + 2R_2)$
3. $\dfrac{6}{R_1}(1+4)$
5. $I(3R_1 + 2R_2 + 5R_3)$
7. $i(iR_1 + iR_2 + 3v)$
9. $\pi\left(r_1^2 - \dfrac{3d^2}{4}\right)$
11. $x^2y^2(y-x)$
13. $(I_1 - I_2)(R_1 + R_2)$
15. $(R_1 + R_2)(i_1 - i_2)$
17. $V\left[t\left(I + \dfrac{V}{R}\right) + Q\right]$
19. $i(iR + V) + P$
21. $(I_1 + I_2)(R_1 + R_2)$
23. $(i+b)(v-a)$
25. $(a+s)(3-t)$
27. $(v+1)(v^2+1)$
29. $(R_1 + R_2 - R_T)(v_1 + v_2)$

7.2 TRINOMIALS

1. (a) $x^2 + 5x + 4$
   (b) $x^2 + 3x - 4$
   (c) $x^2 - 7x + 10$
   (d) $2x^2 - 5x - 3$
   (e) $3i^2 + 9i - 54$
   (f) $6v^2 + 12v + 6$
   (g) $4I^2 + 20I + 25$
   (h) $\dfrac{V^2}{6} - 6$
   (i) $.07a^2 + 1.25a - 3$

7.3 DIFFERENCE OF TWO SQUARES

1. (a) $(x+5)(x-5)$
   (b) $(4+y)(4-y)$

(c) $(4i+3)(4i-3)$
(d) $\left(\dfrac{v^2}{2} + \dfrac{1}{3}\right)\left(\dfrac{v^2}{2} - \dfrac{1}{3}\right)$
(e) $(.1p+.2)(.1p-.2)$
(f) $(3Q-2)(3Q+2)$
(g) $(11x^2 + 7y^2)(11x^2 - 7y^2)$
(h) $\left(4ab^2p^3 + \dfrac{iR}{3}\right)\left(4ab^2p^3 - \dfrac{iR}{3}\right)$
(i) $\left[\dfrac{p^2}{6} + (a-b)^2\right]\left[\dfrac{p^2}{6} - (a-b)^2\right]$
(j) $[(i_1 - i_2)R - 4][(i_1 - i_2)R + 4]$
(k) $R(4i_1 i_2)$
(l) $\left(\dfrac{V_1 - V_2}{R} + \dfrac{V_1 + V_2}{3}\right)$
$\left(\dfrac{V_1 - V_2}{R} - \dfrac{V_1 + V_2}{3}\right)$

8.1 ALGEBRAIC FRACTION

1. (a) $\dfrac{R_2 + 1}{R_2 - 1}$ (b) $\dfrac{i+v}{i-2}$
   (c) $\dfrac{v-2}{v}$ (d) $\dfrac{3i}{2i-3}$
   (e) $\dfrac{R-2}{2R-3}$ (f) $\dfrac{v+2}{3(v-2)}$
   (g) $\dfrac{T+T_0}{20+T_0}$ (h) $\dfrac{s}{3s+2}$
   (i) $\dfrac{t+a}{t-a}$
3. (a) $\dfrac{R_1 I_T}{R_1 + R_2}$
   (b) $(iR - 3v)v$
   (c) $\dfrac{i-2}{v+3}$ (d) $\dfrac{3}{x+3}$
   (f) $\dfrac{10}{14}$

8.2 ALGEBRAIC FRACTIONS— ADDITION AND SUBTRACTION

1. (a) $R_1(R_1 + R_2)$
   (b) $3(i+2)$
   (c) $R_1(R_2 + R_3)$
   (d) $R_1 R_2 R_3$
   (e) $R_1 R_2(R_1 + 2)$
   (f) $(x+1)(x-1)(x+2)$

8.3 MIXED EXPRESSIONS

1. (a) $\dfrac{R+1}{R}$
   (b) $\dfrac{R_1 R_2 + R_2 + R_1}{R_1 R_2}$
   (c) $\dfrac{T+T_0}{T_0}$ (d) $\dfrac{V-V_T}{V_T}$
   (e) $\dfrac{9^0 C + 160}{5}$
   (f) $\dfrac{I_1(R_0 + R_x) + I_T R_0}{R_0 + R_x}$
   (g) $\dfrac{12R_1 R_2 + R_2 V + R_1 V}{R_1 R_2}$
   (h) $\dfrac{(I_1 + I_2)(R_1 + R_2) + V}{R_1 + R_2}$
   (i) $\dfrac{R\omega C + j\omega^2 LC - j}{\omega C}$

8.4 FRACTIONAL EQUATIONS

1. (a) 6 (b) $\dfrac{31}{24}$
   (c) $\dfrac{-4}{3}$ (d) $\dfrac{1}{7}$
   (e) −11 (f) 2
   (g) $\dfrac{16}{3}$ (h) $\dfrac{8}{11}$
   (i) −16 (j) $\dfrac{13}{3}$
   (k) 4 (l) 0
   (m) 667Ω (n) 6.67 kΩ
   (o) 643Ω
3. (a) 1000Ω (b) 1.25 μf
   (c) 400Ω (d) 2 kΩ
   (e) 2 kΩ (f) 11.76
5. (a) 4.12
   (b) 1 min, 30 sec
   (c) 8 hr, 20 min
   (d) 2 hr, 1 min, 12 sec
   (e) 35 min, 1 sec
   (f) 54 min
   (g) 21 hr
   (h) .46

9.1 TWO SIMULTANEOUS EQUATIONS

1. (0,0)
3. $\left(\dfrac{1}{2}, -\dfrac{1}{2}\right)$
5. (−1,4)
7. Inconsistent

9. $(22, -21)$
11. $V_A = 2$, $V_B = 1$
13. $(0,0)$
15. $I_1 = 3$, $I_2 = 1$

9.2   SIMULTANEOUS EQUATION—BY SUBSTITUTION

1. $(0,0)$
3. $\left(-\dfrac{1}{3}, \dfrac{1}{3}\right)$
5. $(0,1)$
7. Inconsistent
9. $(22, -21)$
11. $I_1 = 1$, $I_2 = \dfrac{1}{3}$
13. Indentity
15. $V_1 = 0$, $V_2 = 0$
17. Indentity
19. $I_1 = \dfrac{-5R_3}{R_2^2 - R_1 R_3}$,

   $I_2 = \dfrac{-5R_2}{R_2^2 - R_1 R_3}$

9.3   SIMULTANEOUS EQUATION—BY ADDITION AND SUBTRACTION

1. $(0,0)$ 3. $(-2,5)$
5. $(2, -1)$ 7. $(22, -21)$
9. $V_A = 3$, $V_B = -4$
11. $V_A = 35.56$, $V_B = 8.89$
13. Identity
15. $I_1 = 2.02$, $I_2 = 1.18$
17. $x = \dfrac{b}{ab+1}$, $y = \dfrac{-1}{ab+1}$
19. $I_1 = 2.57$ mA, $I_2 = .86$ mA

11.1   DETERMINANTS

1. $x = y = 0$
3. $x = \dfrac{1}{2}$, $y = -\dfrac{1}{2}$
5. $x = -\dfrac{1}{3}$, $y = -\dfrac{2}{3}$
7. $I_1 = 1$, $I_2 = \dfrac{1}{3}$
9. $V_A = 35.56$, $V_B = 8.9$
11. $I_1 = 2.8$, $I_2 = -1.2$
13. $V_1 = \dfrac{12R_1 R_2}{2R_2 - R_1}$,

   $V_2 = \dfrac{6R_1 R_2}{2R_2 - R_1}$

11.2   DETERMINANTS

1. (a) -1 (b) 24
   (c) -5 (d) $-2a$
   (e) $2A - B$ (f) 0

11.3   DETERMINANTS —EXPANSION BY MINORS

3. (a) -26 (b) 39
   (c) 96   (d) -27
   (e) 54   (f) -3
   (g) -26 (h) -3
   (i) $-2a$ (j) $2A - B$
   (k) -2 (l) 16

11.4   DETERMINANTS

1. (a) 5   (b) -22
   (c) 60   (d) 0
   (e) -10 (f) -4
   (g) 357   (h) 2070
3. (a) Inconsistent (b) Yes
   (c) Dependent   (d) Yes
   (e) Yes

12.1   LAWS OF EXPONENTS

1. x^5 13. $(mn)^7$
3. i^3 15. $\dfrac{v^3}{R^2}$
5. v 17. $\dfrac{aiR}{bv}$
7. -3^5 19. $\dfrac{xyz}{a^2}$
9. -5^3 21. $-ax^3 y^4$
11. $-8x^6$ 23. $a^2 x^5 b^2 y^2$

12.2   FRACTIONAL EXPONENTS

1. (a) $\sqrt[5]{a}$ (b) $\sqrt[6]{y}$
   (c) $\sqrt[3]{z^2}$ (d) $\sqrt[7]{Q^6}$
   (e) $\sqrt[3]{v^4}$ (f) $\sqrt[2]{x^3}$
   (g) $\sqrt[3]{a^4}$ (h) $\sqrt[2]{p^3}$
   (i) $2\sqrt[2]{x}$ (j) $\sqrt[2]{2x}$
   (k) $3\sqrt[3]{y^2}$ (l) $\sqrt[3]{(3y)^2}$
   (m) $4\sqrt[4]{x^5}$ (n) $\sqrt[2]{(i+2)^3}$
   (o) $3\sqrt[3]{(x-y)^2}$
   (p) $3^4 a^4 b^8$

3. (a) 7   (b) 3
   (c) 2   (d) 32
   (e) 8   (f) 16
   (g) $\dfrac{2}{3}$ (h) 1
   (i) .125   (j) -1
   (k) $\dfrac{4}{3}$ (l) $\dfrac{1}{216x^3 y^6}$

12.3   IMAGINARY NUMBERS

1. $j3$ 15. -1
3. $j\dfrac{1}{3}$ 17. 7
5. $-j\dfrac{6}{7}$ 19. $\dfrac{7}{4} x^{3/2}$
7. $j\sqrt{8}\, x$
9. $-j\dfrac{1}{6^{1/2} y^{3/2}}$ 21. $\dfrac{-x^2}{3}\sqrt{\dfrac{5}{6}}$
11. j 23. $\dfrac{2}{3\sqrt{7}}$
13. -1

13.1   EXPONENTIAL EQUATIONS

1. (a) 1, 2, 4, 32
   (b) .5, .13, .03
   (c) 3, 27, 729
   (d) 1, 10, 100, 1000, 10,000, 100,000
   (e) .1, .0000001, 10^{-12}
   (f) .06, .16, .4, 1, 2.5, 6.25, 15.62
   (g) 37.04, 3.33, 1, .3, .03
   (h) 10^6, 100, 1, .01, 10^{-6}
   (i) .4, .58, .83, 1, 1.2, 1.73, 2.49
   (j) $-.13$, $-.25$, $-.5$, -1, -2, -4, -8
   (k) 32, 8, 2, 1, .5, .13, .03
   (l) -7.59, -3.37, -1.5, -1, -1.5, $-.3$, $-.13$
3. (a) $x = 1.71$ (b) 2.71
   (c) 0   (d) -1.7
5. (a) 2   (b) 1.483
   (c) .75   (d) .23
   (e) 1.25

13.2 e^x AND e^{-x}

3. (a) 73.7   (b) .3
   (c) .02   (d) .8
   (e) .6   (f) 22026.5
   (g) 26658331.56
   (h) .0045
   (i) .2194

14.1 QUADRATIC EQUATIONS

1. (a) ± 8 (b) ± 11
   (c) ± 4 (d) ± 3
   (e) ± 2.74 (f) ± 4.58
   (g) $\pm\sqrt{5}$ (h) $x = \dfrac{\pm 5}{2}$
   (i) $\pm\sqrt{8}$ (j) $\pm\dfrac{2}{\sqrt{3}}$
   (k) $\pm\sqrt{2}$ (l) ± 4
   (m) $\pm\sqrt{\dfrac{P}{R}}$ (n) $\pm\sqrt{\dfrac{A}{\pi}}$
   (o) $\pm\sqrt{\dfrac{2s}{g}}$ (p) $\pm 2\sqrt{\dfrac{A}{\pi}}$
   (q) $\pm 2\sqrt{\dfrac{7}{R}}$ (r) $\pm 10^2\sqrt{P}$

   (s) $\pm 7\sqrt{\dfrac{R}{12}}$

   (t) $\pm\sqrt{\dfrac{kM_1M_2}{F}}$

   (u) $\pm\dfrac{yz}{\sqrt{y^2-z^2}}$

14.2 QUADRATIC FORMULA

1. $-1, -4$ (3) $2, 5$
   (5) $3, -6$ (7) $0, -4$
   (9) $4, -6$ (11) 6
   (13) $0, 12$ (15) $.5, 1$
   (17) $.38, -2.62$ (19) $.305, -3.3$
   (21) ± 2.24 (23) $.305, -3.3$
   (25) $\dfrac{1}{2}, -1$ (27) $2.41, -.41$

14.3 COMPLEX ROOTS

1. (a) Yes (b) No
   (c) Yes (d) Yes
   (e) Yes (f) Yes
3. (a) ± 7 (b) $\pm j7$
   (c) ± 5 (d) $0, -3$
   (e) ± 1.25 (f) $\pm j20$
   (g) $\dfrac{3}{2}, 7$ (h) $5, -10$
   (i) $2, \dfrac{-4}{3}$ (j) $.14, -2.41$
   (k) $1, -5$ (l) $\dfrac{1}{4}, \dfrac{-3}{4}$
   (m) $\dfrac{-1}{6}, \dfrac{3}{2}$ (n) $.83, -1.83$
   (o) $-.83, 1.83$ (p) $\dfrac{-3}{2}, \dfrac{5}{3}$
   (q) ± 7 (r) $-1, 1.92$
5. $11.27, 88.73$
7. $.81\Omega, 2\Omega, 4.2\Omega, 10\Omega$
9. $36, 45$

PART III   TRIGONOMETRY—APPLICATIONS

1.1 ANGLES

1. (a) $45°$ (b) $30°$
   (c) $0°$ (d) $-60°$
   (e) $112°$ (f) $-185°$
3. (a) $60°$ (b) $-60°$
   (c) $30°$ (d) $300°$
   (e) $-100°$ (f) $-240°$
5. (a) $1440°$ (b) $630°$
   (c) $1140°$ (d) $-1980°$
   (e) $-1692°$ (f) $3.6\times10^{5°}$

1.2 RADIAN MEASURE

1. (a) 4π (b) 20π
   (c) $\pi/2$ (d) $\pi/4$
   (e) $5\pi/3$ (f) $10\pi/3$
3. (a) 1.114 in (b) 3.501 cm
   (c) $3,000$ ft (d) 1.533 mm
5. 3.71 in.
7. (a) $\pi/6$ (b) $\pi/4$
   (c) $\pi/3$ (d) $\pi/2$
   (e) $2\pi/3$ (f) $3\pi/4$
   (g) $5\pi/6$ (h) π
9. (a) $360°$ (b) $180°$
   (c) $210°$ (d) $300°$
   (e) $315°$ (f) $10°$
   (g) $140°$ (h) $55°$
11. (a) 6.07 (b) 16.5
    (c) 64.2 (d) 344
    (e) 201

13. (a) $\pi/2$ (b) π
    (c) 2π (d) 10π
15. $v = 0.239 r/s$
    $1.047 s$ for $90°$ angle

1.3 TRIANGLES

3. (a) $37°$ (b) $93°$ (c) $86°$
5. (a) $c = 5, \gamma = 36.9°$
   (b) $a = 17.9, \alpha = 58.4°$
   (c) $b = 6.56, \gamma = 46.1°$
   (d) $c = 5.00, \gamma = 60°$
   (e) $c = 9.90, \alpha = 45°$
9. 29.2 ft

1.4 POLAR COORDINATES

3. (a) ± 6.93 (b) ± 4.33
   (c) ± 7.20 (d) ± 3.50
5. -4.33 7. 3.34

2.1 TRIGONOMETRIC FUNCTIONS

1.

	sin	cos	tan
(a)	0.600	0.800	0.750
(b)	0.707	−0.707	−1.000
(c)	−0.707	−0.707	1.000
(d)	−0.707	0.707	−1.000
(e)	−0.555	−0.832	0.667
(f)	0.768	−0.640	−1.200
(g)	−0.530	0.848	−0.625
(h)	0.949	0.316	3.000
(i)	0.949	−0.316	−3.000
(j)	−0.500	0.866	−0.577
(k)	−0.895	−0.447	2.000

1.

	csc	sec	cot
(a)	1.667	1.250	1.333
(b)	1.414	−1.414	−1.000
(c)	−1.414	−1.414	1.000
(d)	−1.414	1.414	−1.000
(e)	−1.802	−1.202	1.500
(f)	1.302	−1.563	−0.833
(g)	−1.887	1.179	−1.600
(h)	1.054	3.165	0.333
(i)	1.054	−3.165	−0.333
(j)	−2.000	1.155	−1.733
(k)	−1.117	−2.237	0.500

3. (a) $0°$ (b) $90°$
   (c) $90°$ (d) $0°$
   (e) $90°$ (f) $0°$

2.2 COMPUTATION OF TRIG FUNCTIONS

1.

	sin	cos	tan
(a)	0.71	0.71	1.00
(b)	0.87	0.50	1.73
(c)	0.39	0.92	0.42
(d)	0.47	−0.88	−0.53
(e)	−0.92	−0.39	2.36
(f)	−0.87	0.50	−1.73
(g)	−0.16	0.99	−0.16
(h)	−0.31	−0.95	0.32
(i)	0.97	−0.26	−3.73
(j)	−0.45	0.89	−0.51

1.

	csc	sec	cot
(a)	1.41	1.41	1.00
(b)	1.15	2.00	0.58
(c)	2.56	1.09	2.36
(d)	2.13	−1.13	−1.88
(e)	−1.09	−2.56	0.42
(f)	−1.15	2.00	−0.58
(g)	−6.39	1.01	−6.31
(h)	−3.24	−1.05	3.08
(i)	1.04	−3.86	−0.27
(j)	−2.20	1.12	−1.96

3. $\sin 0° = 0$ $\cos 0° = 1.00$
$\tan 0° = 0.00$
$\csc 0° = \infty$ $\sec 0° = 1.00$
$\cot 0° = \infty$

5. $\sin 270° = -1.00$
$\cos 270° = 0.00$
$\tan 270° = -\infty$
$\csc 270° = -1.00$
$\sec 270° = \infty$ $\operatorname{ctn} 270° = 0$

2.3 TABLE OF TRIG FUNCTIONS

1. (a) 0.3633   (b) 0.9668
   (c) 0.2053   (d) 0.6211
   (e) 9.8448   (f) 0.8396
   (g) 1.6643   (h) 0.0262
   (i) 0.0279   (j) 0.9999
   (k) 0.9992   (l) 0.0314
3. (a) 2.7529   (b) 1.0343
   (c) 4.8716   (d) 1.6099
   (e) 0.1016   (f) 1.1910

3.1 STANDARD NOTATION

1.

	sin	cos	tan
(a)	0.894	0.447	2.000
(b)	0.555	0.832	0.667
(c)	0.661	0.750	0.882
(d)	0.555	0.832	0.667
(e)	0.894	0.447	2.000
(f)	0.857	0.514	1.667
(g)	0.661	0.750	0.882
(h)	0.527	0.850	0.620

3. $\sin \beta = \dfrac{j}{l}$ $\cos \beta = \dfrac{k}{l}$ $\tan \beta = \dfrac{j}{k}$

5. (a) $\tan \alpha = \dfrac{6.6}{8.8} = 0.750$

$\cot \beta = \dfrac{6.6}{8.8} = 0.750$

$\therefore \tan \alpha = \cot \beta$

(b) $\cot \alpha = \dfrac{8.8}{6.6} = 1.333$

$\tan \beta = \dfrac{8.8}{6.6} = 1.333$

$\therefore \cot \alpha = \tan \beta$

3.2 RIGHT TRIANGLES FOR REFERENCE

1. (a) 1.00   (b) 1.27
   (c) 19.1   (d) 8.66
   (e) 3.54   (f) 3.00
   (g) 2.31   (h) 5.20
   (i) 3.50   (j) 1.00
   (k) 0.00   (l) 1.00
   (m) 0.500   (n) 4.00
   (o) 2.50
3. 38.1 ft
5. 2828 ft

3.3 SOLVING RIGHT TRIANGLES

1. $\Theta = 60.0°$ $X = 4.33 \ \Omega$
3. $\Theta = 45.0°$ $Z = 70.7 \ \Omega$
5. $\Theta = 53.1°$ $X = 4.00 \ \Omega$
7. $\Theta = 90.0°$ $Z = X = 2.7 \ k\Omega$
9. $\Theta = 0.00°$ $R = Z = 500 \ \Omega$
11. $X = 2.86 \ k\Omega$ $R = 1.65 \ k\Omega$
13. $X = 209.8 \ \Omega$ $Z = 326.4 \ \Omega$
15. $R = 56.7 \ \Omega$ $Z = 57.6 \ \Omega$
17. $R = 1787 \ \Omega$ $Z = 2024 \ \Omega$
19. $R = 0.00$ $Z = X = 800 \ \Omega$
21. $R = 65.0 \ \Omega$ $X = 37.5 \ \Omega$
23. $R = 150 \ \Omega$ $X = 260 \ \Omega$
25. $R = 0.00 \ \Omega$ $X = 5 \ k\Omega$
27. $\Theta = 30.0°$ $P = 1.33 \ W$
29. $\Theta = 12.0°$ $P_A = 76.7 \ mVA$

31. $\Theta = 45.0°$ $P_A = 9.55 \ VA$
33. $P = 60.3 \ mW$ $P_Q = 60.3 \ mvar$
35. $P_A = 35.4 \ VA$ $P_Q = 11.3 \ var$
37. $P_A = 3.92 \ VA$ $P = 0.205 \ W$
39. $P_A = 26.8 \ VA$ $P_Q = 26.75 \ var$
41. $P = 159 \ mW$ $P_Q = 0.00$
43. $P = 15.6 \ mW$ $P_Q = 6.07 \ mvar$
45. (a) $\alpha = \beta = 24.9°$
   (b) $\alpha = \beta = 1.01°$
   (c) $\alpha = \beta = 0.867°$
47. (a) $x = 10.6$ $y = 2.85$
   (b) $x = 0.678$ $y = 1.61$
   (c) $x = 24.0$ $y = 25.1$
   (d) $x = 2.98$ $y = 2.20$
49. $\alpha = 29.5°$ $\beta = 60.5°$

4.1 RELATED ANGLES

1. (a) 60°   (b) 30°
   (c) 45°   (d) 61°
   (e) 41°   (f) 5°
   (g) 69°   (h) 72°
   (i) 20°   (j) 60°
   (k) 38°   (l) 80°
3. (a) 0.268   (b) 0.225
   (c) −0.839   (d) 1.225
   (e) −12.86   (f) 5.91
   (g) 118   (h) 33.99
   (i) −0.305   (j) 1.00
   (k) 12.31   (l) 8.74

4.2 TRIG. FUNCTIONS OF NEG. ANGLE

1.

	sin	cos	tan
(a)	−0.1908	0.9816	−0.1944
(b)	−0.9613	0.2756	−3.487
(c)	−0.8572	−0.5150	1.664
(d)	0.7660	0.6428	1.192
(e)	0.8746	−0.4848	−1.804
(f)	0.6157	0.7880	0.7813
(g)	0.3090	−0.9511	−0.3249
(h)	−0.9962	−0.0872	11.43
(i)	−0.7071	0.7071	−1.000

4.3 TRIG. FUNCTIONS IN RADIANS

1. (a) 0.7243   (b) 0.9928
   (c) 2.820   (d) 0.0707
   (e) 0.3994   (f) 0.8957
   (g) 0.9131   (h) 0.3240
   (i) 0.8870

3.

	sin	cos	tan
(a)	0.9982	−0.0592	−16.87
(b)	−0.9860	0.1669	−5.910
(c)	−0.8240	−0.5667	1.454
(d)	−0.2794	0.9602	−0.2910
(e)	0.1411	−0.9900	−0.1425
(f)	−0.7568	−0.6536	1.1578

5.1 $y = \rho \sin \Theta$

5. Peak values:
(a) 1.8   (b) 11.6
(c) 0.18   (d) 17
(e) 31×10^{-3} (f) 167
Peak-to-Peak values:
(a) 3.6   (b) 23.2
(c) 0.36   (d) 34
(e) 62×10^{-3} (f) 334
11. (a) $y = 6.1 \sin \Theta$
(b) $y = 2.9 \sin \Theta$
(c) $y = 0.198 \sin \Theta$
(d) $y = 1.7 \times 10^{-3} \sin \Theta$
13. sine waves with peak values:
(a) 0.8333 A   (b) 1.965 A
(c) 10.67 mA
(d) 0.4468 mA

5.2 $y = \rho \cos \Theta$

5. Peak values:
(a) 1.8   (b) 11.6
(c) 0.18   (d) 17
(e) 31×10^{-3} (f) 167
Peak-to-Peak values:
(a) 3.6   (b) 23.2
(c) 0.36   (d) 34
(e) 62×10^{-3} (f) 334
11. (a) $y = 6.1 \cos \Theta$
(b) $y = 2.9 \cos \Theta$
(c) $y = 0.198 \cos \Theta$
(d) $y = 1.7 \times 10^{-3} \cos \Theta$
13. cosine waves with peak values:
(a) 0.8333 A   (b) 1.965 A
(c) 10.67 mA
(d) 0.4468 mA

6.1   PERIOD AND FREQUENCY

1. (a) 89.01 rad/s (b) 5292°/s
3. 2.094 rad/s
5. 0.4363 rad/s
7. 0.6283s; 0.1571s
9. 7.392 ms; 36.96 ms
11. (a) 1 Hz   (b) 8 Hz

(c) 59.9 Hz   (d) 500 Hz
(e) 172.4 kHz
(f) 4.546 MHz
13. (a) 31.42 rad/s
(b) 377.0 rad/s
(c) 3.142 krad/s
(d) 18.85 krad/s
(e) 5.466×10^5 rad/s
(f) 1.433×10^6 rad/s
(g) 1.131×10^7 rad/s
(h) 1.885×10^8 rad/s

6.2   SINUSOIDAL CURRENTS AND VOLTAGES

1. (a) $i = 6 \sin 220 t$
(b) $v = 12 \sin 314 t$
(c) $v = 167 \sin 377 t$
(d) $i = 0.03 \sin 6283 t$
(e) $i = 0.007 \sin 18,850 t$
(f) $v = 0.038 \sin(12.6 \times 10^6 t)$
5. (a) $i = 5 \cos(80\pi t + \pi/5)$
(b) $v = 15 \cos(120\pi t + 0.937)$
(c) $v = 167 \cos(120\pi t + \pi/4)$
(d) $v = 167 \cos(120\pi t - \pi/4)$
(e) $i = 0.035 \cos(2\pi \times 10^3 t - \pi/2)$
(f) $i = 0.035 \cos(2\pi \times 10^3 t + \pi/2)$
7. They are identical waveforms.
9. (a) $v = 167 \sin 377 t$
(b) $v = 334 \sin 377 t$
(c) $v = 167 \sin 314 t$

6.3   SOME APPLICATIONS

1. $i = 0.170 \sin \omega t$ V
3. (a) $i = 0.175 \sin 10^4 t$ A
(b) $i = 0.0515 \sin 10^4 t$ A
(c) $i = 0.0385 \sin 10^4 t$ A
(d) $i = 23.3 \sin 10^4 t$ mA
(e) $i = 10.61 \sin 10^4 t$ mA
(f) $i = 0.625 \sin 10^4 t$ mA
(g) $i = 15.91 \sin 10^4 t$ μA
5. (a) $v = 170 \sin \omega t$ V
(b) $v = 42.16 \cos \omega t$ V
(c) $v = 5.92 \sin 10^3 t$ V
(d) $V_p = 1.129$ V, $f = 1$ kHz
7. (a) 106.4 V   (b) 24.3 mA
(c) 46.6 mV
9. 12.82 A
11. 0.756 W
13. 1500 W
15. Power dissipated =
$$\frac{V_{eff}^2}{R} = \frac{\left(70.7/\sqrt{2}\right)^2}{100} = 25.0 \text{ W}$$
∴ Not safe

17. (a) 42.5 W   (b) 2.61 W
(c) 51.6 mW   (d) 1.87 mW
19. 70.53 Ω

7.1   PHASORS

1. (a) $V_1 = 6 /0°$ $V_2 = 3 /30°$
(b) $V_1 = 8 \times 10^{-3} /0°$
$V_2 = 22 \times 10^{-3} /17°$
(c) $V_1 = 17 /0°$ $V_2 = 9 /50°$
(d) $V_1 = 71 /0°$ $V_2 = 95 /60°$
3. (a) $V = 7 /16°$
(b) $V = 4.8 \times 10^{-3} /-80°$
(c) $V = 175 /40°$
(d) $V = V_p /-35°$
5.

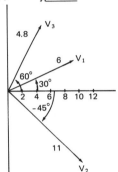

7.2   COMPLEX NUMBERS

3. (a) $2.598 + j1.500$
(b) $3.536 + j3.536$
(c) $9.397 - j3.420$
(d) $-3.857 + j4.596$
(e) $-4.915 + j3.441$
(f) $-6.128 - j5.142$
(g) $1.286 - j1.532$
(h) $2.819 - j1.026$
(i) $0.850 - j1.472$
(j) $5.955 + j3.867$
(k) $-0.660 - j3.742$
(l) $-3.960 - j3.960$
5. (a) $z = 22.36 /63.4°$
(b) $z = 1414 /-45°$
$z = 752 /-85.8°$
(d) $z = 5108 /23.1°$
(e) $z = 36.25 \times 10^3 /-24.4°$
(f) $z = 39.20 \times 10^3 /84.3°$
7. (a) $4.47 e^{j1.11}$
(b) $30.2 e^{-j0.597}$
(c) $2.24 \times 10^3 e^{j1.11}$

(d) $611e^{-j0.878}$

(e) $95.6e^{j0.232}$ (f) $6e^{j1.571}$

(g) $88e^{-j1.571}$ (h) $21.2e^{j0}$

(i) $7.07e^{j0.785}$

9. (a) $v = 21\sin(\omega t + \pi/6)$

    (b) $i = 2.1\cos(\omega t + 1.2)$

    (c) $v = 167\cos(\omega t - 1/5)$

    (d) $i = 0.77\sin(\omega t - 0.1)$

7.3 ADDITION AND SUBTRACTION

1. (a) $3 + j4$ (b) $4 - j2$
   (c) $6 - j6$ (d) $3 + j2$
   (e) $2 + j7$ (f) $2 + j3$
   (g) $32 - j1.1$ (h) $8 + j0.3$
   (i) $8.8 - j6$ (j) $3.12 + j4.12$
   (k) $8.56 + j7.83$ (l) $-j3$
   (m) $j7$ (n) -4
   (o) -6 (p) $1.5 + j3$
   (q) -9
   (r) $5.828 + j0.828$
   (s) $5 + j3$ (t) $6 - j4$
   (u) $1 - j$
3. $\theta = 53.1°$
5. (a) $38 - j7$ (b) $54 - j10$
   (c) $8695 + j10100$
   (d) $890 - j200$
   (e) $1640 - j4300$
7. $664 + j1814$

7.4 MULTIPLICATION

1. (a) $j5$ (b) $8 + j4$
   (c) 16 (d) $3 + j$
   (e) $-12 - j20$ (f) $9 + j6$
   (g) $j6$ (h) -30
   (i) 4.5 (j) $\dfrac{13}{24} + \dfrac{7}{12}j$
   (k) $1.32 - 1.46j$
   (l) $-9 \times 10^6 + 7 \times 10^6 j$
   (m) $4.7 \times 10^5 + 10.3 \times 10^5 j$
   (n) $(a + b) + (a - b)j$

   (o) $(xp + yq) + j(py - qx)$

3. (a) $6.71\underline{/93.4°}$, $-0.398 + j6.698$
   (b) $12.65\underline{/-48.4°}$, $8.40 - j9.46$
   (c) $11.31\underline{/0°}$, $11.31 + j0.000$
   (d) $6.00\underline{/135°}$, $-4.24 + j4.24$
   (e) $30.0\underline{/90°}$, $0.00 + j30.0$
   (f) $14.0\underline{/0°}$, $14.0 + j0.00$
   (g) $4.86\underline{/180°}$, $-4.86 + j0.00$
   (h) $27.0\underline{/0°}$, $27.0 + j0.00$
   (i) $2.76\underline{/90°}$, $0.00 + j2.76$

7.5 DIVISION

1. (a) $0.600\underline{/10°}$ (b) $0.810\underline{/30°}$
   (c) $0.762\underline{/2°}$ (d) $1.176\underline{/27°}$
   (e) $0.600\underline{/180°}$ (f) $1.111\underline{/0°}$
   (g) $2.000\underline{/90°}$ (h) $2.205\underline{/0°}$
   (i) $3.296\underline{/180°}$ (j) $0.370\underline{/-90°}$
   (k) $0.500\underline{/-25°}$ (l) $V/Z\underline{/-\theta°}$

3. (a) $0.8 + j0.6$ (b) $0.5 - j1$
   (c) $-0.5j$ (d) $0.5 - j1.5$
   (e) $-0.75 - j1.25$
   (f) $\dfrac{-9}{13} + j\dfrac{6}{13}$
   (g) -0.50 (h) $\dfrac{-2}{7}j$
   (i) 2.00
5. (a) 3.33Ω
   (b) $1.179\underline{/-45°}\Omega$
   (c) $62.5\underline{/-25.3°}\Omega$
   (d) $2567\underline{/-90°}\Omega$
   (e) $1.25\underline{/-74.5°}k\Omega$
   (f) $297\underline{/25.9°}\Omega$
   (g) $23.5\underline{/23.7°}\Omega$
   (h) $0.868\underline{/14.3°}$ $M\Omega$

8.1 DEFINITION—PROPERTIES

1. (a) 10^6 (b) 2^6
   (c) 3^5 (d) 2.5^2
   (e) 5^6
3. (a) $2^3 = 8$ (b) $8° = 1$
   (c) $3^{-1} = 1/3$ (d) $25^{1/2} = 5$
   (e) $4^1 = 4$ (f) $2^{-3} = 1/8$
5. (a) 81 (b) 64
   (c) 2 (d) -1.0
   (e) 3 (f) $1/36$
   (g) 81 (h) 14.0
   (i) 1.00

8.2 BASE 10

1. (a) 0.2524 (b) 2.4639
   (c) 3.6831 (d) 1.2360
   (e) -0.8576 (f) -2.0841
   (g) 2.5583 (h) 0.8927
   (i) -0.9031 (j) -2.0969
   (k) 4.3979 (l) -6.7471
3. (a) (i) 10.28 (iii) 12.30
       (ii) 22.67 (iv) -3.507
   (b) (i) 28.58 (iii) 29.16
       (ii) 6.021 (iv) -31.06
   (c) (i) 270.2 (iii) 740.7
       (ii) 530.9

8.3 BASE e

1. (a) -0.2107 (b) 0.2700
   (c) 1.1537 (d) 1.7951
   (e) 2.0295 (f) 2.0794
   (g) 2.2116 (h) $?$
3. (a) 8.27
   (b) 5.48×10^{-3}
   (c) 0.8407 (d) 7.57
   (e) 2.851 (f) -1.732
   (g) 1.929 (h) 0.4472
   (i) 1.146

INDEX